MACROSCOPIC AND STATISTICAL THERMODYNAMICS

EXPANDED ENGLISH EDITION

YI-CHEN CHENG

National Taiwan University, Taiwan

NEW JERSEY • LONDON • SINGAPORE • BEIJING • SHANGHAI • HONG KONG • TAIPEI • CHENNAI

Published by

World Scientific Publishing Co. Pte. Ltd.
5 Toh Tuck Link, Singapore 596224
USA office: 27 Warren Street, Suite 401-402, Hackensack, NJ 07601
UK office: 57 Shelton Street, Covent Garden, London WC2H 9HE

British Library Cataloguing-in-Publication Data
A catalogue record for this book is available from the British Library.

MACROSCOPIC AND STATISTICAL THERMODYNAMICS
(Expanded English Edition)

ISBN 981-256-663-5
ISBN 981-256-664-3 (pbk)

Printed by FuIsland Offset Printing (S) Pte Ltd, Singapore

MACROSCOPIC AND STATISTICAL THERMODYNAMICS

Preface

To the Expanded English Edition

After the publication of the Chinese edition of the book, some friends and colleagues encouraged me to translate the book into English so that the book could have a wider readership. This motivated me to do so.

In this expanded English edition, some sections used for the Chinese edition have been rewritten, new sections have been added, and the original Chapter 11 has been expanded and split into two chapters – Chapters 11 and 12. Also, more problems have been added at the end of each chapter so that the reader can have more problem choices. An appendix, added at the end of the text, explains how to derive the approximations and evaluate the integrals used in the text.

More microscopic thermodynamics topics are included in the text. The grand canonical ensemble and the associated grand partition function are introduced in Chapter 10. The grand partition function approach is then applied in Chapter 12 to derive the distribution functions for the quantum ideal gases, which cannot be obtained by using the partition function approach. The topic of ferromagnetism which uses a simple mean field theory Ising model is discussed in Chapter 11. In Chapter 12, thermodynamic properties of ideal quantum gases near absolute zero are studied in more detail.

This expanded English edition was carefully proof-read by both National Central University (Chungli, Taiwan) physics professor J. M. Nester and his wife Debbie Nester. Professor Nester not only refined the English of the text, but in several places also made valuable suggestions about the subject matter.

I appreciate very much their help. Without their effort, the book would not be in the present readable form. Finally, I would like to thank Dr. K. K. Phua of the World Scientific Publishing Company who supported the publication of this edition.

Yi-Chen Cheng
National Taiwan University
March 2006

Preface

To the Chinese Edition

This book is based on lecture notes which the author had written for the undergraduate course "Thermal Physics", which the author has taught during recent years at the National Taiwan University. It is hoped that this book would provide an additional choice as a textbook or a reference book for professors and students who are going to teach or to study the course *Thermal Physics* or *Thermodynamics.*

It is written in Chinese for the convenience of Chinese reading students. However, all the chapter and section titles, and the scientific terminologies are also given in English. This is because at present, and in the foreseeable future, English is the most common language in the physics community, as well as in many other professional communities. Therefore, it is important for the student to be familiar with English terms and terminologies when they are learning a new subject in physics.

There are eleven chapters in the book. Seven of them (Chapters 1, 2, and 4–8) concern macroscopic classical thermodynamics, and the other four (Chapters 3 and 9–11) are about microscopic statistical thermodynamics. In Chapters 1 and 2, the important zeroth, first and second laws of thermodynamics are introduced; they are the basis of thermodynamics.

In these chapters, two quantities, temperature and entropy which play important roles in thermodynamics, are defined. Most students are familiar with the term *temperature* and have little difficulty in understanding its meaning. However, the term *entropy* is rather different. Although most of the students who have learned some thermodynamics may be familiar with the term entropy, many of them have only a vague idea of the physical meaning of entropy.

This is because macroscopic thermodynamics cannot give a clear answer to this question, which can be answered only from a microscopic point of view. Therefore, a microscopic interpretation of entropy is given in Chapter 3 which, hopefully, will help the students understand the physical meaning of entropy. Chapter 4 considers the applications of the combined first and second laws.

Several thermodynamic potentials, which are important in the understanding of the equilibrium condition of a thermodynamic system under different external constraints are introduced and defined in Chapter 5. Chapter 6 introduces methods of cooling and the third law. The conditions for phase equilibrium are studied in Chapter 7, including the coexistence condition for the different phases of single-component matter and the various phase diagrams for a binary mixture. Some applications of thermodynamics are given in Chapter 8. These include the application of thermodynamic relations to non-PVT systems, such as a magnetic system, a surface (thin film) system, and blackbody radiation. Chapter 9 presents the kinetic theoryof gases, in which a gas is considered to be composed of molecules, and classical collision theory is used to study the properties of a dilute gas. These include *non-equilibrium* transport processes. The kinetic theory of gases is a microscopic theory, however this approach cannot be extended to study more condensed matter, such as liquids and solids. Statistical theory is needed to study these more complicated systems.

Chapter 10 studies the basic principles of statistical thermodynamics. Three different kinds of statistics for particles are introduced and their distribution functions are derived. The important elements of statistical mechanics, the concept of an ensemble and the partition function are also introduced. Chapter 11 studies several simple systems whose partition functions are relatively easy to obtain, so their thermodynamic properties can be studied. In this chapter, the properties of ideal quantum gases, including a Fermi and a Bose gas, near the temperature of absolute zero are studied.

This book is intended to be a textbook for a one-semester course. However, it is not quite possible to cover all the material in one semester. Some choices of the topics may be necessary. The following provides suggestions for two possible choices:

(1) If the course emphasizes macroscopic thermodynamics, then Chapters 9–11 may be omitted. Chapter 8 is optional, however, Sections 8.2 and 8.8 are recommended.

(2) If the course emphasizes microscopic statistical thermodynamics, then Chapter 4, Sections 7.6–7.14, and Chapter 8 may be omitted, however, Sections 8.2 and 8.8 are recommended.

Solving problems is an important part of studying any subject in physics. It is recommended that the students would do all, or at least, most of the problems given at the end of each chapter.

Yi-Chen Cheng
National Taiwan University
2003

Contents

Chapter 1

Heat and the First Law

1.1 The Scope of Thermodynamics

Thermodynamics, a branch of science based on the the observations and studies of many scientists over the past three hundred or so years, can be summarized in a few principles and laws (notably the three well-known thermodynamic laws). These laws govern the behavior and the relationships between the thermodynamic state variables of any macroscopic thermodynamic system. Among all the state variables, the most important one is *temperature*. Temperature is defined and derived in thermodynamics. Its nature is quite unique and rather different from other basic physical quantities, such as mass, time, length, electric charge, etc., which are defined in other branches of physics, such as mechanics, electricity and magnetism, etc. We may say that, in general, any phenomenon or theory which involves the physical quantity of temperature belongs to the subject of thermodynamics or its applications. However, the three well-known laws of thermodynamics govern only the thermodynamic behavior in equilibrium states. There are many more, and much more complex phenomena in nature which belong to the category of non-equilibrium thermodynamics. Except for a few sections, this book is confined in the study of equilibrium thermodynamics. Non-equilibrium thermodynamics is beyond its scope.

Any physical system on the earth is not only surrounded by the atmosphere, but also is constantly interacting with its surroundings. The study of the behavior of all these interacting systems belongs to the subject of thermodynamics or its applications. The reason is that the atmosphere is actually a huge macroscopic thermodynamic system; any system in contact with it will have thermodynamic properties. All material systems are composed of atoms; therefore we begin with the simplest system, an atom, to see what thermodynamic properties we can explore. If we put *an atom* in the atmosphere, the atom will be affected by the atmosphere in the following two ways: (1) the motion of the atom as a whole, and (2) the internal structure of the atom. The atmosphere is composed of many molecules and atoms. The *added atom* will become part of the atmosphere, therefore its motion as a whole (the translational motion) will be affected by the atmosphere. This is because the behavior of the atom is influenced by the thermodynamic properties of the atmosphere. However the internal structure of the atom (the orbital electrons) will not be affected. The reason is that the energy gap between the ground state and the first excited state of the orbital electrons is at least several electron volts (eV), while the thermal energy at room temperature is just around 0.025 eV. The orbital electrons will therefore remain in the ground state, unaffected by the atmosphere. This is the reason why, in the study of the electronic structure of an atom, the temperature effect is usually not considered.

In thermodynamics, we often consider an atom, as one of the particles in the atmosphere, a particle without internal structure, but we may also consider an atom as a system (albeit a small system) which is surrounded by the atmosphere (a big system). Now we consider the case where the small system is composed of two atoms (a diatomic molecule). There is not much difference as compared with the case that the small system is a monatomic molecule. The motion of the small system as a whole (translation and rotation) will be affected by the atmosphere,[1] yet the internal structure of the small system is not affected by the atmosphere. It is known that when two atoms attract each other to become a single molecule, each original energy level will be split into two levels, thus the energy gaps will become smaller. However, the energy gap to the first excited state is still on the order of a few eV, which is much greater than the room temperature energy. The orbital electrons will therefore remain in the ground state. In addition to the energy levels of the orbital electrons, the internal structure of a diatomic molecule also includes the relative motion of the two atoms (vibration), but the energy required to

[1] According to quantum mechanics, rotational energy is quantized, however the energy required to excite the rotational motion of a diatomic molecule is usually much smaller than the room temperature thermal energy. Therefore the classical concepts and methods may be used.

excite the vibrational motion is much more than the room temperature energy,[2] therefore we may say that vibrational motion is also not affected by the atmosphere.

However when the number of atoms in the small system is increased to several hundred or more, the situation will be changed significantly. In such a case the mass of the system is much larger than that of a single atom; the motion of the system as a whole is mainly determined by the gravitational force, and therefore its role in thermodynamic properties becomes insignificant. On the other hand, the internal structure now plays the major role in its thermodynamic properties. The energy required to excite the vibrational motion is now decreased to the order of the room temperature thermal energy. Moreover in some materials, the energy required to excite the orbital electrons is also gradually reduced to be close to the room temperature energy. Yet this is still a small system. The vibrational energy levels and the energy levels of the orbital electrons will be changed as the number of the atoms of the system increases, or the geometrical arrangement of the atoms in the system changes. Only when the system contains a very large number of atoms, which are arranged in a normal way (e.g., not in a reduced dimension), will the thermodynamic properties of the system not depend on the number of its atoms, or on the geometrical arrangement of the atoms.

In view of the fact that the thermodynamic properties of a system may depend on the number of atoms in it, we may classify thermodynamic systems, *according to the size of the system*, into the following three classes:

1. A **microscopic system**: the number of atoms in the system is small, so that its internal structure is independent of temperature.

2. A **mesoscopic system**: the number of atoms in the system is much more than that of a microscopic system, but not large enough that the thermodynamic properties of its internal structure depend on the number of the atoms and their geometrical arrangement.

3. A **macroscopic system**: the number of atoms in the system is very large, so that the thermodynamic properties of the system are

[2] According to quantum mechanics, vibrational energy is also quantized. The energy gap between the neighboring levels is equal to the Planck constant times the vibration frequency. The vibrational energy gap of almost of all diatomic molecules is much greater than the room temperature energy.

independent of the number of atoms or their geometrical arrangement.[3]

There are, however, no clear cut distinctions between these three classes of systems, and transition regimes do exist. Here we introduce a concept which is called the **thermodynamic limit**. We consider a thermodynamic system, and double both the number of its atoms and its volume (but keeping the density of the system constant). If the thermodynamic properties of the system remain unchanged, then we say that the system has reached the thermodynamic limit. It is therefore a macroscopic system. A mesoscopic system is not large enough to reach the thermodynamic limit.

According to the above analysis, we may roughly divide the study of equilibrium thermodynamics into two categories. One is to study systems consisting of a very large number of almost independent particles (atoms, molecules) which are confined in the same space (volume). Particle-particle collisions are the main mechanism with which the system maintains its thermodynamic equilibrium. During the collisions, particles exchange their translational and rotational energies, but the internal structures do not change. Between collisions, each particle may be considered as a free particle. We are familiar with this type of systems, they are known as gases. The other is to study systems consisting of a very large number of atoms (molecules) which attract one another strongly and are closely tied together, to become a type of *condensed matter*, in other words a liquid or a solid. In these systems, the main thermodynamic properties come from the relative motion between the particles (vibrations), rather than the translational and rotational motion of each particle. The behavior of the orbital electrons may also play an important role in its thermodynamic properties.

Classical thermodynamics does not look into the microscopic details of the system, and therefore has its limitations. During the latter part of the nineteenth century, scientists began to try to understand thermodynamics from the microscopic point of view. For gases, one may approximate that the only type of interaction that occurs between particles is the *hard core* collision. This theory, known as the **kinetic theory**

[3]If there are two systems composed of the same kind of atoms, but one is macroscopic and the other mesoscopic, then these two systems may have different thermodynamic properties, e.g., magnetic properties, non-equilibrium transport properties, etc. We are familiar with the macroscopic world, but our understanding of the mesoscopic systems is just in the beginning stages. Therefore the study of mesoscopic systems (notably nano-sized systems) is a new and important area. Research in this area may lead us to new and useful discoveries both in science and technology.

of gases, has been very successful in studying the properties of dilute classical gases. Using it, one can also calculate some of the transport coefficients in non-equilibrium thermodynamics, but this theory can not be generalized to study much more complex systems such as liquids and solids. The reason is that in condensed matter, the interactions between the particles are much more complicated than hard core collisions. The theory which successfully treats these complicated systems is called **statistical thermodynamics**, or **statistical mechanics**. In this branch of science, one does not look into the details of the motion of each particle. Instead, one uses the concepts and methods of statistics to treat the complex thermodynamic systems and to simplify the problems. Mean values of thermodynamic variables are taken to represent the thermodynamic properties of the system under consideration. If the condition of the thermodynamic limit is satisfied, the predictions of statistical mechanics (obtained using mean values) can be considered as precise, and the errors (fluctuations) are negligibly small.[4]

1.2 Some Definitions

Thermodynamics is a branch of science which studies the thermodynamic properties of thermodynamic systems. A thermodynamic *system* (for brevity, a system) together with its *surroundings* constitute the universe. There are three types of thermodynamic systems, *according to how the system interacts with its surroundings*:

1. An **open system**: the system can exchange energy and mass with its surroundings.

2. A **closed system**: the system can exchange energy with its surroundings, but not mass.

3. An **isolated system**: the system can exchange neither energy nor mass with its surroundings.

In this book we will confined ourselves, except for a few sections, to the study of equilibrium state thermodynamic properties. A system is said to be in an **equilibrium state** (for brevity, a **state**) if the thermo-

[4]In statistical mechanics, the root mean square fluctuations are usually proportional to the inverse of the square root of the number of particles of the system. The fluctuations are negligibly small if the number of particles of the system becomes very large.

dynamic properties of the system do not change with time.[5] A physical quantity is said to be a **state variable** of the system, if the magnitude of the variable is fixed once the system reaches an equilibrium state. If two systems are identical, and they are in the same equilibrium state, then they will have the same value for every state variable. This implies that if the *state* of a system is specified, then all of the state variables of the system have definite values. Volume, pressure and temperature are known examples of state variables. There are two types of the state variables: **extensive** variables and **intensive** variables. In the same equilibrium state, the value of an extensive variable will be proportional to the size of the system,[6] such as its volume or mass. On the other hand, the value of an intensive variable, such as temperature and pressure, is independent of the size of the system. When an extensive variable is divided by the volume (or mass, or number of kilomoles) of the system, it will become an intensive variable. In this book, if an intensive variable is derived from an extensive variable (e.g., by dividing by volume), then it will be denoted by a *lower case* letter. All the extensive variables are denoted by *upper case* letters. However, it should be noted that the two capital letters P (pressure) and T (temperature) are intensive variables. These two (because of historical reasons) are the only exceptions to the rule.

Bring two isolated systems, A and B, into contact with each other. If we now let the wall which separates A and B becomes thermally conducting (but fixed in position), then A and B can exchange energy (but not mass). A and B separately are no longer isolated, but A+B as a whole is an isolated system. In this situation, we say that A and B are in *thermal contact.* After a period of time, A and B will separately reach their own equilibrium states. When this happens, we say that A and B are in **thermal equilibrium**. If the wall between A and B can move freely (with no friction and no mass exchange) and a final equilibrium state is reached, then we say that A and B are in **thermodynamic equilibrium**. In the latter case, in addition to thermal equilibrium, there is also a mechanical (pressure) equilibrium. The individual volume of A and B may be changed, but the total volume of A+B does not change.

[5]It is also required that the system is not under the action of any external force or field.

[6]This statement is true for macroscopic systems only, but not for mesoscopic systems.

1.3 The Zeroth Law of Thermodynamics

The zeroth law of thermodynamics defines a state variable T called *temperature*. This means that when a system is in an equilibrium state then it has a definite value of T. The physical significance of T in thermodynamics can be understood from the following consideration.

Consider two *isolated* systems A and B which are in their own respective equilibrium states. Bring A and B together and let them to be in thermal contact with each other. Right after A and B make thermal contact, there are two possibilities:

1. A and B reach thermal equilibrium immediately. In this case we say that the temperature of A just before contact T_{A} and the temperature of B just before contact T_{B} are equal to each other, i.e., $T_{\mathrm{A}} = T_{\mathrm{B}}$. Also, after contact, both T_{A} and T_{B} remain unchanged. This means that when two isolated systems have an equal temperature, then their temperatures will remain unchanged when (and after) they are placed in thermal contact.

2. It takes a period of time for A and B to reach thermal equilibrium. This implies that before contact $T_{\mathrm{A}} \neq T_{\mathrm{B}}$. After thermal equilibrium is reached, the temperatures of A and B become T'_{A} and T'_{B}, respectively. Thermal equilibrium requires that $T'_{\mathrm{A}} = T'_{\mathrm{B}}$.

The zeroth law of thermodynamics can be stated as follows. Consider three isolated systems A, B and C, which are in their own respective equilibrium states with temperatures $T_{\mathrm{A}}, T_{\mathrm{B}}$ and T_{C}, respectively. Place A and B in thermal contact. If A and B reach thermal equilibrium right after their contact, then $T_{\mathrm{A}} = T_{\mathrm{B}}$. Remove A from B, and place B and C in thermal contact. If B and C reach thermal equilibrium right after contact, then $T_{\mathrm{B}} = T_{\mathrm{C}}$. If the above situations occur, then one is certain that A and C have the same temperature, i.e., $T_{\mathrm{A}} = T_{\mathrm{C}}$. A and C will reach thermal equilibrium immediately if they are placed in thermal contact.

The above consideration and the zeroth law have the following implications. Consider two systems which are in their respective equilibrium states; place them in thermal contact. If the systems have the same temperature before contact, then after contact they will remain in their equilibrium states with the same temperature. If the systems have different temperatures before contact, then after contact, the systems will experience a period of non-equilibrium. The temperatures of the two systems will eventually become equal to each other and the systems will reach their new respective equilibrium states. The temperature in the final equilibrium states will be different from the temperatures before contact. However if the size (such as the mass) of one of the systems is much larger than that of the other, then the temperature change of the large system will be very small, and can be neglected. The final temperature of the small system will be equal to that of the large system. An example of the above case is the use of a thermometer. The thermometer is the small system, which is used to measure the temperature of the large system.

Now we would like to introduce the concept of a heat reservoir. A **heat reservoir** is a very large system which makes thermal contact with much smaller systems. In the contact it can exchange thermal energy with the smaller systems, without changing its own temperature. A constant temperature process usually needs the help of a heat reservoir.

Finally, it may be helpful to give *temperature* a more precise physical meaning. We may say that temperature is a measure of the average energy per particle in a system due to the *random motions* of the particles. We will derive this result microscopically in Chapters 3, 9 and 11.

1.4 Equilibrium State

We have stated that if a system is in an equilibrium state then its thermodynamic properties will not change with time. The thermodynamic properties mean the behavior of the state variables of the system. The values of the state variables depend on the state of the system only, and are independent of the past history of the system. The most familiar state variables are *pressure*, *volume* (quantities defined in mechanics) and *temperature*. Temperature is a physical quantity defined in thermodynamics (not in other disciplines of physics), and therefore it may be considered as the most important state variable in thermodynamics. For magnetic systems, the main concern is the magnetic properties, therefore instead of pressure and volume, the basic state variables are the magnetic field, the magnetic moment (quantities defined in electromagnetism), and the temperature. Besides the above well-known state variables, we will introduce more state variables in the latter chapters and sections of this book. These additional state variables will be useful in studying thermodynamic properties under different situations.

For most of the thermodynamic systems we consider, if the values of *any two* of the state variables are fixed, then the values of all other state

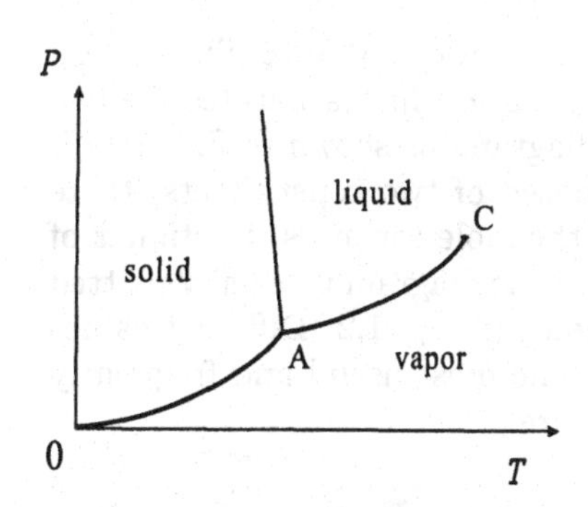

Figure 1.1: A sketched phase diagram of H_2O. A is the triple point; C is the critical point. Note that the slope of the solid-liquid coexistence curve is negative.

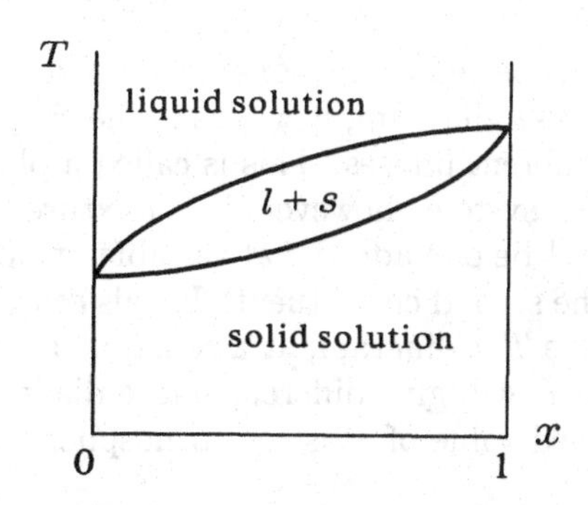

Figure 1.2: A sketched phase diagram (at a constant P) of the alloy $A_{1-x}B_x$, where x is the fraction of B by weight. The area labeled by $l + s$ is the liquid-solid coexistence region.

variables will also be fixed. This means that there are only two independent variables for most of the thermodynamic systems. For ordinary (non-magnetic) systems, the most frequently used variables are pressure P, volume V, and temperature T (known as PVT systems). However only two of them are independent, therefore there is a relation between P, V and T, which can be written as

$$f(P, V, T) = 0. \tag{1.1}$$

This is called the equation of state. The most well-known is the equation of state for an ideal gas $f(P, V, T) = PV - nRT$. In magnetic systems, the equation of state can be written as $f'(B, M, T) = 0$, where B and M are the magnetic field and the magnetic moment, respectively. We see that an equation of state always involves the state variable T. This is one of the reasons why we say that temperature plays quite a unique role in thermodynamics.

If we need two equations of state to describe two different states of the same system, then these two different states (denoted as states 1 and 2) belong to different *phases.* When the system changes from state 1 to state 2 (or vice versa), there will be a process of *phase transition.* For a phase transition to occur, the system must satisfy an additional phase equilibrium condition. Therefore the number of independent variables will be reduced to one. If we plot the condition of phase equilibrium on a P-T diagram, it will take the shape of a curve which separates the two different phases. This is called a phase diagram, as shown in Fig. 1.1. If the system, however, is a mixture composed of two constituents, there will be one additional variable, which is the mole (or mass) fraction x of the second constituent. In this case, the phase diagram is usually plotted as a T-x diagram, at a constant P, as shown in Fig. 1.2. Different values of P will give different phase diagrams. The most useful and frequently used value of P is one atmospheric pressure.

1.5 The First Law of Thermodynamics

Before we state the first law of thermodynamics, we would like to introduce two important concepts: (1) the internal energy [7] of the system U is a state variable; (2) heat is a form of energy, which is also called thermal energy.[8]

The first law can be stated as follows. When a system changes from an initial state to a final state, the change in its internal energy ΔU is equal to the heat it absorbs from the surroundings ΔQ minus the work done by the system ΔW. This can be expressed as

$$\Delta U = \Delta Q - \Delta W. \tag{1.2}$$

If the changes of both of the two independent variables are very small, then the changing process is called an *infinitesimal process.* In this case, the first law is written as

$$dU = đQ - đW, \tag{1.3}$$

where dU, $đQ$ and $đW$ indicate that in the process, the changes in the internal energy U, the heat absorbed Q, and the work done W are all very small quantities. From the above equations, we may say that the first law is just a statement of the conservation of energy. Here we should note that in Eq. (1.3) we use two different notations d and $đ$ to denote

[7]Internal energy is the sum of the kinetic and potential energy of all particles in the system, but excluding the kinetic energy for the system moving as a whole.

[8]The first scientist who expressed heat in terms of the units of energy was James Joule (1840). The latest data on the mechanical equivalent of heat is 1 cal=4.186 J. One calorie is the energy required to increase the temperature of 1 gram of water from 14.5°C to 15.5°C. James P. Joule, British physicist (1818–1889).

small quantities. The notation dU means that the internal energy U is a state variable. The function U exists if the state of the system is specified. Whereas $đQ$ (or $đW$) implies that Q (or W) is not a state variable. This means that the function which represents Q (or W) does not exist. Therefore the value of ΔU in Eq. (1.2) depends only on the initial and the final states, but the value of ΔQ (or ΔW) will depend on the *process*, i.e., the path by which the system changes from the initial to the final state. In mathematics, we call dU an *exact* differential and $đQ$ (or $đW$) an *inexact* differential.

1.6 Exact and Inexact Differentials

Consider a system with two independent variables x and y. An infinitesimal quantity dF can be expressed as

$$dF = a(x,y)\,dx + b(x,y)\,dy. \tag{1.4}$$

If dF is an exact differential, then the function $F = F(x,y)$ exists. From calculus we find

$$dF = \left(\frac{\partial F}{\partial x}\right)_y dx + \left(\frac{\partial F}{\partial y}\right)_x dy. \tag{1.5}$$

Therefore (by comparing Eq. (1.4) and (1.5)) we have

$$a(x,y) = \left(\frac{\partial F}{\partial x}\right)_y, \quad b(x,y) = \left(\frac{\partial F}{\partial y}\right)_x, \tag{1.6}$$

and

$$\left(\frac{\partial a}{\partial y}\right)_x = \left(\frac{\partial b}{\partial x}\right)_y. \tag{1.7}$$

Equation (1.7) is the necessary and sufficient condition for the infinitesimal quantity $dF = a(x,y)dx + b(x,y)dy$ to be an exact differential. In this case $F(x,y)$ is a well-defined function. But if the condition Eq. (1.7) is not satisfied, then the function $F(x,y)$ does not exist. In that case we rewrite dF as $đF'$ and call it an inexact differential (for the convenience of discussion we replace F by F').

The main difference between an exact differential dF and an inexact differential $đF'$ is the following. When the independent variables (x,y) change from the initial (x_1,y_1) to the final (x_2,y_2), the change of F is

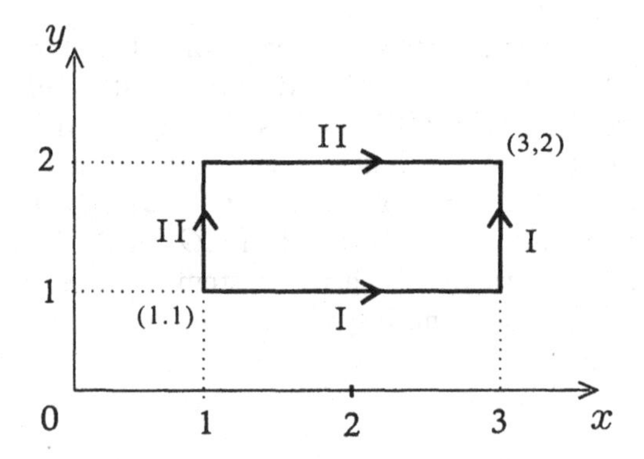

Figure 1.3: Two different integration paths.

a fixed value. It is simple to get $\Delta F = \int dF = F(x_2, y_2) - F(x_1, y_1)$, but the change of F' depends on the path, i.e., $\Delta F' = \int đF'$ depends on the path. A path is a curve that connects the initial point (x_1, y_1) and the final point (x_2, y_2). Since $F'(x, y)$ does not exist, we can not write $\Delta F'$ as $\Delta F' = F'(x_2, y_2) - F'(x_1, y_1)$; $F'(x_1, y_1)$ and $F'(x_2, y_2)$ are not defined. We give a simple example in the following. If $dF = ydx + xdy$, then $a(x, y) = y$ and $b(x, y) = x$, therefore Eq. (1.7) is satisfied,

$$\left(\frac{\partial a}{\partial y}\right)_x = \left(\frac{\partial b}{\partial x}\right)_y = 1. \tag{1.8}$$

From calculus we easily get

$$dF = ydx + xdy = d(xy + c), \tag{1.9}$$

thus $F = xy + c$ (c is a constant). Therefore $\Delta F = \int dF = F(x_2, y_2) - F(x_1, y_1) = x_2y_2 - x_1y_1$, independent of the path. However if $đF' = ydx - xdy$, then $a(x, y) = y$ and $b(x, y) = -x$, therefore

$$\left(\frac{\partial a}{\partial y}\right)_x = 1, \quad \left(\frac{\partial b}{\partial x}\right)_y = -1, \tag{1.10}$$

Eq. (1.7) is not satisfied. The function $F'(x, y)$ does not exist and $\Delta F'$ will depend on the path. If we take $(x_1, y_1) = (1, 1)$ and $(x_2, y_2) = (3, 2)$, then $\Delta F = x_2y_2 - x_1y_1 = 5$ is a fixed value and independent of the path, but $\Delta F'$ will depend on the path. Consider path I in Fig. 1.3. We have

$\Delta F' = -1$. However if we take path II, we get $\Delta F' = 3$. Thus exact and inexact differentials have quite different properties, that should not be mixed up. An exact differential is the same as a differential in calculus, but an inexact differential is not a differential (as defined in calculus), it is just an infinitesimal quantity. In thermodynamics, both $đQ$ and $đW$ are inexact differentials, which mean that *heat* Q and *work* W are *not* state variables.

1.7 Work in Thermodynamics

In mechanics, work equals force times the displacement. In thermodynamics, work equals pressure times the volume change. We get this result because pressure is force divided by area, and the volume change is the displacement times the area. In an infinitesimal process, we can write

$$đW = PdV, \tag{1.11}$$

where P is the pressure and dV is the infinitesimal change of volume. The work defined by Eq. (1.11), is usually called **configuration work.** This type of work is the result due to a change of the configuration of the system (here the volume). The pressure P can be considered as a generalized force X, and the volume change dV a generalized displacement dY. The general form of a configuration work can be written as

$$đW_{\text{conf}} = XdY. \tag{1.12}$$

In thermodynamics, different systems may have different appearances for the configuration work. Some are listed in Table 1.1.

Table 1.1 Examples of configuration work

System	X (force)	Y (displacement)	$đW$
PVT system	P(pressure)	V(volume)	PdV
magnetic material	B(magnetic field)	M(magnetization)	$-BdM$
dielectric material	E(electric field)	$\mathcal{P}$(polarization)	$-Ed\mathcal{P}$
thin film	σ(surface tension)	A(area)	$-\sigma dA$
string	J(tension)	L(length)	$-JdL$

One of the main characteristics of configuration work is that it may be positive or negative. When the value of X makes an infinitesimal change (keeping the sign unchanged), it is possible that dY may change from positive to negative (or vice versa), thus $đW_{\mathrm{conf}}$ changes its sign. This means that the system can do work on its surroundings and the surroundings can also do work on the system. We call this ***reversible work***. There is another type of work, which can do work in one direction only. We call this *irreversible work*. For example, the work done by a resistor is always on its surroundings (by producing heat). It is impossible, by an infinitesimal change (or by reversing the direction) of the current, to change the direction of the work (i.e., so that the resistor absorbs heat from the surroundings). This type of work is called *dissipative work*, because it can not be recovered. Any work produced by a frictional force is categorized as this type of work. The result of this type of work is to generate heat; it is not related to a change in the configuration of the system. It is therefore quite different from configuration work.

In the following, we use $đW = PdV$ as an example to explain the fact that work depends on the path. Consider a system which changes from state 1 (P_1, V_1) to state 2 (P_2, V_2); the work done by the system during the process is

$$\Delta W = \int_{V_1}^{V_2} PdV. \tag{1.13}$$

From calculus we know that ΔW is equal to an area in Fig. 1.4. The magnitude of the area depends on the shape of the curve connecting points 1 and 2. The connecting curve is the *path* mentioned above. In the figure, we plot two paths, I and II. It is obvious that the magnitude of $\Delta W_{\mathrm{II}} - \Delta W_{\mathrm{I}}$ equals the area enclosed by the paths I and II. Except for when there are two identical paths, the enclosed area will not be equal to zero. Therefore ΔW depends on the path.

1.8 Enthalpy

In a constant pressure process (*isobaric* process) , if the volume changes from V_1 to V_2, the work done by the system is $\Delta W = P(V_2 - V_1)$. In this case the first law can be written as

$$\Delta U = \Delta Q_P - P(V_2 - V_1), \tag{1.14}$$

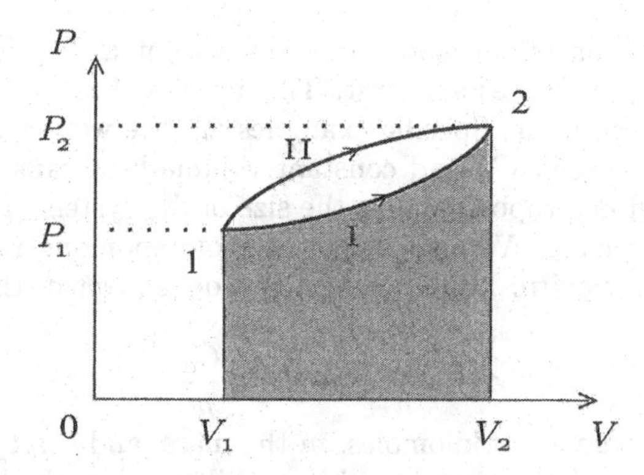

Figure 1.4: Work depends on the path. The difference between the work done along paths I and II is equal to the area (in white) enclosed by the paths I and II.

where ΔQ_P means that heat is absorbed in a constant P process. Since $\Delta U = U_2 - U_1$, Eq. (1.14) can be written as

$$\Delta H = H_2 - H_1 = \Delta Q_P, \tag{1.15}$$

where we have defined a new state variable the **enthalpy**:

$$H = U + PV. \tag{1.16}$$

From Eq. (1.15) we see that in a constant P process, the heat absorbed by the system is equal to the change of the enthalpy of the system. This is very useful in the study of phase transitions, because during a phase transition the pressure is usually kept constant. Therefore, when two phases coexist, the difference of the enthalpy between the two phases is the well-known *latent heat* in the phase transition.

1.9 Heat Capacity

When a system absorbs a certain amount of heat ΔQ, its temperature T will increase by ΔT. From this we can define a physical quantity known as the *heat capacity* C:

$$C = \lim_{\Delta T \to 0} \frac{\Delta Q}{\Delta T} = \frac{đQ}{dT}. \tag{1.17}$$

The last expression of the above equation is not a differentiation, because the function Q does not exist. This implies that the magnitude of C depends on the path. Specific examples are the well-known constant pressure heat capacity C_P and constant volume heat capacity C_V. Apparently, C will be proportional to the size of the system. It is therefore an extensive variable. We can define the corresponding intensive variable called the **specific heat** c, which is independent of the size of the system:

$$c_\alpha = \frac{C_\alpha}{n}, \quad \text{or } c_\alpha = \frac{C_\alpha}{m}, \tag{1.18}$$

where n is the number of kilomoles, m the mass, and α may be P or V, etc. Therefore, c_α is the heat capacity per kilomole, or the heat capacity per unit mass and is known as the specific heat.

That the heat capacity C depends on the path in PVT systems is well-known. This means that the constant volume C_V is not equal to the constant pressure C_P. In the following we derive the formula for $C_P - C_V$ for a PVT system. In an *isochoric* (i.e., constant V) process $đW = 0$, $đQ_V = dU$ (Eq. (1.3)), and in an *isobaric* (i.e., constant P) process $đQ_P = dH$ (Eq. (1.15)), thus we get from Eq. (1.17)

$$C_V = \left(\frac{\partial U}{\partial T}\right)_V \quad \text{and} \quad C_P = \left(\frac{\partial H}{\partial T}\right)_P. \tag{1.19}$$

From the above equations, we have

$$\begin{aligned} C_P - C_V &= \left(\frac{\partial H}{\partial T}\right)_P - \left(\frac{\partial U}{\partial T}\right)_V \\ &= \left(\frac{\partial U}{\partial T}\right)_P + P\left(\frac{\partial V}{\partial T}\right)_P - \left(\frac{\partial U}{\partial T}\right)_V \\ &= \left(\frac{\partial V}{\partial T}\right)_P \left[P + \left(\frac{\partial U}{\partial V}\right)_T\right]. \end{aligned} \tag{1.20}$$

Here in the second equality, we have used $dH = dU + PdV + VdP$, and in the last equality we have used

$$\left(\frac{\partial U}{\partial T}\right)_P = \left(\frac{\partial U}{\partial T}\right)_V + \left(\frac{\partial U}{\partial V}\right)_T \left(\frac{\partial V}{\partial T}\right)_P, \tag{1.21}$$

which is obtained from the following equation divided by dT term by term, and keeping P constant:

$$dU = \left(\frac{\partial U}{\partial T}\right)_V dT + \left(\frac{\partial U}{\partial V}\right)_T dV. \tag{1.22}$$

1.10 $C_P - C_V$ for an Ideal Gas

We use the ideal gas as an example to evaluate $C_P - C_V$. The equation of state of an ideal gas is

$$PV = nRT, \tag{1.23}$$

where n is the number of kilomoles (or moles) and R is the **gas constant.** The temperature T in this equation is called the **ideal gas temperature**, which is also known as the absolute temperature.[9] The unit of the absolute temperature is K (kelvin). The SI units of P and V are $\mathrm{N/m^2} \equiv \mathrm{Pa}$ (pascals) and $\mathrm{m^3}$, respectively. The value of the gas constant R is then

$$R = 8.314 \times 10^3 \,\mathrm{J\,kilomole^{-1}\,K^{-1}} = 8.314 \,\mathrm{J\,mole^{-1}\,K^{-1}}. \tag{1.24}$$

Some people use one atmospheric pressure (atm) as the unit for pressure and liters as the unit of volume; the value of R is then[10]

$$R = 8.206 \times 10^{-2} \text{ liter atm mole}^{-1}\,\mathrm{K}^{-1}. \tag{1.25}$$

We will show in Chapter 2 that for an ideal gas $(\partial U/\partial V)_T = 0$, which implies that the internal energy U is a function of T only, independent of V. Therefore, Eq. (1.20) becomes

$$C_P - C_V = P\left(\frac{\partial V}{\partial T}\right)_P = nR, \tag{1.26}$$

which implies that $C_P - C_V > 0$ for an ideal gas. In Chapter 4 we will show that $C_P - C_V \geq 0$ for all systems. This means that C_P is always greater than, or at least equal to, C_V. This can easily be understood, because in an isochoric process the system does no work. The heat absorbed is entirely used to increase the temperature. In an isobaric process, however, part of the absorbed heat is used to do work, and the system needs to absorb more heat to increase the temperature by the same amount as in the isochoric case.

[9] We will introduce another temperature, the thermodynamic temperature, in Chapter 2. The origins by which the ideal gas and the thermodynamic temperatures are defined are quite different, but we will show in Chapter 2 that these two temperatures can be chosen to be identical to each other. This new temperature scale is known as the absolute temperature.

[10] For the units conversion: 1 atm=1.013×10^5 Pa; 1 liter=$10^{-3}\mathrm{m}^3$.

1.11 Reversible Adiabatic Process for an Ideal Gas

In thermodynamics, we frequently encounter both reversible and irreversible processes. These are important concepts, and we will come back to this subject and give a more detailed explanation in Chapter 2. Here we give simple criteria for a reversible process. A process is called reversible, if the following two conditions are satisfied:

1. There is no dissipative work involved (i.e., frictional force is negligible).

2. The process is a *quasi-static* process.

A process is said to be quasi-static if it proceeds very slowly, such that at any moment in the process the system can be considered as almost being in an equilibrium state. Next we shall consider an adiabatic process. A process is said to be adiabatic if, during the process, the system absorbs or rejects no heat, that is $đQ = 0$. In a reversible adiabatic process we have $dU = -PdV$.

As an example, we consider an ideal gas. The internal energy U is a function of T only, it is independent of V. Therefore

$$dU = \left(\frac{\partial U}{\partial T}\right)_V dT + \left(\frac{\partial U}{\partial V}\right)_T dV = C_V dT \text{ (ideal gas)}, \tag{1.27}$$

and

$$C_V dT = -PdV = -\frac{nRT}{V}dV, \tag{1.28}$$

$$c_v \frac{dT}{T} = -R\frac{dV}{V}, \quad c_v = \frac{C_V}{n}. \tag{1.29}$$

Integrating both sides from (T_1, V_1) to (T_2, V_2) we obtain

$$\frac{T_2}{T_1} = \left(\frac{V_1}{V_2}\right)^{R/c_v}. \tag{1.30}$$

From the equation of state and $R = c_p - c_v$, we can rewrite Eq. (1.30) as

$$P_1 V_1^\gamma = P_2 V_2^\gamma, \quad \gamma \equiv \frac{c_P}{c_v}. \tag{1.31}$$

Equation (1.31) is equivalent to

$$PV^\gamma = \text{constant}. \tag{1.32}$$

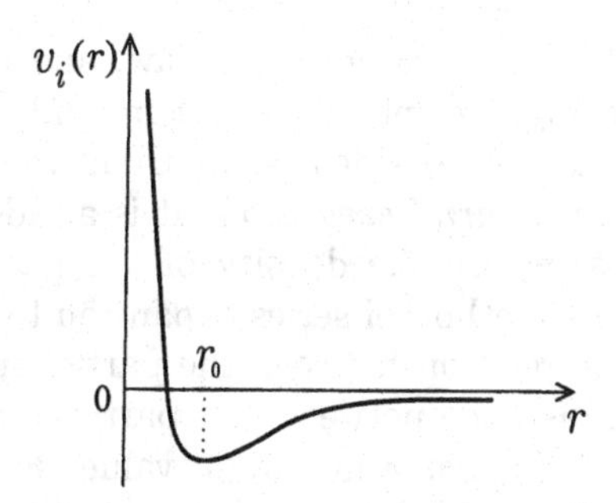

Figure 1.5: The relation between the interaction energy $v_i(r)$ and the distance between molecules r, where $v_i(r)$ is minimum at r_0.

This is the well-known equation which an ideal gas must obey in a reversible adiabatic process. For a monatomic ideal gas, the molar specific heat $c_v = 3R/2$ and $c_p = 5R/2$ by Eq. (1.26), therefore $\gamma = 5/3 \approx 1.67$. For a diatomic ideal gas [11] $c_v = 5R/2$ and $c_p = 7R/2$, thus $\gamma = 7/5 = 1.40$. Note that in this process a monatomic and a diatomic ideal gas obey different equations.

1.12 Virial Expansion

In thermodynamics, the most frequently discussed system is the ideal gas, which has the equation of state $PV = nRT$. The ideal gas is an idealized system which is based on the following two assumptions:

1. There is no interaction between the molecules, except for collisions.
2. The molecular volume is approximately zero.

These two conditions can be met only when the density of the gas is very small.[12] When the density is not sufficiently low, the behavior

[11] For a diatomic molecule, in addition to the translational energy, there is also rotational energy, which also makes a contribution to the heat capacity. We will give a more detailed discussion on this subject in Chapters 9 and 11.

[12] Here we are talking about the *classical ideal gas*. In addition to a low density, one also needs the condition of high temperature, in order that quantum effects can be neglected. The condition under which a gas can be considered as a classical ideal gas is $\lambda_{\rm th} \ll \overline{d}$, where $\lambda_{\rm th} = h/\sqrt{2\pi mkT}$ is the thermal wavelength (m is the particle mass, h and k are the Planck constant and the Boltzmann constant) and $\overline{d} = (V/N)^{1/3}$ is the mean distance between two particles. This condition will be derived in Chapter 10.

of the gas may deviate from the ideal gas law, and corrections may be needed. In the following, we introduce a commonly used approximate method which includes the non-ideal gas part in the equation of state. The method is called the *virial expansion.* It is an ideal method for the study of a dilute gas, where the density of the gas is still low. This approximation uses the method of series expansion to treat the problem. In Fig. 1.5 we plot the relation between the particle-particle interaction energy $v_i(r)$ and the distance between two particles r. In the figure, r_0 is the distance where $v_i(r)$ has a minimum value. For $r > r_0$, the slope $dv_i/dr > 0$, and particles attract each other. When r increases, the attraction decreases, going to zero for large r. For $r < r_0$, $dv_i/dr < 0$, and the particles repel each other. The repulsive force increases very rapidly as r decreases, which is rather similar to a hard core.

Consider a system with n kilomoles of particles moving in a container of volume V. The condition under which the density can be considered as dilute is

$$\frac{v_0}{v} \ll 1; \quad v_0 \equiv N_A r_0^3, \;\; v \equiv \frac{V}{n}, \tag{1.33}$$

where $N_A = 6.02 \times 10^{26}$ molecules kilomole^{-1} is **Avogadro's** [13] **number.** The virial expansion method is to express the equation of state of a gas as a power series in v_0/v:

$$Pv = RT\left[1 + \frac{B_2(T)}{v} + \frac{B_3(T)}{v^2} + \frac{B_4(T)}{v^3} + \cdots\right]. \tag{1.34}$$

In the above equation, we let the constant v_0 be included in the expansion coefficients B_i, thus v_0 does not appear in the equation. We also replace B_1 by 1, because it is equal to 1 identically. The condition for an ideal gas is that $v_0/v \to 0$, i.e., $B_i/v^{i-1} = 0$ for $i \geq 2$. The coefficients of expansion $B_2(T)$, $B_3(T)$ $\cdots$ etc. are called the **virial coefficients.**

In addition to the ideal gas law, the *van der Waals*[14] *equation of state* is the other most frequently studied equation of state for a gas system. The equation is

$$\left(P + \frac{a}{v^2}\right)(v - b) = RT; \quad a > 0, \;\; b > 0. \tag{1.35}$$

In the equation, a and b are constants, which may have different values for different gases. In many applications, we may consider this as the

[13] Count A. Avogadro, Italian physicist (1776–1856).

[14] Johannes D. van der Waals, Dutch physicist (1837–1923), Nobel prize laureate in physics in 1910.

equation of state for a real gas. It can be used to study the phenomenon of gas-liquid coexistence in phase transitions. Equation (1.35) may be considered as being obtained from the virial expansion. If we take $B_2 = b - a/RT$, $B_3 = 0$, $B_4 = ab^2/RT$, and $B_i = 0$ for $i \geq 5$, and substitute them into Eq. (1.34), we have

$$\begin{aligned} Pv &= RT\left[1 + \left(b - \frac{a}{RT}\right)\frac{1}{v} + \left(\frac{ab^2}{RT}\right)\frac{1}{v^3}\right] \\ &= RT\left(1 - \frac{a}{RT}\frac{1}{v} + \frac{ab}{RT}\frac{1}{v^2}\right)\left(1 + \frac{b}{v}\right). \end{aligned} \tag{1.36}$$

This can be rewritten as

$$Pv\left(1 + \frac{b}{v}\right)^{-1} = RT - \frac{a}{v} + \frac{ab}{v^2} = RT - \frac{a}{v}\left(1 - \frac{b}{v}\right). \tag{1.37}$$

Under the condition $0 < b \ll v$, $(1+b/v)^{-1} \approx 1 - b/v$, after a rearrangement of the terms, Eq. (1.37) becomes the van der Waals equation of state (1.35). If we define the two new variables, P' and v', as

$$P' = P + \frac{a}{v^2}, \quad v' = v - b, \tag{1.38}$$

then Eq. (1.35) becomes

$$P'v' = RT. \tag{1.39}$$

If we interpret P' and v' as the new pressure and the new molar volume, respectively, then P' and v' satisfy the ideal gas equation of state. Therefore we may consider that P' and v' are the pressure and the volume of an ideal gas, which is derived from the *original* real gas system. We note that the *real* pressure P is smaller than P' $(a > 0)$. This is understandable since attractions between the molecules make the real gas pressure smaller than the corresponding ideal gas pressure. Since $b > 0$, this implies that v' (the volume available for the ideal gas) is smaller than v (volume available for the real gas). The reason for this is that in the real gas, the *free* volume available for a molecule to move without colliding with other molecules is smaller than the volume of the container V $(V = nv)$. Part of the volume V is occupied by the hard cores of the molecules. From the above analysis, we can see that the constant a in the van der Waals equation of state is related to the attractions between the molecules, and the other constant b represents the size of the hard core of a molecule.

1.13 Expansion Coefficient and Compressibility

One can rewrite the equation of state of a system $f(P,V,T) = 0$ as $V = V(T,P)$, i.e., V is a function of T and P. The temperature T and pressure P are controllable variables in the laboratory. Therefore, in principle, the function $V(T,P)$ can be determined experimentally. However in practice, it is more useful to measure experimentally the *expansion coefficient* (also called *expansivity*) and the *compressibility*. These two quantities determine the rate of change of the volume V as the temperature T or the pressure P changes.

From $V = V(T,P)$ we have

$$dV = \left(\frac{\partial V}{\partial T}\right)_P dT + \left(\frac{\partial V}{\partial P}\right)_T dP. \tag{1.40}$$

The **expansion coefficient** (or **expansivity**) β determines, when P is kept constant ($dP = 0$), the rate of change of V as T changes,

$$\beta = \frac{1}{V}\left(\frac{\partial V}{\partial T}\right)_P. \tag{1.41}$$

On the other hand, the **compressibility** κ determines, when T is kept constant ($dT = 0$), the rate of change of V as P changes,

$$\kappa = -\frac{1}{V}\left(\frac{\partial V}{\partial P}\right)_T. \tag{1.42}$$

Therefore, Eq. (1.40) can be written as

$$dV = \beta V dT - \kappa V dP. \tag{1.43}$$

There are two points in Eq. (1.42) which are worthwhile mentioning. (1) There is a *minus* sign in front of the partial differentiation, which makes κ a positive quantity. The reason is that for a thermodynamic system to be stable, the volume of the system should decrease as the pressure increases.[15] (2) The compressibility defined in Eq. (1.42) under the condition of constant temperature is called the *isothermal compressibility*. However, the compressibility may also be measured adiabatically; this is called *adiabatic compressibility*, which is denoted as κ_S. This means that

$$\kappa_S = -\frac{1}{V}\left(\frac{\partial V}{\partial P}\right)_S, \tag{1.44}$$

[15] We will prove this statement in Chapter 5.

where S denotes the entropy, which will be defined in Chapter 2. The subscript S of the partial differentiation means that during the differentiation S is kept constant, which is the condition for an adiabatic process. There is a relation between the isothermal compressibility κ and the adiabatic compressibility κ_S, which will be discussed in Chapter 4.

When the equation of state for a system is known, for example the ideal gas, we can use Eqs. (1.41)–(1.42) to calculate the theoretical expansion coefficient and the isothermal compressibilty. From the ideal gas equation of state $V = nRT/P$, we obtain the expansion coefficient

$$\beta = \frac{1}{V}\left(\frac{nR}{P}\right) = \frac{1}{T}, \tag{1.45}$$

and the isothermal compressibility

$$\kappa = -\frac{1}{V}\left(\frac{-nRT}{P^2}\right) = \frac{1}{P}. \tag{1.46}$$

However for systems for which the equation of state is unknown, we can use the experimental values of β and κ to determine the equation of state. For example, for some dilute real gases, the equation of state can be expressed as the form of a virial expansion. Usually only a small number of the expansion terms are non-negligible, and higher order terms may be neglected. The lower order virial coefficients may be determined by the experimental values of β and κ and some other properties of the system. The equation of state may then be determined.

For liquids and solids, it is not easy to obtain the complete form of $V = V(T, P)$. In general both β and κ are functions of T and P, thus if one wants to obtain the complete form of $V = V(T, P)$ by integrating Eq. (1.43), one needs a very large amount of experimental data for β and κ. It is not practical to do this. In practice, one usually only needs to know the equation of state for the small range of T and P of interest. If the range is small enough, then we may consider that both β and κ are constants, independent of T and P. If one is interested in the properties of the system near $T = T_0$ and $P = P_0$, then by integrating Eq. (1.43), one obtains

$$V = \int dV = V_0[1 + \beta(T - T_0) - \kappa(P - P_0)], \tag{1.47}$$

where $V_0 = V(T_0, P_0)$, and β and κ are considered as constants. This may be considered as an approximate equation of state for a liquid or a solid for temperatures T that do not deviate too much from T_0, and pressures P that do not deviate too much from P_0.

1.14 Problems

1.1. A cylindrical container is divided into two parts A and B by a wall between them. Both A and B contain gas molecules. Describe what kind of the dividing wall can make the gases in A and B maintain the following situations: (a) in mechanical but not thermal equilibrium; (b) in thermal but not mechanical equilibrium; (c) in thermodynamic equilibrium.

1.2. The initial state of n kilomoles of an ideal gas is at temperature T_1 and pressure P_1. The gas is then compressed reversibly against a piston to a volume equal to one-half of its original volume. The temperature of the gas is varied during the compression so that at each instant the relation $P = AV$ is satisfied; A is a constant.

(a) Express A in terms of P_1, T_1, n and the gas constant R.

(b) Draw a diagram of the process in the P-V plane.

(c) Find the final temperature T_2 in terms of T_1.

(d) Find the work done on the gas in terms of n, R, and T_1.

1.3. The temperature of an ideal gas at an initial pressure P_1 and volume V_1 is increased at constant volume until the pressure is doubled. The gas is then expanded isothermally until the pressure drops to its original value, where it is compressed at constant pressure until the volume returns to its initial value.

(a) Sketch these processes in the P-V plane and the P-T plane.

(b) Compute the work in each process and the net work done in the cycle if $n = 2$ moles, $P_1 = 1$ atm and $V_1 = 0.04$ m^3.

1.4. An ideal gas is contained in a cylindrical container with a movable piston. Initially, the volume of the gas is 0.1 m^3, the pressure is 2×10^6 N m^{-2}, and the temperature is 300 K.

(a) The gas is expanded isothermally until the final pressure is reduced to 1×10^6 N m^{-2}. Find the final volume of the gas.

(b) The piston is held fixed at its initial position and the pressure is reduced to 1×10^6 N m^{-2}. Find the final temperature.

(c) The system starts from its initial conditions, expands isothermally until the pressure is 1.5×10^6 N m^{-2}, and then it is cooled at a constant volume until the pressure is 1×10^6 N m^{-2}. Find the final temperature and volume of the gas.

(d) The system starts from its initial conditions, is cooled at a constant volume until the pressure is 1.5×10^6 N m^{-2}, and this is followed by an isothermal expansion until the pressure is 1×10^6 N m^{-2}. Find the final temperature and volume of the gas.

(e) Plot each of the above processes on a T-V diagram.

1.5. The initial state of a monatomic ideal gas is $P_0 = 1$ atm, $T_0 = 273$ K, and $V_0 = 10^{-3}$ m^3. The gas is isothermally expanded to $V_1 = 2\,V_0$ and is then cooled at constant pressure to a volume V. The volume V is such that a reversible adiabatic compression to the pressure P_0 returns the system to its initial state. All of the changes are conducted reversibly. Calculate the value of V and the total work done on or by the gas.

1.6. One mole of N_2 gas is contained at $T = 273$ K and $P = 1$ atm. The addition of 3000 J of heat to the gas at constant pressure causes 832 J of work to be done during the expansion. Calculate

(a) the final state of the gas;

(b) the values of ΔU and ΔH for the change of state; and

(c) the values of c_v and c_p for N_2.

Assume that nitrogen behaves as an ideal gas, and that the above changes of state is conducted reversibly.

1.7. For a diatomic ideal gas near room temperature, what fraction of the heat absorbed is available for doing external work if the gas expands at a constant pressure? What is the result if the gas expands at a constant temperature?

1.8. Show that the work done by a gas in an arbitrary process can be expressed as

$$đW = PV\beta\,dT - PV\kappa\,dP.$$

What is this expression for an ideal gas?

1.9. The equation of state of a certain gas is $(P+b)v = RT$, its specific internal energy is given by $u = aT + bv + u_0$. (a) Find c_v. (b) Show that $c_P - c_v = R$.

1.10. Suppose the molar internal energy of a van der Waals gas is given by $u = cT - a/v + u_0$, where a is one of the constants in the equation of state, and c and u_0 are constants. Calculate the molar specific heat capacities c_v and c_P.

1.11. The constant pressure molar specific heat of most substances (except at very low temperatures) can be satisfactorily expressed by the empirical formula

$$c_P = a + 2bT - cT^{-2},$$

where a, b and c are constants and T is measured in kelvin.

(a) Find the heat required to raise the temperature of n moles of the substance at constant pressure from T_1 to T_2. Express your answer in terms of a, b, c, T_1, and T_2.

(b) Find the mean heat capacity per mole between T_1 and T_2.

1.12. Show that $\left(\frac{\partial h}{\partial P}\right)_T = -c_P \left(\frac{\partial T}{\partial P}\right)_h$.

1.13. Show that $\left(\frac{\partial u}{\partial T}\right)_P = c_P - P\beta v$.

1.14. Show that $\left(\frac{\partial u}{\partial P}\right)_T = P\kappa v - (c_P - c_v)\frac{\kappa}{\beta}$.

1.15. Show that the following relations hold for a reversible adiabatic expansion of an ideal gas:
(a) $TV^{\gamma-1} = \text{constant}$, (b) $TP^{(1/\gamma)-1} = \text{constant}$.

1.16. One liter of air at a pressure of 1 atm is pumped into a bicycle tire. The final pressure of the tire is 7 atm. Consider this process as a reversible adiabatic process and the air molecules as diatomic.

(a) What is the final volume of the air in the tire?

(b) How much work is done in compressing the air?

(c) If the temperature of the air is initially 300 K, what is the temperature after compression?

1.17. Two identical bubbles of gas, A and B, are formed at the bottom of a lake and then rise to the surface. Suppose the pressure at the bottom of the lake is twice that of the surface, and the temperature of the water in the lake is a constant independent of the depth. Consider the situation where bubble A rises so rapidly that no heat is exchanged between it and the water, while bubble B rises so slowly that it always remains in thermodynamic equilibrium with the water. Find the ratio of the final volume (just beneath the surface) of A and B, assuming that A undergoes a reversible adiabatic process and B undergoes a reversible isothermal process. The gas molecules are assumed to be diatomic.

1.18. (a) Show that the expansion coefficient β can be expressed as

$$\beta = -\frac{1}{\rho}\left(\frac{\partial \rho}{\partial T}\right)_P,$$

where $\rho = 1/v$ is the density (v is the molar volume).

(b) Show that the isothermal compressibility κ can be written as

$$\kappa = \frac{1}{\rho}\left(\frac{\partial \rho}{\partial P}\right)_T.$$

1.19. Show that the expansion coefficient of a van der Waals gas is

$$\beta = \frac{Rv^2(v-b)}{RTv^3 - 2a(v-b)^2}.$$

Show that this reduces to the ideal gas result when $a = b = 0$.

1.20. Show that the isothermal compressibility of a van der Waals gas is

$$\kappa = \frac{v^2(v-b)^2}{RTv^3 - 2a(v-b)^2}.$$

Show that this reduces to the ideal gas result when $a = b = 0$.

1.21. A hypothetical substance has an expansivity of $\beta = 2bT/v$ and an isothermal compressibility of $\kappa = a/v$, where a and b are constants. Show that the equation of state is $v - bT^2 + aP =$constant.

1.22. A hypothetical substance has an expansivity of $\beta = bT^2/P$ and an isothermal compressibility of $\kappa = aT^3/P^2$, where a and b are constants. Find the equation of state of the substance and the ratio a/b.

1.23. Show that, in general,

$$\left(\frac{\partial \beta}{\partial P}\right)_T + \left(\frac{\partial \kappa}{\partial T}\right)_P = 0,$$

where β and κ are the expansivity and the isothermal compressibility, respectively.

1.24. If the equation of state is given in the form $P = P(T, V)$, show that the following relation holds:

$$\left(\frac{\partial P}{\partial T}\right)_V = \frac{\beta}{\kappa},$$

where β and κ are the expansion coefficient and the isothermal compressibility, respectively. From this, show that

$$dP = \frac{1}{\kappa}\left(\beta\, dT - \frac{dV}{V}\right).$$

Chapter 2

Entropy and the Second Law

2.1 Introduction

In Chapter 1 we introduced the zeroth and the first law of thermodynamics. The zeroth law defines a state variable the *temperature*, the most important physical quantity in thermodynamics. The first law tells us that *heat* is a form of energy; therefore the first law is a statement of the conservation of energy. Experiences tell us, however, that the conservation of energy is not the only law which governs the behavior of all the thermodynamic systems. There are many processes which would conserve energy but they never happen in nature. For example, if we put a glass of hot water in the colder atmosphere, after a few moments, the temperature of the water will decrease, eventually it will become equal to that of the air. The water loses heat to the air, and energy is conserved. However we have never seen the reverse process (the water in the glass gets hotter than the air) happen without extra effort (e.g., by externally heating the water). Even though the reverse process (water absorbs heat from the air *spontaneously*) does not violate the energy conservation law, it never happens. There are many phenomena like this in nature. In these phenomena an isolated system (air plus water in the above example) starts from an *initial* state and undergoes a process to reach a *final* state, while the reverse process, the system beginning from the *final* state and returning to the *initial* state, with the same external conditions, never happens. Such a phenomenon is called an *irreversible process*. The phenomena of irreversible processes are very diversified and complex. It is hard to imagine that there could exist a simple principle to tell whether a process is reversible or irreversible. The second law of thermodynamics tells us that such a simple principle does exist; we will proceed to discuss this subject in this chapter.

The second law of thermodynamics defines a new state variable, known as *entropy*. Many students who have studied thermodynamics do not quite understand the role entropy plays. Some of the most puzzling questions are: Why entropy should be defined in thermodynamics? What is the physical meaning of entropy? However, if one tries to understand the necessity of introducing a new state variable (i.e., entropy) from the mathematical point of view, the reason is very clear. In the first law we have seen that the infinitesimal heat $đQ$ is an inexact differential. In mathematics it can be proved that any inexact differential with two independent variables can always be converted to an exact differential by multiplying by a suitable integrating factor. The inexact differential $đQ$ thus can be converted to an exact differential, which defines a new function (i.e., a state variable) now known as *entropy*.

In this chapter we first define the state variable entropy mathematically, the procedure also defines a *thermodynamic temperature*. The thermodynamic temperature defined in this way is just an assumption, it is not easy to find its relation with the practical temperature scales which are used in physics or daily life. Therefore we need to derive the entropy function *thermodynamically*, which will enable us to prove that the thermodynamic temperature is identical to the ideal gas temperature. One of the important features of these two temperature scales is that they have the *same zero point*, this fact plays a very important role in thermodynamics. A new name, the *absolute temperature* is given to represent these two temperatures.

It is understandable that it is not easy to understand the physical meaning of entropy from the macroscopic point of view, as its existence is due to a mathematical theorem. Most of the physical quantities we know are defined physically, which means that these quantities can be observed and/or measured by physical means, therefore there is little difficulty in understanding their physical meanings. Thus we may say that entropy is defined mathematically but not physically. We need to give a physical interpretation of this term, otherwise we may just know its mathematical definition, but still do not understand its physical content. In order to fill this gap, which can not be found from macroscopic thermodynamics, we will give a microscopic interpretation of entropy in the next chapter.

2.2 Entropy

It can be proved in mathematics that for an inexact differential with two independent variables one can always find an integrating factor which converts the inexact differential to an exact differential. Multiplying the inexact differential by the integrating factor, the product will be an exact differential. For example, the differential $ydx - xdy = đg$, mentioned in Chapter 1, is an inexact differential. The function $g(x, y)$ does not exist. However if we multiply $đg$ by $1/y^2$, then we have

$$\frac{1}{y^2}\, đg = \frac{ydx - xdy}{y^2} = d\left(\frac{x}{y}\right), \tag{2.1}$$

which is an exact differential. Therefore $1/y^2$ is an integrating factor for $đg$.

We apply this theorem to $đQ$ and $đW$ in thermodynamics. In PVT systems, $đW = PdV$, therefore $1/P$ is the integrating factor for $đW$. Thus $đW/P = dV$ is an exact differential, i.e., V is a state variable. Due to the fact that both P and V are frequently used state variables in thermodynamics, there is nothing new required to converting $đW$ to an exact differential. The next problem is how to convert $đQ$ to an exact differential. Apparently one has to define a new state variable which corresponds to the exact differential derived from $đQ$. Another problem is what is the integrating factor? If we choose P and T as the two independent variables, it is reasonable to assume that the integrating factor is a function of T only. This is reasonable since $đQ$ is *thermal* in nature, and the integrating factor should not depend on the *mechanical* variable P. This is just like $đW$ being *mechanical* in nature, and its integrating factor is $1/P$, which is independent of T. Therefore we may assume that the integrating factor for $đQ$ is of the form $1/F(T)$, which is a function of T only. Here we may choose T as the ideal gas temperature, while $F(T)$ is an unknown function of T. For simplicity, we may define a **thermodynamic temperature**[1] $T^{\rm th} \equiv F(T)$. After multiplying by the integrating factor, $đQ$ becomes an exact differential dS, which defines a new state variable *entropy* S. The mathematical formulation

[1]Because $T^{\rm th} = F(T)$ is a formula for transforming from one temperature scale T to another $T^{\rm th}$, therefore $F(T)$ is required to be a single-valued function, which satisfies $F(T_2) > F(T_1)$, if $T_2 > T_1$. This means that for each value of T, there corresponds only one $F(T)$, and the relative magnitude of T_1 and T_2 will not change after the transformation.

can be expressed as[2]

$$dS = \frac{đQ}{T^{\text{th}}} \text{ (reversible process).} \tag{2.2}$$

This is the definition of the new state variable **entropy** S. Note that this definition applies to *reversible processes* only. The reason for this restriction is that, in a reversible process, the quantity $đQ$ is well defined, and all the state variables are also well defined in the process (the system remains in the equilibrium state). For irreversible processes, either $đQ$ or the state variables are not well defined. Therefore Eq. (2.2) does not hold. However, for an irreversible process which is quasi-static but with dissipative work, Eq. (2.2) may be applicable, if the dissipative heat involved is known. In this case, if the dissipative heat is included in $đQ$, then Eq. (2.2) still holds. This is because the system cannot distinguish the heat rejected by its surroundings and the dissipative heat involved in the process. Note, however, that only when the system absorbs heat should the dissipative heat be included in $đQ$; if the system rejects heat then the dissipative heat should not be included.[3] We will come back to this point in Sec. 2.14. For *non-quasi-static irreversible processes*, Eq. (2.2) does not hold. The reason is quite simple. In this case, the system undergoes a period of non-equilibrium states; all the state variables are not well defined in the process. It is meaningless to define a new state variable in this situation. For a *finite reversible isothermal* process, the entropy change for a system is

$$\Delta S = \frac{\Delta Q}{T^{\text{th}}}. \tag{2.3}$$

The above equation may also hold for a *quasi-static but irreversible isothermal process, if the dissipative heat is properly included.*

In Eq. (2.2) above we defined a new state variable entropy S and a new temperature, the thermodynamic temperature T^{th}. We obtained this result by a purely mathematical consideration. However it is not

[2]For thermodynamic systems with more than two independent variables, it can be shown that an integrating factor for $đQ$ always exists if the condition of *adiabatic inaccessibility* is satisfied. The entropy function S can then be defined as in Eq. (2.2) with T^{th} being connected with Kelvin temperature. This is known as the principle of Carathéodory. For an explanation of the principle of Carathéodory, the reader is referred to the book "*Heat and Thermodynamics*" 7th ed., by M. W. Zemansky and R. H. Dittman, McGraw-Hill, New York, 1997.

[3]The reason is that the dissipative heat is not rejected by the system which rejects heat, but it is absorbed by the system which absorbs heat.

easy to know how the thermodynamic temperature T^{th} is related to the ideal gas temperature T if no thermodynamics is involved. Therefore in the following sections of the present chapter, one of the main tasks is to define entropy *thermodynamically*. Only when we can do this, will we be able to find the relation between the thermodynamic temperature T^{th} and the ideal gas temperature T. We will show in Sec. 2.9 that we can choose T^{th} to be equal to T, thus there is no distinction between T^{th} and T. However in the following sections, before T^{th} is identified, the symbol T denotes the ideal gas temperature. After T^{th} is proved to be identical to T, the symbol T will be given a new name, **the absolute temperature.**

2.3 Spontaneous or Irreversible Processes

We consider an isolated system [4] and let it alone without any external disturbance for a sufficiently long period of time. There are two possibilities:

1. The state of the system remains in its initial state.
2. The system evolves to a *new* equilibrium state.

Case 1 indicates that the initial state of the system is an *equilibrium* state. Therefore the state of the system will not change with time, which is the definition of an equilibrium state. Case 2 is the situation we call a spontaneous process, which is an irreversible process. Here we give a simple definition for an irreversible process.

Consider a system which starts from an initial state and undergoes a process to reach a *different* final state. If, *by only infinitesimal changes of the external conditions*, the system cannot reverse the process and return to its initial state, then the process is called an irreversible process.

Here, the term *by only infinitesimal changes of the external conditions*, means that the external conditions are exactly the same as the process which brings the system from the initial to the final state, except that there may be some *infinitesimal* changes. All spontaneous processes are irreversible processes. Examples are: the mixing of two (or more) different kinds of gases; the free expansion of a gas; the thermal contact of two bodies which initially have different temperatures. In all these spontaneous irreversible processes, there is a common and

[4] An isolated system may consist of two open or closed systems, which can exchange energy and/or mass with each other, but the combined system is isolated.

important feature. If a system spontaneously goes from an initial state to a *different* final state, it must experience a period of *non-equilibrium* processes. This implies that the initial state is a non-equilibrium state, otherwise the system would remain in the *equilibrium* initial state without changing. In most cases, the initial state involves the thermal (or thermodynamic) contact of two isolated systems, which are in their respective equilibrium states. *The combined system is an isolated system, but after contact the two subsystems are no longer isolated and can exchange energy and/or mass with each other spontaneously.* We have mentioned in Sec. 1.11 that a reversible process must meet the following two requirements: (1) the process proceeds very slowly so that at any moment the system can be considered as being in equilibrium; this is known as a quasi-static process; and (2) no dissipative work is involved. In addition to spontaneous processes, most of the non-spontaneous processes are also irreversible. A non-spontaneous process is a process which involves external forces. Generally external forces are accompanied by friction and dissipative work. Therefore we can see that most of the processes in nature are irreversible. Reversible processes constitute only a small part of the processes in the universe.[5]

2.4 Heat Engines and Refrigerators

A device which can convert heat into mechanical energy (which can do work on the surroundings) is called a *heat engine*, or just an engine. Consider an engine M as shown in Fig. 2.1. After a complete cycle of the operation, M absorbs heat Q from the heat source at temperature T_1 and then does a total amount of work W to an object. No other systems are involved in this process. From the conservation of energy we have $W = Q$, because after a complete cycle, the energy of M does not change, and no other systems are involved. This result indicates that the engine M absorbs heat Q from a heat source and converts 100% of it into work W. We call this a *perfect engine.* However a perfect engine does not exist, because the work W can be done on another heat source at temperature T_2. Since work can be 100% converted into heat, a thermal energy $Q = W$ is added to the heat source T_2. The net effect of the

[5]One of the most familiar examples of a reversible processes is the phase transition. For example, ice can melt to water by absorbing heat from the surroundings. The reverse process is possible if there is an infinitesimal decrease in the temperature of the surroundings, and water will reject the same amount of heat to the surroundings to condense to ice.

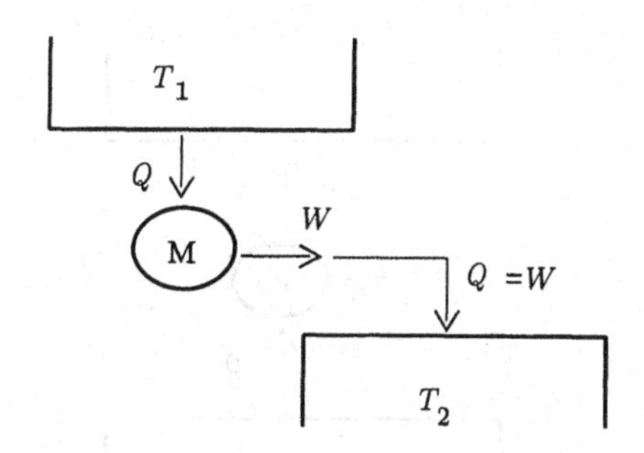

Figure 2.1: A perfect engine M absorbs heat Q from the heat reservoir at T_1, and converts Q completely into work W. Work W can then be completely converted into heat Q, which can be put into any heat reservoir at temperature T_2.

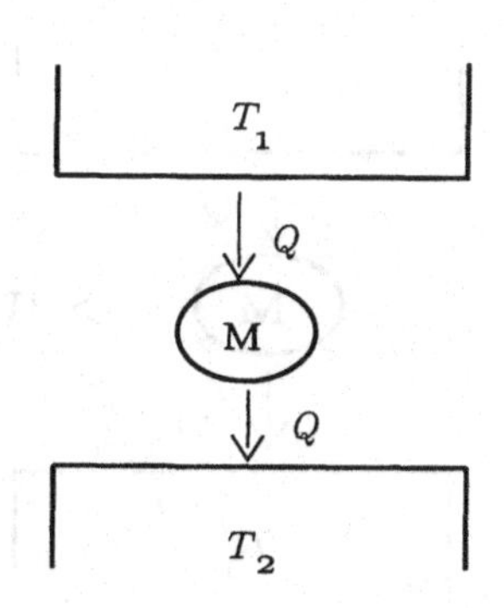

Figure 2.2: An engine M absorbs heat Q from a lower temperature reservoir at T_1 and puts it into a higher temperature reservoir at T_2.

engine M is therefore to remove an amount of heat Q from a heat source at temperature T_1 and put it into another heat source at temperature T_2, as shown in Fig. 2.2. Suppose $T_2 > T_1$, then the above result shows that it is possible for heat Q ($Q > 0$) to flow from a lower temperature T_1 to a higher temperature T_2 with no changes of the surroundings.[6] This violates the well-known phenomenon that heat always flows from a place with a higher temperature to a place with a lower temperature. Therefore, for a realistic engine we need two heat sources, one source at a higher temperature T_2, which supplies heat, and the other at a lower temperature T_1 ($T_1 < T_2$), which absorbs the unused heat, as shown in Fig. 2.3. From the conservation of energy we have

$$W = Q_2 - Q_1 \quad (Q_2 > 0,\ Q_1 > 0). \tag{2.4}$$

We are considering an engine which does *positive* work on an object outside of the engine, therefore we should have $W > 0$ and $Q_2 > Q_1 > 0$. The efficiency of the engine η, defined below, is always smaller than 1,

$$\eta \equiv \frac{W}{Q_2} = 1 - \frac{Q_1}{Q_2} < 1. \tag{2.5}$$

[6]After a complete cycle, the engine M returns to its original state, and no other systems are involved in the process.

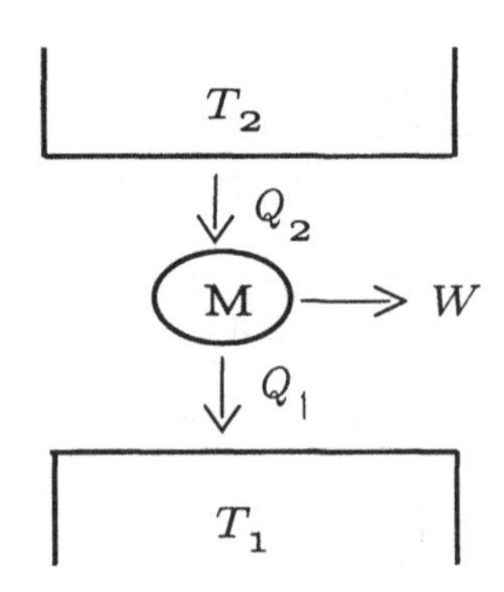

Figure 2.3: A realistic engine M, which absorbs heat Q_2 from a heat reservoir at T_2, does work $W = Q_2 - Q_1$ on an object, and rejects the unused heat Q_1 to a heat reservoir T_1 ($T_2 > T_1$).

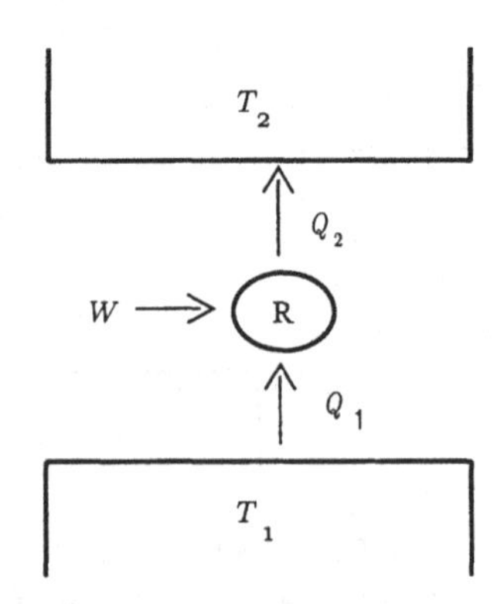

Figure 2.4: A refrigerator R absorbs heat Q_1 from a lower temperature reservoir at T_1 and puts heat Q_2 into a higher temperature reservoir at T_2. An external work $W = Q_2 - Q_1$ is done on R.

For a perfect engine, as mentioned above, the efficiency is $\eta = 1$, because $Q_1 = 0$. A perfect engine is an engine which can convert 100% of its heat into mechanical work.

A *refrigerator* is a machine which may be thought of as the inverse operation of a heat engine. An external work W is *done on* the refrigerator R, which removes heat Q_1 from a lower temperature reservoir T_1, and an amount of heat Q_2 is put into a higher temperature reservoir T_2 ($T_2 > T_1$), as shown in Fig. 2.4. Here we assume that W, Q_1, and Q_2 are all positive quantities. From the conservation of energy we have $Q_2 = Q_1 + W$, which implies that the heat absorbed by the higher temperature reservoir is greater than that extracted from the lower temperature reservoir. A refrigerator is said to have a better efficiency when the ratio of the heat removed Q_1 to the input work W is larger. However, unlike an engine whose efficiency is always less than 1, this ratio may be greater than 1. We call this ratio the *coefficient of performance* c (rather than the efficiency), which is defined as

$$c = \frac{Q_1}{W} = \frac{Q_1}{Q_2 - Q_1}. \tag{2.6}$$

When the heat absorbed by the higher temperature reservoir Q_2 is not too much larger than the heat removed from the lower temperature reservoir Q_1, the coefficient of performance c of the refrigerator will be large. And, in fact, it may be much larger than 100%. Note that when $W = 0$, $c \to \infty$, but this is impossible, because if $W = 0$, the net effect of the refrigerator is that heat flows from a colder body to a hotter body. Therefore, c is a positive quantity; it may be smaller, or greater than 1, but it is always finite.

2.5 The Second Law of Thermodynamics

In terms of the definition of a heat engine, we have the following two statements for the second law of thermodynamics:

1. The **Clausius**[7] **statement:**

   It is impossible to construct a device that operates in a complete cycle and whose sole effect is to transfer heat from a place at lower temperature to a place at higher temperature.

2. The **Kelvin-Planck**[8] **statement:**

   It is impossible to construct a perfect heat engine, which can completely convert heat into mechanical energy with no other effect.

We have proved in the preceding section that if the Kelvin-Planck statement were false (i.e., a perfect engine does exist), then the Clausius statement would also be false (i.e., heat can flow from a lower temperature to a higher temperature). Now we want to prove that if the Clausius statement were false, then the Kelvin-Planck statement would also be false. This means that these two statements are equivalent, and it is sufficient to keep just one of them. There is nothing new in the other statement. As shown in Fig. 2.5(a), we have two engines M and M′ which are operating between the same two heat reservoirs T_1 and T_2, with $T_2 > T_1$. Suppose the engine M is an engine which violates the Clausius statement; thus after one cycle, it can transfer heat $Q_1 > 0$ from the colder reservoir T_1 and put it in the hotter reservoir T_2, with no other effect (that is $Q_2 = Q_1$). The engine M′ is an ordinary engine

[7]Rudolph J. E. Clausius, German physicist (1822–1888).

[8]Lord Kelvin, originally William Thomson, British physicist (1824–1907); Max K. E. L. Planck, German physicist (1858–1947), Nobel prize laureate in physics in 1918.

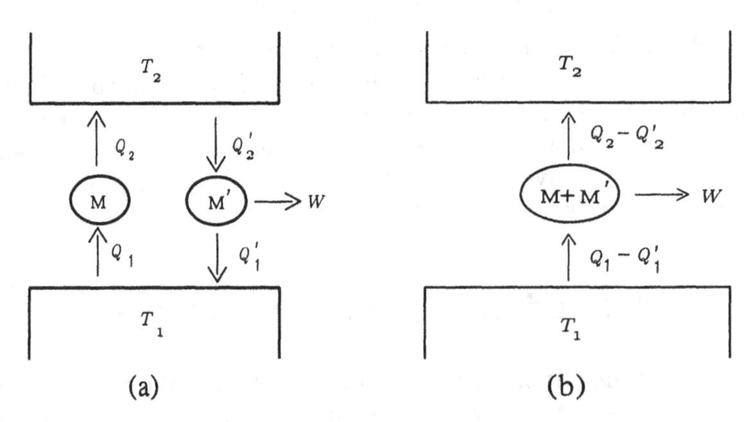

Figure 2.5: An ordinary engine M′ and an engine M which violates the Clausius statement are combined to operate between two heat reservoirs T_1 and T_2, with $T_2 > T_1$. (a) M and M′ operate separately. (b) M and M′ operate jointly.

which, after one cycle, absorbs heat Q_2' from the hotter reservoir T_2, and does work $W' = Q_2' - Q_1'$ on an object, rejecting the unused heat Q_1' to the colder reservoir T_1. We can adjust the sizes of M and M′, such that $Q_2 = Q_2'$. We may consider the joint operations of M and M′ as the operation of a single engine M+M′, as shown in Fig. 2.5(b). We easily see that the combined engine M+M′ and the reservoir T_2 have no heat exchange or transfer, because $Q_2 - Q_2' = 0$. Only the reservoir T_1 transfer a positive heat[9] $Q_1 - Q_1' > 0$ to the engine M+M′. The engine M+M′ converts 100% of this heat into mechanical work $W = Q_1 - Q_1' > 0$. This violates the Kelvin-Planck statement.

In the above, we have proved that the Clausius statement and the Kelvin-Planck statement of the second law of thermodynamics are equivalent. This means that, if one of the statements were false, then the other one would also be false. Or we can say it another way that, if one of the statements is true, then the other one is also true. However, these two statements concern only the restrictions on the efficiency of an engine and the direction of heat flow, which cover only a very limited type of phenomena in the universe. In the following sections, we will generalize the statement of the second law to a very general form, which can be

[9]That $Q_1 - Q_1' > 0$ is due to $Q_1 = Q_2 = Q_2'$ and $Q_2' - Q_1' = W' > 0$, therefore $Q_1 - Q_1' = Q_2' - Q_1' > 0$.

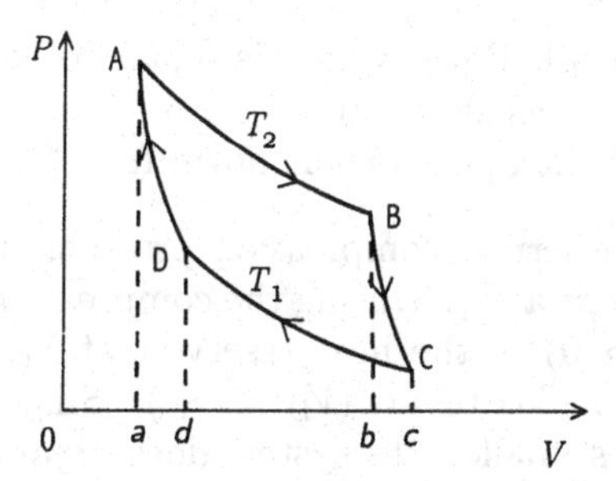

Figure 2.6: A Carnot cycle.

applied to almost all phenomena in the universe. This is possible only after the new state variable, *entropy*, is derived thermodynamically.

2.6 A Carnot Cycle and Carnot Engine

An engine which is operated and does work according to the **Carnot**[10] **cycle** is called a *Carnot engine.* The most important characteristic of a Carnot cycle is that every process in the cycle is *reversible.* We use a *PVT* system as an example to illustrate the cycle. It consists of *two reversible isothermal processes* and *two reversible adiabatic processes*, as shown in Fig. 2.6. The system starts from an initial state A, then undergoes the above mentioned processes to states B, C, and D, and then returns back to A to complete a Carnot cycle. We describe the processes in detail in the following. Note that every process is *reversible.*

1. A→B is a process of isothermal expansion. The temperature[11] of the system is kept at T_2, and the volume expands from V_A to V_B ($V_B > V_A$). During the expansion, the system absorbs heat Q_2 ($Q_2 > 0$) from a heat reservoir[12] and does positive work W_{AB}. The magnitude of W_{AB} equals the area enclosed by the curves ABba, as shown in Fig. 2.6.

2. B→C is a process of adiabatic expansion. The system does not absorb (reject) heat during the expansion, thus $\Delta Q_{BC} = 0$, but the volume expands from V_B to V_C ($V_C > V_B$). Therefore the system

[10]N. L. Sadi Carnot, French engineer (1796–1832).

[11]Any temperature scale may be used, but here we use the ideal gas temperature.

[12]The system must be in thermal contact with a heat reservoir in order that the temperature can be kept constant.

does positive work $W_{\rm BC}$, which is equal to the area enclosed by the curves BCcb, as shown in Fig. 2.6. During the expansion the temperature of the system cools down from T_2 to T_1 ($T_1 < T_2$).

3. C→D is an isothermal compression process. The temperature of the system is kept at T_1. During the compression the system rejects heat Q_1 ($Q_1 > 0$) to the heat reservoir at T_1, and the volume is compressed from $V_{\rm C}$ to $V_{\rm D}$ ($V_{\rm D} < V_{\rm C}$). Since the volume of the system becomes smaller, the system does *negative* work (work done to the system) $W_{\rm CD}$. The magnitude of $W_{\rm CD}$ is equal to the area enclosed by CcdD, as shown in Fig. 2.6.

4. D→A is an adiabatic compression process. The system does not absorb (reject) heat during the compression, thus $\Delta Q_{\rm DA} = 0$, but the volume is compressed from $V_{\rm D}$ to $V_{\rm A}$ ($V_{\rm A} < V_{\rm D}$). The system returns to its initial state A to form a complete cycle. During the last process, the system does *negative* work $W_{\rm DA}$, whose magnitude is equal to the area enclosed by DdaA, as shown in Fig. 2.6.

When a system undergoes a complete Carnot cycle, (A→B→C→D →A), the net heat absorbed by the system is equal to $\Delta Q = Q_2 - Q_1$; the net work done *by* the system ΔW is given by (cf. Fig. 2.6)

$$\Delta W = W_{\rm AB} + W_{\rm BC} - |W_{\rm CD}| - |W_{\rm DA}| = \text{area ABCD}. \tag{2.7}$$

After a complete cycle, the system returns to its initial state A, therefore there is no change of the internal energy U, i.e., $\Delta U = 0$. From the first law of thermodynamics, Eq. (1.2), we have $\Delta Q = \Delta W \equiv W$, i.e., $W = Q_2 - Q_1$. The area enclosed by the curve ABCD is a positive area, therefore $Q_2 > Q_1 > 0$. The efficiency of the engine η is given by

$$\eta = \frac{W}{Q_2} = \frac{Q_2 - Q_1}{Q_2} = 1 - \frac{Q_1}{Q_2} < 1. \tag{2.8}$$

Here we would like to emphasize that *a Carnot cycle is a reversible cycle*, therefore the heat rejected (or absorbed) by the heat reservoir *is equal to* the heat absorbed (or rejected) by the system. However *for engines which operate in irreversible processes*, the heat rejected (or absorbed) by the heat reservoir *is not necessarily equal to* the heat absorbed (or rejected) by the system, because dissipative heat may be involved in the processes. This distinction is a key point in the derivation of the *Clausius inequality* which will be given in Sec. 2.12.

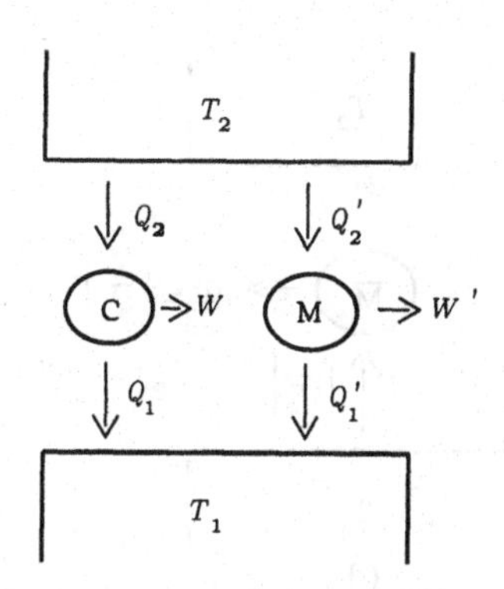

Figure 2.7: A Carnot engine C and some other engine M operating between two heat reservoirs T_1 and T_2.

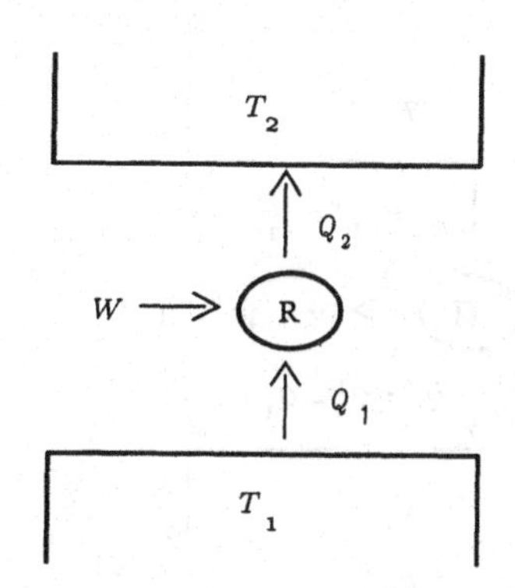

Figure 2.8: The operation of the Carnot engine C, shown in Fig. 2.7, is reversed to become a Carnot refrigerator R.

2.7 Carnot's Theorem

Carnot's theorem can be stated as follows:

Operating between the same two heat reservoirs, no engine can be more efficient than a Carnot engine.

We now give the proof. As shown in Fig. 2.7, we consider two engines C and M, both of which are operating between the same two heat reservoirs, T_1 and T_2 ($T_2 > T_1$). The engine C is a Carnot engine, and M is any other engine. The operation of C is such that it absorbs heat Q_2 from the heat reservoir T_2, does work $W = Q_2 - Q_1$, and then rejects the unused heat Q_1 to the reservoir T_1. Here, we take both Q_1 and Q_2 to be positive, and $Q_2 > Q_1 > 0$. The operation of M is exactly the same as that of C, except that the amount of heat and work are *primed* quantities. Thus it absorbs heat Q_2' from T_2, does work $W' = Q_2' - Q_1'$, and rejects heat Q_1' to T_1. We also have $Q_2' > Q_1' > 0$. Now we assume that the efficiency of M is greater than that of C, i.e., $\eta_M > \eta_C$. We will prove that this will lead to a result which violates the second law.

Since a Carnot engine is reversible, we can reverse its operation, then the Carnot engine C becomes a Carnot refrigerator R, as shown in Fig. 2.8. External work $|W_R|$ is done *on* R (i.e., $W_R = -W < 0$), which enables R to remove heat Q_1 ($Q_1 > 0$) from the lower temperature reservoir T_1 to the higher temperature reservoir T_2. Since the external

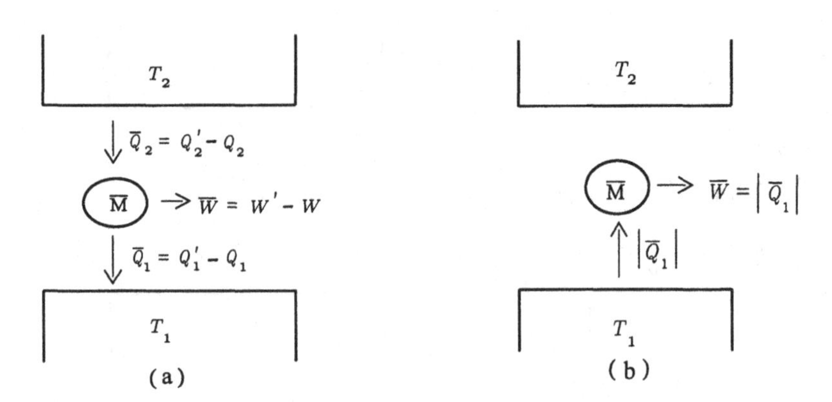

Figure 2.9: (a) A Carnot refrigerator R and an engine M operating jointly (R+M=$\overline{\mathrm{M}}$) between the heat reservoirs T_1 and T_2. (b) The condition $Q_2 = Q_2'$ is chosen for the combined engine $\overline{\mathrm{M}}$.

work is done *on* R, the heat absorbed by T_2 is greater than that removed from T_1, $Q_2 = Q_1 - W_{\mathrm{R}} = Q_1 + |W_{\mathrm{R}}| > Q_1$. We combine the operations of R and M together, which is equivalent to the operation of a single engine $\overline{\mathrm{M}}$, as shown in Fig. 2.9(a). The engine $\overline{\mathrm{M}}$ operates between the heat reservoirs T_1 and T_2 ($T_2 > T_1$). It absorbs heat $\overline{Q}_2 = Q_2' - Q_2$ from the reservoir T_2, does work $\overline{W} = W' + W_{\mathrm{R}} = W' - W$, and then rejects the unused heat $\overline{Q}_1 = Q_1' - Q_1$ to the reservoir T_1. We can adjust the relative sizes of R and M, such that $Q_2 = Q_2'$ and $\overline{Q}_2 = 0$. This implies that the engine $\overline{\mathrm{M}}$ and the reservoir T_2 are not related to each other. Also we have[13] $\overline{W} = -\overline{Q}_1$. Now we want to prove that, if $\eta_{\mathrm{M}} > \eta_{\mathrm{C}}$, then $\overline{W} > 0$ and $\overline{Q}_1 < 0$, as shown in Fig. 2.9(b). This figure shows that the engine $\overline{\mathrm{M}}$ absorbs heat $|\overline{Q}_1| > 0$ from the heat reservoir T_1, and converts it completely into mechanical work $\overline{W}(= |\overline{Q}_1|)$. This violates the Kelvin-Planck statement of the second law. The proof that $\overline{Q}_1 < 0$ (or $\overline{W} > 0$) is given in the following. From the condition

$$\eta_{\mathrm{C}} = 1 - \frac{Q_1}{Q_2} < \eta_{\mathrm{M}} = 1 - \frac{Q_1'}{Q_2'}, \tag{2.9}$$

together with the condition $Q_2 = Q_2'$ (by adjusting the relative sizes of R and M), gives

$$Q_1 > Q_1' > 0, \quad \text{and} \quad \overline{Q}_1 = Q_1' - Q_1 < 0. \tag{2.10}$$

[13] $\overline{W} = W' - W = Q_2' - Q_1' - (Q_2 - Q_1) = Q_1 - Q_1' = -\overline{Q}_1$.

This result implies that if $\eta_{\rm M} > \eta_{\rm C}$, then we will get an engine $\overline{\rm M}$, as shown in Fig. 2.9(b). However this engine can not exist because it violates the Kelvin-Planck statement. Therefore, among all engines, the Carnot engine is the most efficient one. The efficiency of any other engine M must be less than or at most equal to that of a Carnot engine, i.e., $\eta_{\rm M} \leq \eta_{\rm C}$.

If M is a reversible engine, then we can reverse the operation of M to become a refrigerator $\rm R_M$. Consider the joint operations of $\rm R_M$ and C. By the same argument as given above, we can prove that $\eta_{\rm C} \leq \eta_{\rm M}$. Therefore we get the following conclusion: Operating between the same two heat reservoirs, **all reversible engines have the same efficiency.** Moreover **the efficiency of any irreversible engine is less than that of a Carnot engine.**[14] Therefore any reversible engine may be considered as a Carnot engine.

2.8 Thermodynamic Temperature

In the previous section we have proved that, when operating between the same two heat reservoirs, all Carnot engines have the same efficiency. This implies that the efficiency of a Carnot engine is independent of the *working substance* of the engine. The efficiency depends only on the temperatures of the reservoirs T_1 and T_2. Therefore we have

$$\eta = \frac{Q_2 - Q_1}{Q_2} = 1 - \frac{Q_1}{Q_2} \equiv f'(T_1, T_2).$$

If we define $f(T_1, T_2) = 1 - f'(T_1, T_2)$, then the above equation can be rewritten as

$$\frac{Q_1}{Q_2} = f(T_1, T_2). \tag{2.11}$$

Now we consider three Carnot engines which use the same working substance and operate between the respective heat reservoirs (T_1, T_2), (T_2, T_3), and (T_1, T_3). Each Carnot engine consists of two isothermal processes, which needs two heat reservoirs, and two adiabatic processes, with an expansion process and a compression process. However the above three engines have only three reservoirs (at T_1, T_2, and T_3, respectively), which means that each of these reservoirs must serve for two of the engines. If we plot the three Carnot cycles on the same P-V diagram,

[14]If there is an engine M which has an efficiency equal to that of a Carnot engine, then by reversing the Carnot engine we can show that M must be reversible.

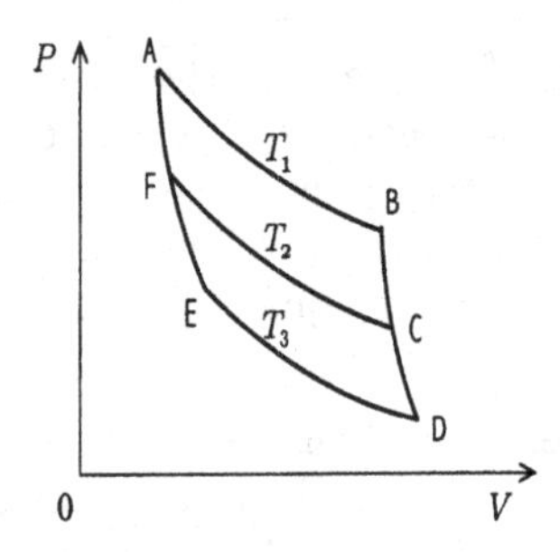

Figure 2.10: Three Carnot engines which operate between the respective heat reservoirs (T_1, T_2), (T_2, T_3) and (T_1, T_3).

there will be only three isothermal paths. Furthermore, if we let the three adiabatic expansion (compression) processes have the same path on the P-V diagram, then we will get the diagram shown in Fig. 2.10. In this figure, only six states of the working substance (A, B, C, D, E, F) are needed to specify the three Carnot cycles.[15] When the system undergoes the process A→B (constant temperature at T_1), it absorbs heat Q_1; in the process C→F (constant temperature at T_2) it rejects heat Q_2; and in the process D→E (constant temperature at T_3) it rejects heat Q_3. It should be noted that *here we take all the Q values to be positive.* From Eq. (2.11) we get

$$\frac{Q_1}{Q_2} = f(T_1, T_2); \quad \frac{Q_2}{Q_3} = f(T_2, T_3); \quad \frac{Q_1}{Q_3} = f(T_1, T_3), \tag{2.12}$$

therefore

$$\frac{Q_1}{Q_2} = \frac{Q_1}{Q_3} \times \frac{Q_3}{Q_2} = \frac{f(T_1, T_3)}{f(T_2, T_3)} = f(T_1, T_2). \tag{2.13}$$

From the last equality of the above equation, it is easy to see that the two variables T_1 and T_2 in the function $f(T_1, T_2)$ must be separable, thus

$$f(T_1, T_2) = F(T_1)F'(T_2) = \frac{F(T_1)}{F(T_2)} = \frac{Q_1}{Q_2}. \tag{2.14}$$

In the second equality, both the numerator and the denominator must have the same functional form for F, otherwise Eq. (2.13) will not hold.

[15] A Carnot cycle needs four states to specify the processes, therefore for three Carnot cycles one may need a maximum of twelve states.

Kelvin defined $F(T)$ as the thermodynamic temperature $T^{\rm th}$, i.e., $F(T) \equiv T^{\rm th}$, where T is the ideal gas temperature. Therefore Eq. (2.11) becomes

$$\frac{Q_1}{Q_2} = \frac{T_1^{\rm th}}{T_2^{\rm th}}. \tag{2.15}$$

The efficiency of a Carnot engine is then given by

$$\eta = 1 - \frac{Q_1}{Q_2} = 1 - \frac{T_1^{\rm th}}{T_2^{\rm th}}. \tag{2.16}$$

2.9 Ideal Gas Temperature vs. Thermodynamic Temperature

In the previous section we defined a thermodynamic temperature $T^{\rm th} = F(T)$, where T is the ideal gas temperature, but we do not know the functional form of $F(T)$. In this section we will show that $F(T) = cT$, where c is a constant. If we take $c = 1$, then $T^{\rm th} = T$, which means that the thermodynamic temperature and the ideal gas temperature are identical. Therefore we need use *only one* temperature scale T in thermodynamics. Moreover, no matter what value of c is taken, $T^{\rm th}$ and T have the same zero point, i.e., $T^{\rm th} = 0$ and $T = 0$ denote the same temperature, even when $c \neq 1$. We call $T = 0$ **absolute zero.** For convenience, we take $c = 1$ to make $T^{\rm th} = T$, and give this temperature scale T a new name: **absolute temperature.** The unit of the absolute temperature is K (kelvin). In terms of the *Celsius*[16] *temperature* scale (also known as the *centigrade temperature* scale), absolute zero is $-273.15°$C. Therefore the relation between these two temperature scales is

$$T(°\mathrm{C}) = T(K) + 273.15. \tag{2.17}$$

The proof of $F(T) = cT$ is given in the following. We use an ideal gas as the working substance of a Carnot engine. A complete Carnot cycle consists of A→B→C→D→A, as shown in Fig. 2.11.

1. A→B is a reversible isothermal process. During the process, the temperature is kept constant at T_2, and the volume expands from

[16] Anders Celsius, Swedish astronomer (1701–1744).

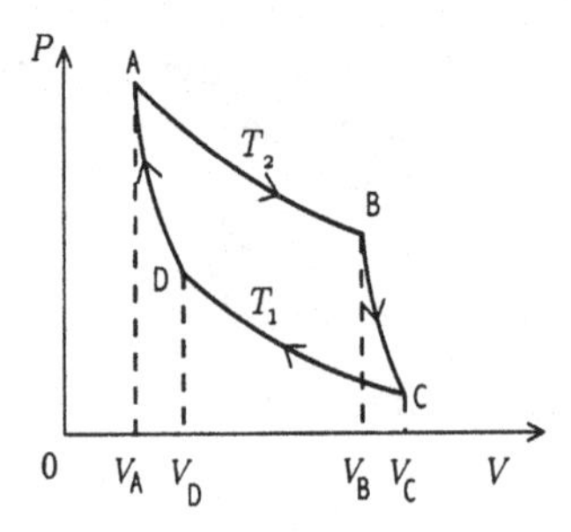

Figure 2.11: A Carnot cycle for an ideal gas.

V_A to V_B. Since[17] $\Delta U = 0$, $\Delta Q = \Delta W$; the heat Q_2 absorbed from the heat reservoir T_2 can be found from

$$Q_2 = W_{AB} = \int_{V_A}^{V_B} PdV = nRT_2 \int_{V_A}^{V_B} \frac{dV}{V} = nRT_2 \ln \frac{V_B}{V_A}. \quad (2.18)$$

2. B→C is a reversible adiabatic process, thus $\Delta Q = 0$. The volume expands from V_B to V_C and the temperature decreases from T_2 to T_1. The work done by the system W_{BC} is

$$W_{BC} = \Delta Q - \Delta U = -\Delta U = -n \int_{T_2}^{T_1} c_v dT. \quad (2.19)$$

3. C→D is a reversible isothermal process. The temperature is kept constant at T_1 and the volume is compressed from V_C to V_D. The system absorbs heat Q_1 from the heat reservoir T_1. Note that here we use the convention that if the system rejects heat to the reservoir $Q < 0$, and if the system absorbs heat from the reservoir $Q > 0$.

$$Q_1 = W_{CD} = \int_{V_C}^{V_D} PdV = nRT_1 \ln \frac{V_D}{V_C}. \quad (2.20)$$

4. D→A is a reversible adiabatic process, thus $\Delta Q = 0$. The volume is compressed from V_D to V_A, and the temperature increases from

[17]The internal energy of an ideal gas depends on the temperature only, therefore $dU = nc_v dT = 0$, if $dT = 0$.

T_1 to T_2. The work done by the system W_{DA} ($W_{\mathrm{DA}} < 0$ implies work is done *on* the system by external forces) is given by

$$W_{\mathrm{DA}} = \Delta Q - \Delta U = -\Delta U = -n \int_{T_1}^{T_2} c_v dT. \tag{2.21}$$

After a complete cycle, the work done *by* the system is

$$\begin{aligned} W_{\mathrm{total}} &= W_{\mathrm{AB}} + W_{\mathrm{BC}} + W_{\mathrm{CD}} + W_{\mathrm{DA}} \\ &= nRT_2 \ln \frac{V_{\mathrm{B}}}{V_{\mathrm{A}}} - n \int_{T_2}^{T_1} c_v dT + nRT_1 \ln \frac{V_{\mathrm{D}}}{V_{\mathrm{C}}} - n \int_{T_1}^{T_2} c_v dT \\ &= nRT_2 \ln \frac{V_{\mathrm{B}}}{V_{\mathrm{A}}} + nRT_1 \ln \frac{V_{\mathrm{D}}}{V_{\mathrm{C}}} \\ &= nR(T_2 - T_1) \ln \frac{V_{\mathrm{B}}}{V_{\mathrm{A}}}. \end{aligned} \tag{2.22}$$

Here we have used the relation $V_{\mathrm{C}}/V_{\mathrm{D}} = V_{\mathrm{B}}/V_{\mathrm{A}}$ which holds for an ideal gas operating in a Carnot cycle. The proof of this relation will be given as an exercise (Problem 2.4). Therefore the efficiency for a Carnot engine is

$$\begin{aligned} \eta &= \frac{W_{\mathrm{total}}}{Q_2} = \frac{nR(T_2 - T_1) \ln(V_{\mathrm{B}}/V_{\mathrm{A}})}{nRT_2 \ln(V_{\mathrm{B}}/V_{\mathrm{A}})} \\ &= \frac{nR(T_2 - T_1)}{nRT_2} = 1 - \frac{T_1}{T_2}. \end{aligned} \tag{2.23}$$

Comparing Eqs. (2.23) and (2.16) we get the following relation:

$$T^{\mathrm{th}} = cT, \quad c = \text{constant}. \tag{2.24}$$

By taking $c = 1$, we obtain the relation $T^{\mathrm{th}} = T$, which implies that the thermodynamic temperature is exactly the same as the ideal gas temperature.

2.10 Entropy Derived From the Second Law

At the beginning of this chapter we used a mathematical argument to assert that there exists a new state variable which is called *entropy*. In this section we will use a thermodynamic argument to show that a *new state variable* entropy does exist. We consider any reversible cyclic process which is plotted on the T-V diagram, as shown in Fig. 2.12. It is

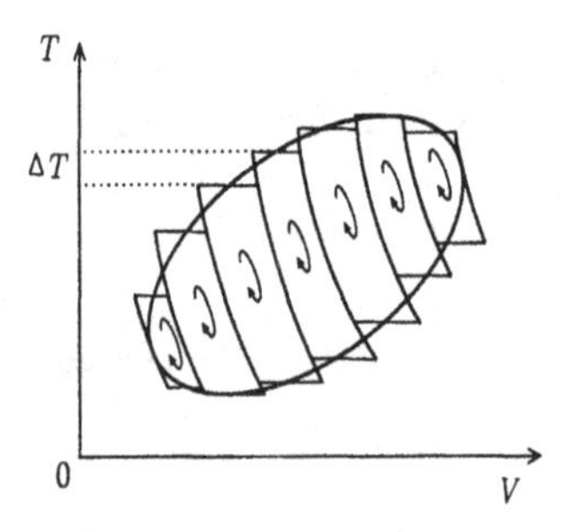

Figure 2.12: Any arbitrary reversible cyclic process can be approximated by a very large number of infinitesimal Carnot cycles.

a closed curve because for a cyclic process the end point must return to its starting point. For the convenience of discussion, we use T as the vertical axis, because we will consider many isothermal processes which will be horizontal lines on the diagram. This cyclic process can be approximated by a very large number of infinitesimal Carnot cycles as illustrated in Fig. 2.12. Every Carnot cycle consists of two isothermal and two adiabatic processes. We let the isothermal processes be as close to the curve of the real process as possible. In the limit where the temperature difference ΔT between the isothermal processes of any two neighboring Carnot cycles approaches zero,[18] we may say that the combination of all the Carnot cycles is a good approximation to the real cyclic process. This is because the adiabatic processes of the neighboring Carnot cycles will cancel each other (the directions of the processes are opposite to each other). Only isothermal processes will remain intact. Of course when ΔT is small but finite, the cancellation of the adiabatic processes is not perfect, the approximate path is therefore not as smooth as the real process in Fig. 2.12. However in the limit $\Delta T \to 0$, the approximate path will be very close to the real process.

We have proved that for any Carnot engine operating between two reservoirs T_1 and T_2, Eq. (2.15), i.e., $Q_1/Q_2 = T_1/T_2$ ($T^{\rm th} = T$), holds. In this relation we have assumed that both Q_1 and Q_2 are positive quantities. This, however, is not convenient in the following discussion. It is better to let Q itself decide whether it is positive or negative. We use the convention that $Q > 0$ if it is *rejected by the reservoir*, and $Q < 0$

[18]ΔT represents the temperature difference between the hotter (or the colder) heat reservoirs of two neighboring Carnot cycles.

if it is *absorbed by the reservoir.*[19] With this sign convention, Eq. (2.15) becomes

$$\frac{Q_1}{T_1} + \frac{Q_2}{T_2} = 0. \tag{2.25}$$

This is due to the fact that Q_1 and Q_2 have opposite signs, one of them is positive and the other one is negative. For the combination of a very large number of infinitesimal Carnot cycles, we can write

$$\sum_i \frac{Q_i}{T_i} = 0 \quad \text{(reversible process)}, \tag{2.26}$$

where i denotes the ith reservoir. When the temperature difference between any two neighboring reservoirs approaches zero, i.e., $\Delta T_i = |T_{i+1} - T_i| \to 0$ (for all i), then each Q_i becomes an infinitesimal quantity $đQ$, and the summation in the above equation becomes an integral, actually a cyclic integral,

$$\oint \frac{đQ}{T} = 0 \quad \text{(reversible process)}. \tag{2.27}$$

This equation holds for any reversible cyclic process. Therefore we can consider that for any system which starts from an initial point (state), and then undergoes any reversible process and returns to the initial state, the contour integral of this reversible cyclic process is always equal to zero. If this conclusion holds for any reversible cyclic process, then the integrand $đQ/T$ under the integral sign must be an exact differential. Otherwise the above conclusion will not be true. We define the exact differential as dS, the function S exists and is a state variable. If we recall the *entropy* we have defined in Eq. (2.2), we see that the function S defined above is identical to *entropy* we have defined mathematically. Of course we also obtained the answer $T^{\text{th}} = T$, which can not be obtained by a purely mathematical consideration. We may also write Eq. (2.2) as

$$đQ = TdS \quad \text{(reversible process)}. \tag{2.28}$$

Note that this equation also holds for a quasi-static irreversible process, if the dissipative heat is properly taken into account.

[19]When the reservoir rejects heat, this implies that the system absorbs heat, therefore $Q > 0$. When the reservoir absorbs heat, this implies that the system rejects heat, thus $Q < 0$. Therefore, this convention is the same as the one we used in the first law.

2.11 Entropy Change for a Reversible Process

We may use Eq. (2.28) to calculate the entropy change of a system for any reversible process. However entropy S is a state variable, therefore the entropy change depends only on the initial state and the final state, and is independent of the process. Therefore, if we want to calculate the entropy difference between state 2 (the final state) and state 1 (the initial state) $\Delta S = S_2 - S_1$, we may choose any reversible process which would make the system to change from state 1 to state 2. From Eq. (2.28) we get

$$\Delta S = S_2 - S_1 = \int_1^2 \frac{đQ}{T} \quad \text{(reversible process).} \tag{2.29}$$

There are three reversible processes which are frequently considered:

1. An isothermal process (T=constant),

$$\Delta S = S_2 - S_1 = \int_1^2 \frac{đQ}{T} = \frac{\Delta Q}{T}. \tag{2.30}$$

2. An isochoric process (V =constant),

$$\begin{aligned} \Delta S &= S_2 - S_1 = \int_1^2 \frac{đQ}{T} = \int_{T_1}^{T_2} \frac{C_V}{T}\, dT \\ &= C_V \ln \frac{T_2}{T_1}, \quad \text{if } C_V = \text{constant.} \end{aligned} \tag{2.31}$$

3. An isobaric process (P =constant),

$$\begin{aligned} \Delta S &= S_2 - S_1 = \int_1^2 \frac{đQ}{T} = \int_{T_1}^{T_2} \frac{C_P}{T}\, dT \\ &= C_P \ln \frac{T_2}{T_1}, \quad \text{if } C_P = \text{constant.} \end{aligned} \tag{2.32}$$

These are just simple examples. We will come back to the problem of calculating the entropy change in Chapter 4, where more general cases and examples will be given.

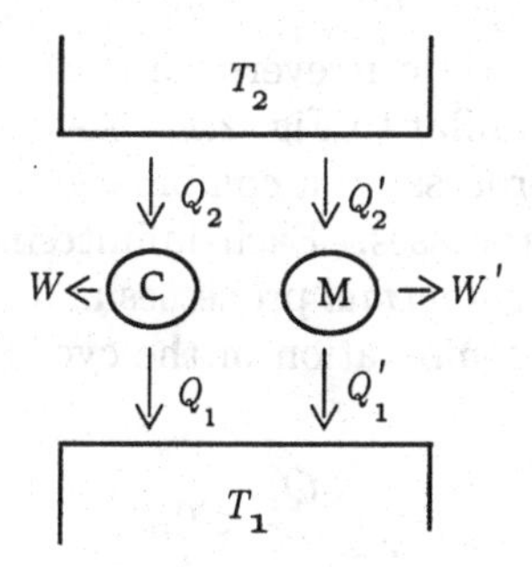

Figure 2.13: A Carnot engine C and an irreversible engine M operating between the same two heat reservoirs T_1 and T_2.

2.12 The Clausius Inequality

We have shown in Sec. 2.7 that operating between the same two heat reservoirs, no engine can be more efficient than a Carnot engine. In particular, the efficiency of any irreversible engine must be less than that of a Carnot engine. We consider a Carnot engine C and an irreversible engine M operating between the same two reservoirs T_1 and T_2 ($T_1 < T_2$), as shown in Fig. 2.13. We have[20]

$$\eta_{\mathrm{C}} = 1 - \frac{Q_1}{Q_2} = 1 - \frac{T_1}{T_2} > \eta_{\mathrm{M}} = 1 - \frac{Q_1'}{Q_2'}, \tag{2.33}$$

which implies $Q_1'/Q_2' > T_1/T_2$. Here we have assumed that both Q_1' and Q_2' are positive quantities. This assumption is, however, not ideal and we will let the sign of Q_i' ($i = 1, 2$) be determined by thermodynamics. As in Sec. 2.9, if Q is rejected by the reservoir $Q > 0$, and if Q is absorbed by the reservoir $Q < 0$.[21] With this sign convention, and for simplicity of notation replacing Q_i' by Q_i ($i = 1, 2$), the inequality $|Q_1'/Q_2'| > T_1/T_2$ can be rewritten as

$$\frac{Q_1}{T_1} + \frac{Q_2}{T_2} < 0 \quad \text{(irreversible process)}. \tag{2.34}$$

Here Q_i is the heat *rejected by the reservoir* T_i (if $Q_i < 0$, it implies that T_i is the colder reservoir).

[20]We may let $Q_2 = Q_2'$, which implies $Q_1' > Q_1$. This is due to $Q_1' = Q_1 + Q_{\mathrm{dis}} > Q_1$, where Q_{dis} is the dissipative heat produced per cycle by the irreversible engine.

[21]See also footnote 19 in p. 49.

We consider a quasi-static irreversible cyclic process, which can be plotted in a T-V plane similar to Fig. 2.12. As discussed in Sec. 2.10, we may approximate this process by a combination of a very large number of infinitesimal cyclic processes. Each infinitesimal cyclic process consists of two *irreversible isothermal* processes and two *reversible adiabatic* processes. Thus for the combination of the cyclic processes, we have

$$\sum_i \frac{Q_i}{T_i} < 0. \tag{2.35}$$

Any two neighboring infinitesimal cyclic processes will share the same adiabatic path, but operate in opposite directions. As the adiabatic processes are reversible, the opposite operations will cancel each other, with only small residual paths (which are not cancelled) near the isothermal processes (cf. Fig. 2.12). When the temperature difference between the hotter (colder) reservoirs of any two neighboring infinitesimal cyclic processes approaches zero, $\Delta T \to 0$, the errors introduced by the approximation will be negligible. Therefore we can replace the real cyclic process by the combination of a very large number of infinitesimal cyclic processes. In the limit $\Delta T \to 0$, $Q_i \to đQ$, and inequality (2.35) becomes

$$\oint \frac{đQ}{T} < 0 \quad \text{(irreversible process).} \tag{2.36}$$

Combining Eq. (2.27) and inequality (2.36), we get the Clausius inequality

$$\oint \frac{đQ}{T} \leq 0, \tag{2.37}$$

where the *equal sign* applies to reversible processes only. We note that the physical meaning of the term $đQ$ in the integrand is *the heat rejected by the heat reservoir* T, which is in thermal contact with the system. We recall that when the reservoir rejects heat $đQ > 0$, which is the case where the system absorbs heat. However, *in an irreversible process, the heat absorbed by the system is not necessarily equal to the heat rejected by the reservoir.* The reason is that dissipative heat may be involved in the irreversible process. This is the reason why there is a distinction between a reversible and an irreversible process in the Clausius inequality (2.37).

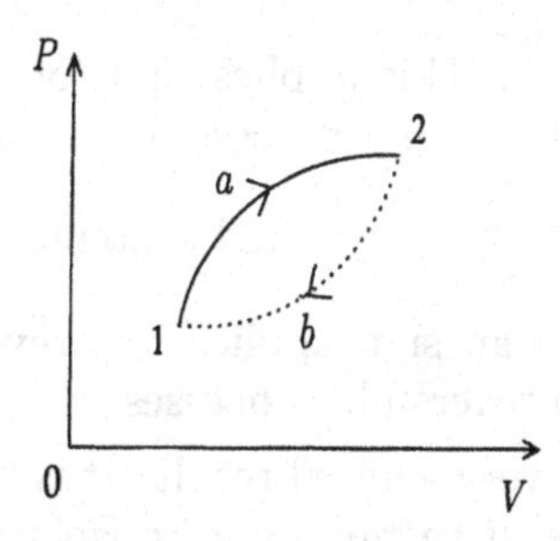

Figure 2.14: Process a may be reversible or irreversible, but process b is reversible. The combination of the processes a and b may be a reversible or an irreversible cyclic process.

2.13 The Principle of Non-Decrease of Entropy and the Second Law

We consider an isolated system which undergoes process a $(1 \to 2)$, as shown in Fig. 2.14. Process a may be reversible or irreversible. We imagine another reversible process b $(2 \to 1)$, which brings the system from state 2 back to state 1 (Fig. 2.14). Therefore the combination of a and b is a cyclic process, which is either reversible or irreversible. From inequality (2.37) we get

$$\oint \frac{dQ}{T} = \int_a \frac{dQ}{T} + \int_b \frac{dQ}{T} \leq 0. \tag{2.38}$$

The *equal sign* holds only when a is a *reversible process.* We know that process b is reversible, therefore the integrand dQ/T for the integration along path b is an exact differential dS, so we can rewrite the above inequality as

$$\int_a \frac{dQ}{T} \leq S_2 - S_1 \equiv \Delta S. \tag{2.39}$$

For an infinitesimal process, we may write ΔS as dS and remove the integral sign to obtain

$$dS \geq \frac{dQ}{T}, \tag{2.40}$$

where the *equal sign holds for reversible processes and the greater than sign holds for irreversible processes.* Applying the above formula to an isolated system we get $dS \geq 0$ for an infinitesimal process, because for an isolated system $đQ = 0$. This implies that for an *isolated* system the following inequality holds for any process, infinitesimal or finite:

$$\Delta S \geq 0 \quad \text{(isolated system).} \tag{2.41}$$

Here again the greater than sign applies to irreversible processes, and the equal sign applies to reversible processes.

Inequality (2.41) is a very general result. It is often referred to as the *principle of non-decrease of entropy.* For an isolated system, its entropy can only increase (for irreversible processes) or remain constant (for reversible processes). The total change in entropy of an isolated system never decreases. In fact, this principle is often referred to as the second law of thermodynamics. This is a very general principle; it includes not only the Clausius and the Kelvin-Planck statements, but also all the other kinds of processes. Therefore the most general statement of the second law is the following:

The entropy change of an isolated system never decreases. It remains constant in reversible processes, and it increases in irreversible processes.

Now we want to prove that both the Clausius and Kelvin-Planck statements are included in the above principle.

1. The Clausius statement:

   We consider a device M which violates the Clausius statement, as shown in Fig. 2.15. After a complete cycle, the device M removes heat Q from the reservoir T_1 and the same amount of heat is put in the reservoir T_2, with $T_2 > T_1$. After a complete cycle, M returns to its initial state, therefore the entropy change for M is $\Delta S_{\mathrm{M}} = 0$. For the heat reservoir T_1 this is a constant temperature process, thus the entropy change is $\Delta S_1 = -Q/T_1$ $(Q > 0)$. Similarly for T_2, the entropy change is $\Delta S_2 = Q/T_2$. Therefore, after a complete cycle, the total entropy change for the isolated system (M, T_1, and T_2) is

$$\begin{aligned} \Delta S_{\text{total}} &= \Delta S_1 + \Delta S_2 + \Delta S_{\mathrm{M}} \\ &= -\frac{Q}{T_1} + \frac{Q}{T_2} + 0 = Q\left(\frac{1}{T_2} - \frac{1}{T_1}\right) \\ &< 0, \text{ if } T_2 > T_1. \end{aligned} \tag{2.42}$$

   Therefore a violation of the Clausius statement (heat can flow from

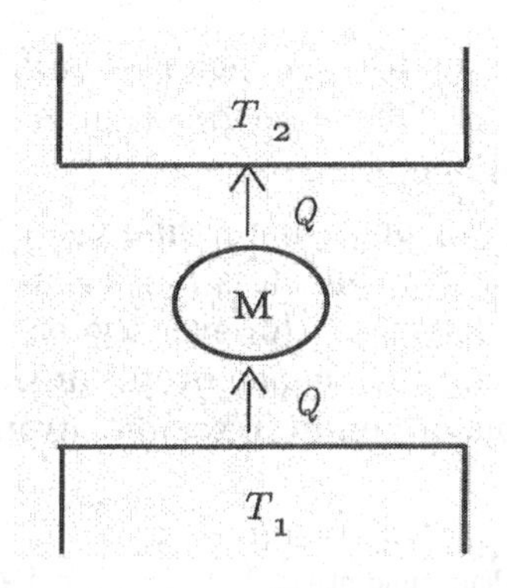

Figure 2.15: An engine M, which violates the Clausius statement, operates between heat reservoirs T_1 and T_2.

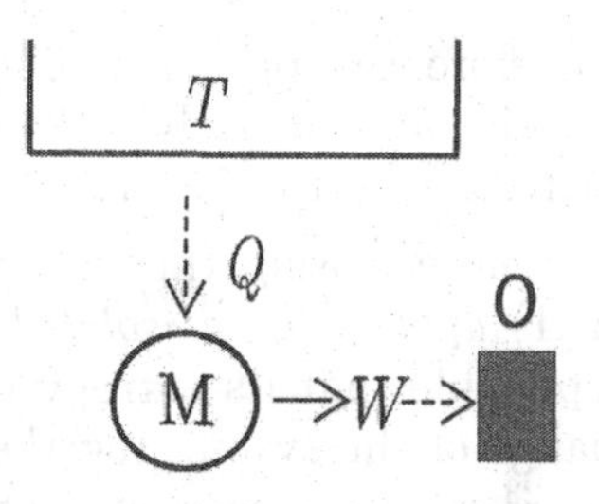

Figure 2.16: An engine M, which violates the Kelvin-Planck statement, absorbs heat Q from a heat reservoir T and does work $W = Q$ on an object O.

a lower temperature to a higher temperature) will lead to a violation of the principle of non-decrease of entropy.

2. The Kelvin-Planck statement:

   Consider a device M which violates the Kelvin-Planck statement, as shown in Fig. 2.16. After a complete cycle, device M absorbs an amount of heat Q from the heat reservoir T, and converts the heat 100% into mechanical work W ($W = Q$), which is used to lift an object O against the gravitational force. As in the above case, after a complete cycle, the entropy change for M is zero. The entropy change of the object O is also zero, because the potential energy of O may be increased, but the microscopic states will not change.[22] The entropy change for the reservoir is $-Q/T$ ($Q > 0$). Therefore for the isolated system (M, O, and T) the total entropy change is

$$\begin{aligned} \Delta S_{\text{total}} &= \Delta S_T + \Delta S_{\text{M}} + \Delta S_{\text{O}} = -\frac{Q}{T} + 0 + 0 \\ &= -\frac{Q}{T} < 0. \end{aligned} \tag{2.43}$$

   Therefore a violation of the Kelvin-Planck statement (such that heat can be 100% converted to work) will also lead to a violation of the principle of non-decrease of entropy.

[22] All molecules (atoms) in O will be increased by the *same amount* of potential energy, which will not affect the microscopic states of the molecules.

Therefore the principle of non-decrease of entropy includes both the Clausius and the Kelvin-Planck statements. This principle is quite often referred to as the second law of thermodynamics.

Before closing this section, we mention once more that inequality (2.41) *applies to an isolated system only.* If a system is not isolated, it is possible that its entropy change may decrease. However the entropy change of the system together with the entropy change of its surroundings[23] must increase, so that the principle of non-decrease of entropy is not violated. This implies that

$$\Delta S_{\text{total}} = \Delta S_{\text{system}} + \Delta S_{\text{surroundings}} \geq 0. \tag{2.44}$$

From this we know that the entropy change of a system ΔS_{system} may be less than zero, but the total entropy change ΔS_{total} will always be greater than or equal to zero.

2.14 Entropy Change for an Irreversible Process

For any reversible process, we can use Eq. (2.29) to calculate the entropy change, but for an irreversible process, how can we calculate the entropy change? We have said that entropy is a state variable, therefore the entropy change depends on the initial state and the final state, independent of the path. If the initial state is denoted as 1, and the final state as 2, then the entropy change is given by $\Delta S_{\text{system}} = \Delta S = S_2 - S_1$, independent of the process, even though it may be an irreversible one. Therefore, in order to calculate the entropy change ΔS, we may choose any *arbitrary reversible process* which will change the system from state 1 to state 2. When a process is chosen, Eq. (2.29) may be used to calculate ΔS.

We have mentioned earlier that there are two types of irreversible processes: (1) Those involving non-equilibrium states during the process. (2) A quasi-static process which accompanies dissipative heat. For type 1 irreversible processes, it is not easy to describe it using the equilibrium state variables. If we want to calculate the change of any state variable between the initial and the final states, then we can choose an arbitrary reversible process which will make the system change from the initial state to the final state. The chosen reversible process is then used to calculate the change of the state variable, and there is no need to

[23]The system plus its surroundings is considered as an isolated system.

think about the irreversible process. An example of this kind of calculation is the use of Eq. (2.29) to calculate the change in entropy ΔS. Therefore, for a system which undergoes a non-equilibrium irreversible process, one can still use the equilibrium reversible processes to study its properties. For type 2 irreversible processes, it is relatively simple. At any moment, the system can be considered as being in an equilibrium state (quasi-static), and we can use the equilibrium state variables to describe the process. Equation (2.29) may be used to calculate the change in entropy, but there is dissipative heat in the process, and it must also be properly included in the calculation. Therefore, as we have discussed in Sec. 2.2, for *the system which absorbs heat* during the process, the change in entropy is

$$\Delta S = \int_1^2 \frac{đQ_{\text{surr}} + đQ_{\text{dis}}}{T} \quad \text{(system absorbs heat).} \tag{2.45}$$

Here $đQ_{\text{surr}} + đQ_{\text{dis}}$ is the heat absorbed by the system under consideration, in which $đQ_{\text{surr}}$ is the heat rejected by the surroundings (i.e., other systems), and $đQ_{\text{dis}}$ is the dissipative heat ($đQ_{\text{dis}} > 0$). However for *the system which rejects heat* in the process, the change in entropy is

$$\Delta S = \int_1^2 \frac{đQ_{\text{system}}}{T} \quad \text{(system rejects heat),} \tag{2.46}$$

where $đQ_{\text{system}}$ is the heat rejected by the system under consideration to the surroundings (thus $đQ_{\text{system}} < 0$). Here we emphasize again that in a quasi-static irreversible process, *only for a system which absorbs heat should the dissipative heat be included in the entropy calculation. For a system which rejects heat, the dissipative heat should not be included.*

2.15 An Example of Reversible vs. Irreversible Heat

In Fig. 2.14, if we choose path a as an irreversible process, and path b is a reversible process, then inequality (2.38) may be written as

$$\oint \frac{đQ}{T} = \int_1^2 \frac{(đQ^{(\text{irr})} - đQ^{(\text{r})})}{T} < 0, \tag{2.47}$$

where $đQ^{(\text{irr})}$ represents the heat rejected by the heat reservoir T in an infinitesimal *irreversible* process, and $đQ^{(\text{r})}$ the heat rejected by the reservoir T in an infinitesimal *reversible* process. When the process $1 \to 2$ becomes an infinitesimal one, we get

$$đQ^{(\mathrm{irr})} < đQ^{(\mathrm{r})}. \tag{2.48}$$

Since a finite process is the sum of a very large number of infinitesimal processes, and the above inequality is valid for all infinitesimal processes, for a finite process we have

$$Q^{(\mathrm{irr})} < Q^{(\mathrm{r})} \quad \text{(finite process)}. \tag{2.49}$$

Inequality (2.49) tells us that when a system changes from state 1 to state 2, the amount of heat rejected by the heat reservoir will depend on whether the process is reversible or irreversible. The amount of heat rejected in a reversible process is always greater than that of an irreversible process. In the following, we give a practical example in order to help the reader gain more insight into the above statement. We consider a system consisting of water which is in coexistence with its vapor. The system (water plus its vapor) is contained in a cylindrical container, which is in thermal contact with a heat reservoir at temperature T, as shown in Fig. 2.17. There is a movable piston on the upper side of the container. An adjustable weight, which exerts a pressure P_{ext} on the vapor, is placed on top of the piston. When equilibrium is reached, the external pressure P_{ext} is equal to the vapor pressure $P_{\mathrm{vap}}(T)$, i.e., $P_{\mathrm{ext}} = P_{\mathrm{vap}}(T)$. Note that the vapor pressure $P_{\mathrm{vap}}(T)$ is a function of T only. Now we will consider the situation where the external pressure is slightly different from the equilibrium value. Will the piston move? If so how? What thermodynamic problems do we have to consider? Two different situations will be discussed in the following.

1. There is no friction between the piston and the container:

   We let the external pressure have a small change,

$$P_{\mathrm{ext}} = P_{\mathrm{vap}}(T) + \delta P, \tag{2.50}$$

where δP is a very small pressure. There are two cases depending on the sign of δP.

   (a) When $\delta P > 0$, the piston will move downward. We would like to consider a reversible process, which requires the process to be quasi-static, therefore the speed of the piston must be very small

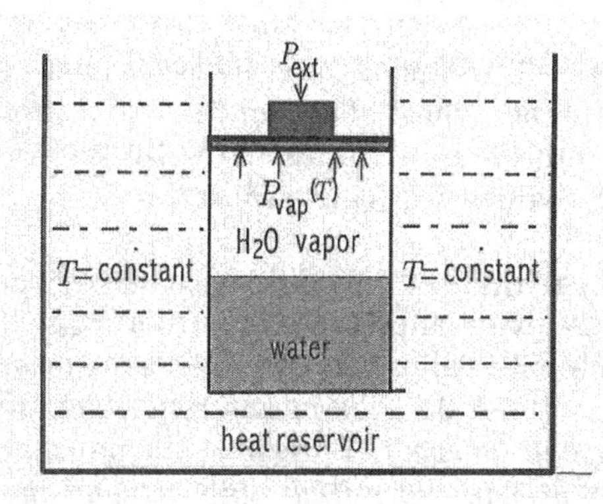

Figure 2.17: Water and its vapor are contained in a cylindrical container, which is in thermal contact with a heat reservoir at temperature T. There is a movable piston at the upper side of the container, with an adjustable external pressure on the vapor.

(almost zero). This implies that δP is almost zero (but positive). We write the condition as $\delta P = 0^+$. Because T is kept constant, the vapor pressure $P_{\text{vap}}(T)$ is also a constant. When the piston moves downward, some of the vapor has to condense into water to keep $P_{\text{vap}}(T)$ constant. In the condensation process, the system will deliver latent heat of condensation $L(T)$ to the reservoir. Suppose that when x kilomoles of vapor have condensed to water, the extra external pressure δP is removed. The piston and the system is again in equilibrium, and the piston stops moving. We have $L(T) = xl(T)$, where $l(T)$ is the latent heat of condensation per kilomole. In this process the temperature T is always kept constant, therefore the total entropy change for the system and the reservoir is

$$\begin{aligned}\Delta S_{\text{total}} &= \Delta S_{\text{reservoir}} + \Delta S_{\text{system}} \\ &= +\frac{xl(T)}{T} - \frac{xl(T)}{T} = 0. \end{aligned} \tag{2.51}$$

This is a reversible process, therefore $\Delta S_{\text{total}} = 0$, as expected.

(b) After the process described in (a) is completed, the external pressure is then changed to $\delta P < 0$, and the piston will move upward. If δP is very small, such that $\delta P = 0^-$, the speed of the piston will be very small (almost zero) and in the reverse direction. Therefore this process is the reverse of case (a). In order to keep the vapor pressure $P_{\text{vap}}(T)$ constant, some water molecules will evaporate to vapor. When x kilomoles of water molecules have evaporated, the system must absorb a heat of evaporation $xl(T)$ from the reservoir. The system returns to its initial state. In the whole process, the total entropy change ΔS_{total} is also zero.

We emphasize here that the only difference between cases (a) and (b) is that we change the external pressure P_{ext} from $P_{\text{ext}} + 0^+$ to $P_{\text{ext}} + 0^-$. The difference of the external pressure in these two cases is $(P_{\text{ext}} + 0^+) - (P_{\text{ext}} + 0^-) = 0^+ - 0^- \to 0$. This means that when *we make only an infinitesimal change of the external pressure*, the process can be reversed and *the system returns to the initial state without any change of the surroundings.* This is the criterion for a reversible process, and it can be used to check whether a process is reversible or not.

2. There is friction between the piston and the container:

Suppose the maximum frictional force between the piston and the container is $F^{\text{M}}_{\text{fric}}$; we define a maximum frictional pressure $P^{\text{M}}_{\text{fric}} = F^{\text{M}}_{\text{fric}}/A$, where A is the area of the piston (the area of the cross section of the container). We take $P^{\text{M}}_{\text{fric}}$ to be a positive quantity. Apparently, when the magnitude of the external pressure P_{ext} is in the following range,

$$P_{\text{vap}}(T) - P^{\text{M}}_{\text{fric}} < P_{\text{ext}} < P_{\text{vap}}(T) + P^{\text{M}}_{\text{fric}}, \tag{2.52}$$

the piston will not move. Since $P^{\text{M}}_{\text{fric}} \neq 0$, we can not make an infinitesimal change in the external pressure to reverse the direction of motion of the piston. Therefore this is an irreversible process. As in the above, we discuss the two cases where the piston moves downward and upward, respectively.

(a) When $P_{\text{ext}} = P_{\text{vap}}(T) + P^{\text{M}}_{\text{fric}} + 0^+$, the piston will move downward very slowly (quasi-statically). During the process some of the vapor condenses into water, which keeps the vapor pressure constant. When x kilomoles of vapor have condensed into water, an amount of latent heat of condensation $xl(T)$ is delivered by the vapor which is absorbed by the reservoir. However, in addition to the latent heat, some dissipative heat (due to friction) Q_f ($Q_f > 0$) will be produced, which is also absorbed by the reservoir. By using Eqs. (2.45)–(2.46) we get the total change in entropy

$$\Delta S_{\text{total}} = +\frac{xl(T) + Q_f}{T} - \frac{xl(T)}{T} = \frac{Q_f}{T} > 0. \tag{2.53}$$

The total change in the entropy is greater than zero, therefore this is an irreversible process, as expected. The irreversible heat (which is the heat absorbed or delivered by the heat reservoir) in this process is[24]

$$Q^{(\mathrm{irr})} = -(xl(T) + Q_f) < -xl(T) = Q^{(\mathrm{r})}, \qquad (2.54)$$

which satisfies inequality (2.49).

(b) When the external pressure $P_{\mathrm{ext}} = P_{\mathrm{vap}}(T) - P_{\mathrm{fric}}^{\mathrm{M}} + 0^-$, the piston will move upward very slowly (quasi-statically) and some of the water molecules will evaporate to vapor, in order to keep the vapor pressure constant. When x kilomoles of water molecules have evaporated to vapor, the system returns to its initial state. During this process, the system absorbs an amount of heat of evaporation $xl(T)$, part of which comes from the dissipative heat Q_f ($Q_f > 0$) produced during this process (due to friction). The heat reservoir provides only $xl(T) - Q_f$ of heat to the system. From Eqs. (2.45)–(2.46) we get the total change in entropy as

$$\Delta S_{\mathrm{total}} = -\frac{xl(T) - Q_f}{T} + \frac{xl(T)}{T} = \frac{Q_f}{T} > 0. \qquad (2.55)$$

Since $\Delta S_{\mathrm{total}} > 0$, this is an irreversible process, as expected. As mentioned above, we may also use the criterion for a reversible process to see why this is an irreversible process. The process considered in (b) is not *really the reverse process* of (a), although it does bring the system back to its initial state. The reason is that the difference between the external pressure of (b) and (a) is equal to $2P_{\mathrm{fric}}^{\mathrm{M}}$, which is not an infinitesimal quantity, as required in a reversible process. The irreversible heat (the heat absorbed or delivered by the heat reservoir) in this process is

$$Q^{(\mathrm{irr})} = xl(T) - Q_f < xl(T) = Q^{(\mathrm{r})}, \qquad (2.56)$$

which also satisfies inequality (2.49).

2.16 The Natural Variables for *U* and *S*

For a reversible process $dQ = TdS$, and $dW = PdV$ (reversible work), consequently the first law Eq. (1.3) can be written as

$$dU = TdS - PdV. \qquad (2.57)$$

[24]See footnote 19 in p. 49 for the sign convention of Q.

As the quantity on the left hand side of Eq. (2.57) depends only on the state of the system, independent of the path, therefore the quantity on the right hand side should be also independent of the path. This means that Eq. (2.57) can also be applied to irreversible processes. The physical quantities in this equation, U, T, S, P, and V are all state variables; their values depend on the state of the system, independent of the history of the system. Therefore we may consider Eq. (2.57) as a thermodynamic relation, involving no paths or processes. A PVT system has two independent variables. From Eq. (2.57) we see that the *natural* independent variables for U are S and V. Therefore we may write $U = U(S, V)$, and from Eq. (2.57)

$$\left(\frac{\partial U}{\partial S}\right)_V = T, \quad \left(\frac{\partial U}{\partial V}\right)_S = -P. \tag{2.58}$$

Equation (2.57) can also be written as

$$dS = \frac{1}{T}\,dU + \frac{P}{T}\,dV, \tag{2.59}$$

which implies that the *natural* independent variables for S are U and V; thus $S = S(U, V)$, and

$$\left(\frac{\partial S}{\partial U}\right)_V = \frac{1}{T}, \quad \left(\frac{\partial S}{\partial V}\right)_U = \frac{P}{T}. \tag{2.60}$$

These relations will be useful in the latter chapters and sections.

2.17 Increase of Entropy and Degradation of Energy

Equation (2.57) is derived from the differential form of the first law Eq. (1.3) by replacing $đQ$ by TdS and $đW$ by PdV, which is valid for reversible processes. However for irreversible processes $đQ \neq TdS$, we may therefore ask whether Eq. (2.57) still holds for irreversible processes? The answer is yes, because as we have said, Eq. (2.57) is independent of the path. However we have to examine the contents of Eq. (2.57) more carefully. Since this equation is independent of the path, we may write

$$dU = đQ^{(\mathrm{r})} - đW^{(\mathrm{r})} = đQ^{(\mathrm{irr})} - đW^{(\mathrm{irr})}. \tag{2.61}$$

Here the superscript r (irr) represents the heat or work in a *reversible* (*irreversible*) process. Usually we interpret the quantity dQ in the first law as the heat *absorbed by the system*. However if we re-interpret it as the heat *rejected by the heat reservoir* (or its surroundings), we can gain more physical information. These two interpretation are equivalent in a reversible process,[25] but they are different in an irreversible process. The latter case also involves dissipative heat which does not belong to the heat reservoir (or the surroundings). We recall that in deriving the Clausius inequality, we emphasized that the quantity dQ in the integrand is the heat *rejected or absorbed by the reservoir* (if $dQ > 0$ which implies the heat is *rejected* by the reservoir). Therefore, the two dQ in Eq. (2.61) satisfy the inequality $dQ^{(\mathrm{irr})} < dQ^{(\mathrm{r})}$, which is just inequality (2.48). We may write

$$dQ^{(\mathrm{irr})} = dQ^{(\mathrm{r})} - \delta q, \tag{2.62}$$

where $\delta q > 0$ is the dissipative heat produced in an irreversible process. From Eq. (2.61) we find

$$dW^{(\mathrm{irr})} = dW^{(\mathrm{r})} - \delta q < dW^{(\mathrm{r})}. \tag{2.63}$$

Therefore the infinitesimal work done by a system in an infinitesimal irreversible process is always smaller than that of an infinitesimal reversible process.[26] The reason for this is that there is always dissipative heat produced in an irreversible process.

A finite process is the sum of a very large number of infinitesimal processes. The above inequality holds for every infinitesimal process, therefore for a finite process we have

$$W^{(\mathrm{irr})} < W^{(\mathrm{r})} \quad \text{(finite process)}. \tag{2.64}$$

Hence we get the conclusion: *A system undergoes a change from one equilibrium state to another equilibrium state. The process may be either reversible, or irreversible, and the system may do work during the process. We know for certain that the work done by the system in a reversible process is always greater than that in any irreversible process.* This means that when a system changes from one state to another, the maximum work the system can do is via a reversible process. We have seen this

[25]In a reversible process, the heat absorbed by the system is equal to the heat rejected by the heat reservoir or its surroundings.

[26]Of course, we assume that these two processes have the same initial and final states.

kind of conclusion before: the efficiency of a Carnot engine (a reversible one) is always greater than that of an irreversible engine.

We know that the free expansion of an ideal gas is an irreversible process. In this process the system absorbs no heat and does no work, and the temperature remains constant (the internal energy remains constant). However if the system undergoes a reversible isothermal process (with the same initial and final states as the free expansion), the system does work W in the process

$$W = \int_{V_1}^{V_2} PdV = nRT \int_{V_1}^{V_2} \frac{dV}{V} = nRT \ln \frac{V_2}{V_1}. \tag{2.65}$$

We have $W > 0$, because $V_2 > V_1$ (expansion). This is consistent with the above conclusion, because in the irreversible process (free expansion) $W = 0$. Of course, in the reversible isothermal process, the system absorbs heat Q from a heat reservoir which converts to work. However in this consideration $W = Q$ (because $\Delta U = 0$), which is the maximum work that the amount of heat Q can do.[27]

From the above result, we get an important conclusion. Although in any thermodynamic process, conservation of energy always holds, but if the process is irreversible some of the energy is *wasted.* We say that the energy is wasted because, if the process is reversible then this energy can do work, but it does no work in an irreversible process. The wasted energy may be called a *degraded energy*, which means that, although the energy is not lost, it has lost the chance to do work (due to irreversible processes). The Kelvin-Planck statement of the second law tells us that we can not construct a perfect engine, that part of the heat (energy) obtained from the hotter reservoir (energy source) is always lost to the colder reservoir as *wasted* heat. This may be part of the reason that we are unable to use energy as efficiently as we want. However the main reason why the problem of energy efficiency is difficult is due to the occurrence of irreversible processes in our daily life. If we can construct a reversible engine, such as a Carnot engine, then its efficiency could be better than any irreversible engine, and therefore would have less wasted energy. Moreover, we can reverse the operation of the engine, then it can *recover the wasted energy* from the colder reservoir and put it back into the hotter reservoir without any extra expense. Therefore the wasted energy in a reversible process can be recovered, it is not really lost.

[27]The result that $W = Q$ does not violate the second law, because the system does not return to its initial state. It is not an engine.

For an irreversible engine, the efficiency is smaller, because dissipative heat is always present. Dissipative heat can not be recovered and is usually considered as degraded. However, not all irreversible processes will produce dissipative heat, for example, the free expansion of an ideal gas, a glass of hot water in air cooling down, etc. Although there is no dissipative heat in these processes, they are irreversible processes. In these processes there is still some energy that is degraded. For the free expansion case, the gas does no work. However we have shown in the above that in an equivalent reversible process the gas can do a maximum work that an amount of heat can do. For the case of a glass of hot water, it can do several things, such as make a hot cup of instant coffee or tea. But after the water cools down it can no longer be used for the above purposes. All these phenomena may help us to understand the physical meaning or significance of *entropy*. An irreversible process will increase the entropy of an isolated system, and some of the energy becomes degraded. This may indicate that entropy is a measure of the *randomness* of a system. The reasoning is partly due to the belief that if a system becomes more random (disordered), then its ability to do work may be weakened. However this is only a very vague idea, it may have little practical use. To get a concrete and deeper understanding of the physical meaning of entropy, a microscopic (atomic level) definition of entropy is necessary. This is what we will do in the next chapter.

2.18 Problems

2.1. Given 1.0 kg of water at 100°C and a very large block of ice at 0°C. A reversible heat engine absorbs heat from the water and expels heat to the ice, until work can no longer be extracted from the system. At the completion of the process:

   (a) What is the temperature of the water?
   (b) How much ice has been melted? The heat of fusion of ice is 80 kcal kg^{-1} at 0°C.
   (c) How much work has been done by the engine?

2.2. Sketch a Carnot cycle which uses an ideal gas as the working substance:

   (a) on a U-V diagram;
   (b) on a U-T diagram;

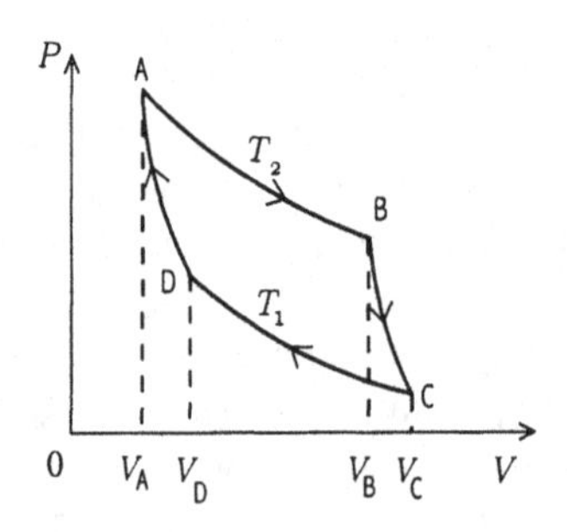

Figure 2.18: Problems 2.4 and 2.5.

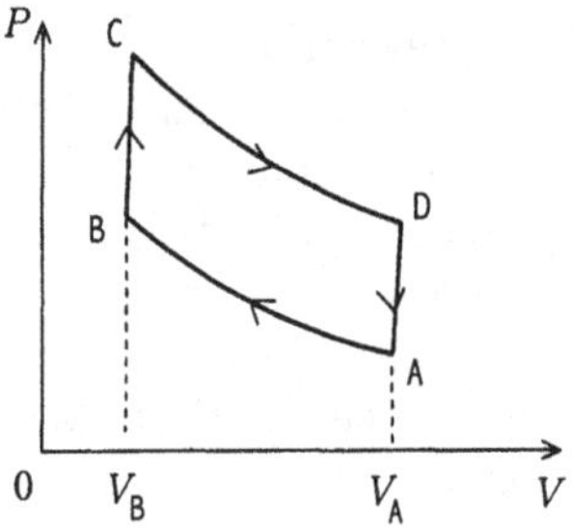

Figure 2.19: Problem 2.8.

(c) on a U-H diagram; and

(d) on a P-T diagram.

2.3. A Carnot engine is operated between two heat reservoirs at temperatures of 600 K and 300 K.

(a) If the engine receives 8×10^6 J of heat from the 600 K reservoir in each cycle, how much heat per cycle does it deliver to the 300 K reservoir?

(b) If the engine is operated as a refrigerator and receives from the 300 K reservoir 2×10^6 J of heat in each cycle, how much heat per cycle does it deliver to the 600 K reservoir?

(c) How much work is done per cycle by the engine? How much work is done per cycle on the refrigerator?

(d) What is the efficiency of the engine in (a) and the coefficient of performance of the refrigerator in (b)?

2.4. Consider a Carnot cycle ABCDA plotted on the P-V diagram in Fig. 2.18. Show that for an ideal gas

$$\frac{V_{\rm B}}{V_{\rm A}} = \frac{V_{\rm C}}{V_{\rm D}}.$$

2.5. Consider a Carnot engine which uses one kilomole of an ideal gas as the working substance. The Carnot cycle operated by the engine is shown in Fig. 2.18. Suppose $T_2 = 2T_1$ and $V_{\rm C} = 8V_{\rm A}$. Is there any difference if the ideal gas is monatomic or diatomic? If the work done by a monatomic gas per cycle is $W_{\rm mon}$ and that for a diatomic gas is $W_{\rm dia}$, compute the ratio $W_{\rm mon}/W_{\rm dia}$. If the ratio is not equal to 1, can you still expect that they have the same efficiency? Explain your reasoning.

2.6. Consider a Carnot engine which uses one kilomole of a monatomic ideal gas as the working substance. The system undergoes the following reversible cyclic processes: A→B (isothermal), B→C (adiabatic), C→D (isothermal), and D→A (adiabatic). Suppose $V_{\rm B} = 2V_{\rm A}$, $V_{\rm C} = 8V_{\rm B}$, and $W = 5.185 \times 10^6$ J is the total work done in each circle. Find the temperatures of the heat reservoirs $T_{\rm A}$ and $T_{\rm C}$.

2.7. A reversible heat engine works between *three* heat reservoirs at temperatures T_1, T_2, and T_3. After an integral number of cycles, the engine absorbs heat Q_3 from T_3, absorbs heat Q_2 from T_2, and rejects heat Q_1 to T_1. The total work done by the engine in these cycles is W. Suppose T_1=300 K, T_2=450 K, T_3=600 K, W=800 J, and Q_3=1200 J, Find Q_1 and Q_2. What is the efficiency of the engine?

2.8. Consider the Otto cycle ABCDA plotted on the P-V diagram in Fig. 2.19. On the diagram AB and CD are reversible adiabatic processes, and BC and DA are reversible constant volume processes. Show that, for an ideal gas, the efficiency is given by

$$\eta = 1 - \left(\frac{V_{\rm B}}{V_{\rm A}}\right)^{\gamma-1},$$

where $V_{\rm A}/V_{\rm B}$ is known as the compression ratio and $\gamma = c_p/c_v$.

2.9. Consider the Joule cycle ABCDA plotted on the P-V diagram in Fig. 2.20. On the diagram AB and CD are reversible adiabatic processes, and BC and DA are reversible constant pressure processes. Show that, for an ideal gas, the efficiency is given by

$$\eta = 1 - \left(\frac{P_{\rm B}}{P_{\rm A}}\right)^{(\gamma-1)/\gamma},$$

where $P_{\rm B} < P_{\rm A}$ and $\gamma = c_p/c_v$.

2.10. Consider the Diesel cycle ABCDA plotted on the P-V diagram in Fig. 2.21. On the diagram AB and CD are reversible adiabatic

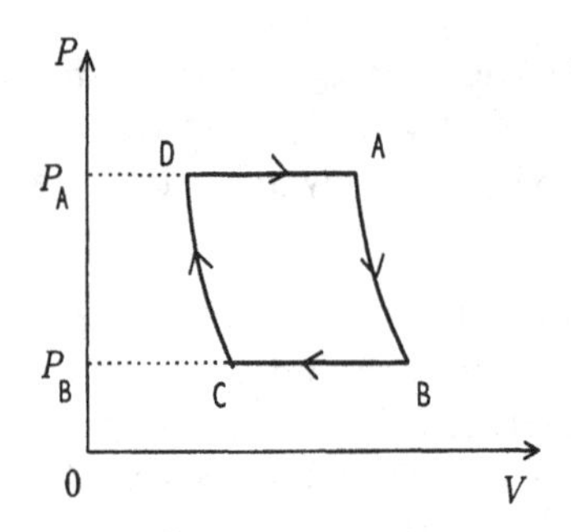

Figure 2.20: Problem 2.9.

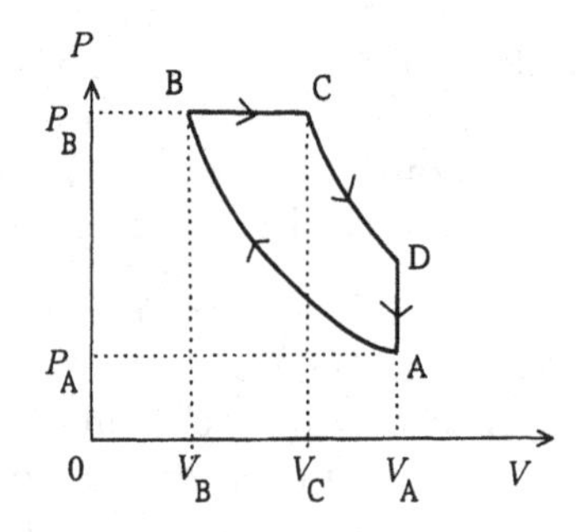

Figure 2.21: Problem 2.10.

processes, BC is a reversible constant pressure process, and DA is a reversible constant volume processes. Show that, for an ideal gas, the efficiency is given by

$$\eta = 1 - \frac{1}{\gamma}\frac{(V_C/V_A)^\gamma - (V_B/V_A)^\gamma}{(V_C/V_A) - (V_B/V_A)}.$$

where $V_B < V_C < V_A$ and $\gamma = c_p/c_v$.

2.11. Find the change in entropy of the system during the following processes:

(a) 1 kg of ice at $T = 0°C$ and $P = 1$ atm melts at the same T and P. The latent heat of fusion is 3.34×10^5 J kg^{-1}.

(b) 1 kg of steam at $T = 100°C$ and $P = 1$ atm condenses into water at the same T and P. The latent heat of vaporization is 2.26×10^6 J kg^{-1}.

2.12. (a) One kilogram of water at 0°C is brought into thermal contact with a large heat reservoir at 100°C. When the water has reached 100°C, what has been the change in entropy of the water, of the heat reservoir, and of the universe?

(b) If the water had been heated from 0°C to 100°C by first bringing it into thermal contact with a heat reservoir at 50°C and then with a reservoir at 100°C, what would have been the change in the entropy of the universe?

(c) Explain how the water might be heated from 0°C to 100°C with no change in the entropy of the universe.

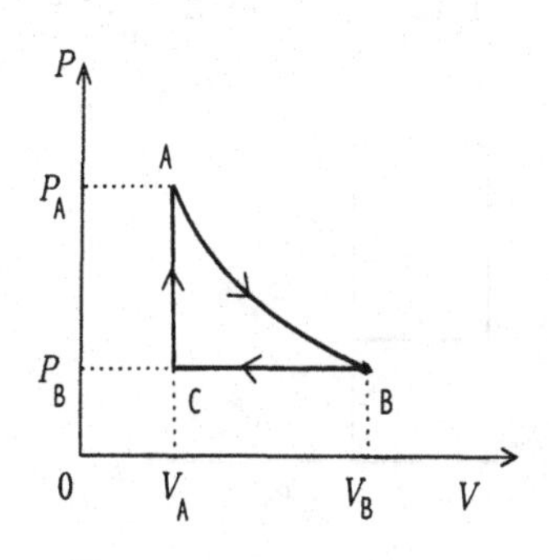

Figure 2.22: Problem 2.13.

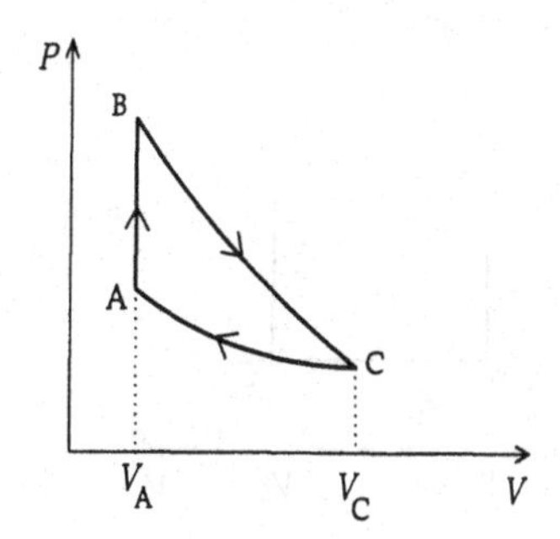

Figure 2.23: Problem 2.14.

2.13. The initial state of one kilomole of a monatomic ideal gas is pressure $P_A = 16$ atm and volume $V_A = 1$ liter, which is the state A shown in Fig. 2.22. The system then undergoes the following processes: (1) a reversible adiabatic process from A to B, with volume $V_B = 8$ liter, (2) a reversible isobaric process from B to C, where $V_C = V_A$, and then (3) a reversible isochoric process from C to A. This completes a closed cycle ABCA. Calculate the change in entropy in each process in the cycle. What is the total change of entropy for the complete cycle?

2.14. The initial state of an ideal gas is $P = 10^3$ Pa, $V = 0.2$ m^3, and $T = 200$ K, which is the state A in Fig. 2.23. The gas undergoes the following reversible processes: (1) a constant V process from A to B, with $T_B = 288$ K, (2) an adiabatic process from B to C, with $T_C = 200$ K, and (3) an isothermal process from C back to the initial state A. The constant volume specific heat of the gas c_v is $3R/2$.

(a) Calculate the heat absorbed, and the work done by the system in each process.

(b) Calculate the entropy change of the system in each process.

(c) What is the efficiency of the cycle, if it is considered as a heat engine?

2.15. A system is taken *reversibly* around the cycle ABCDA on the *P*-*V* diagram of Fig. 2.24, where A→B and C→D are isobaric processes; and B→C and D→A are isochoric processes. The temperature (in the Celsius scale) in each state is $T_A = 326.85°$C, $T_B = 1526.85°$C,

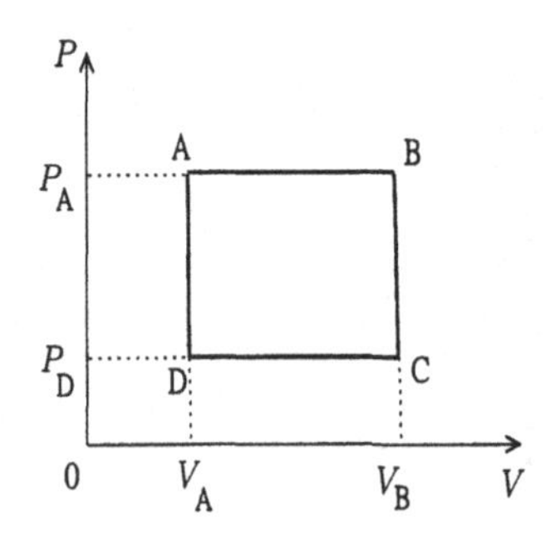

Figure 2.24: Problem 2.15.

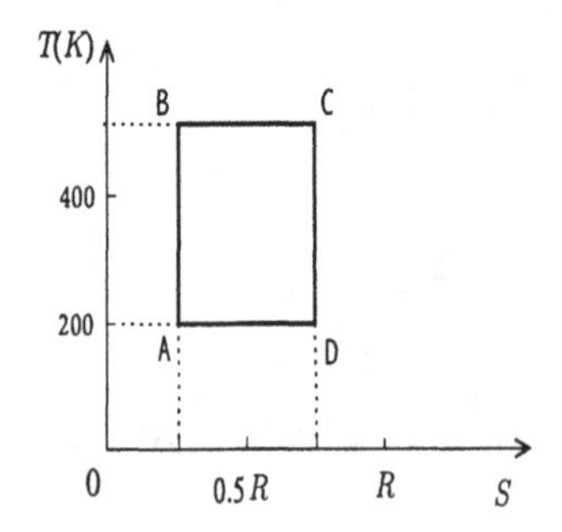

Figure 2.25: Problem 2.16.

$T_C = 626.85°C$, and $T_D = 26.85°C$. Assume that the heat capacities are constants with $C_V = 8\ \mathrm{Jk}^{-1}$ and $C_P = 10\ \mathrm{Jk}^{-1}$. It is also known that $V_B = 3\,V_A = 3 \times 10^{-2}\ \mathrm{m}^3$.

(a) Calculate the heat flow into the system in each process.

(b) Calculate the pressure difference $P_A - P_D$.

(c) Calculate the entropy change in each process.

2.16. A system is taken reversibly around the cycle ABCDA on the T-S diagram of Fig. 2.25, in which $T_A = T_D$=200 K, $T_B = T_C$=500 K, $S_A = S_B = 0.25\,R$, and $S_C = S_D = 0.75\,R$.

(a) Does the cycle ABCDA operate as an engine or a refrigerator?

(b) Calculate the heat transferred in each process.

(c) Find the efficiency of this cycle operating as an engine, graphically as well as by direct calculation.

(d) What is the coefficient of performance of this cycle, when operating as a refrigerator?

2.17. A Carnot engine operates on 1 kg of methane (CH_4), which is considered to be an ideal gas with $c_p/c_v = 1.35$. If the ratio of the maximum volume to the minimum volume is 4 and the cycle efficiency is 25%, find the entropy increase of the methane during the isothermal expansion.

2.18. Two identical finite systems of constant heat capacity C_P, are initially at temperatures T_1 and T_2, respectively, where $T_2 > T_1$. These systems are used as the heat reservoirs of a Carnot engine which does an infinitesimal amount of work in each cycle.

(a) Why is it that the work done in each cycle can only be an infinitesimally small amount?

(b) Show that the final equilibrium temperature of the reservoirs is $(T_1T_2)^{1/2}$.

(c) Show that the total amount work done is $C_P(T_2^{1/2} - T_1^{1/2})^2$.

2.19. A home air conditioner is operating on a reversible Carnot cycle between T_2 (temperature inside the house) and T_1 (temperature outside) ($T_1 > T_2$). The air conditioner consumes power W J/s, and absorbs heat Q_2 J/s from the house and rejects heat Q_1 J/s to the outside. Assume that the amount of heat that leaks into the house from outside per second can be described by the formula $Q = \alpha(T_1 - T_2)$ ($\alpha > 0$). We may take both T_1 and T_2 to be the absolute temperatures.

(a) What is the coefficient of performance for the air conditioner?

(b) In the steady state, both T_1 and T_2 are constants, derive a formula for the room temperature T_2 in terms of T_1, W, and α.

(c) In the winter, the cycle is reversed and the device becomes a heat pump which absorbs heat from outside and rejects heat into the house. Find the lowest outside temperature in °C for which it can maintain 20°C inside the house, if $W/\alpha = 1.1$.

2.20. In the water behind a high power dam (110 m high) the temperature difference between the surface and bottom may be 10°C. Compare the possible energy extraction from the thermal energy of 1 kg of water with that generated by allowing the water to flow over the dam through turbines in the conventional way. You may assume that the surface temperature of the water is 300 K.

2.21. The following two relations are given for a given gas: $PV = f(\theta)$ and $(\partial U/\partial V)_\theta = 0$, where $f(\theta)$ is some function of θ which is the temperature on some temperature scale. Here P, V, and U are the pressure, volume, and internal energy of the system. Show that $f(\theta)$ has the significance of a thermodynamic temperature (or an absolute temperature).

2.22. When operating between two heat reservoirs at temperatures T_1 and T_2, with $T_2 > T_1$, show that no refrigerator can have the coefficient of performance c better than the ideal value $c_{\text{ideal}} = T_1/(T_2 - T_1)$, which is the c value for a Carnot refrigerator.

2.23. A heat engine is operating between a body of finite mass and a heat reservoir at a constant temperature T_2. The initial temperature of the body is T_1, which is higher than T_2. After many cycles the temperature of the body is lowered from T_1 to T_2. Prove that the maximum work obtainable from the engine is $Q + T_2(S_2 - S_1)$, where Q is the heat flow out of the body and $S_2 - S_1$ is the entropy change of the body.

2.24. Two identical bodies of constant heat capacity C_P have the same initial temperature T_i. If a refrigerator working between the two bodies cools one down to temperature T_f, show that the minimum work required to do this is

$$W_{\text{min}} = C_P \left(\frac{T_i^2}{T_f} + T_f - 2T_i \right) .$$

2.25. If T_{max} is the maximum temperature of the heat reservoirs from which a heat engine receives heat, and T_{min} is the minimum temperature of the heat reservoirs to which it gives up heat, show that the efficiency of the engine never exceeds $1 - (T_{\text{min}}/T_{\text{max}})$.

2.26. Consider all the possible processes by which a body, at temperature T_2 with a heat capacity C, can lose heat and cool down to the temperature T_1 of a heat reservoir, where $T_1 < T_2$. How one can make the work done in this process be maximum? What is the maximum value? For simplicity, assume C to be a constant.

Chapter 3

Microscopic Interpretation of Entropy

3.1 Introduction

In order to understand the physical meaning of entropy, a microscopic definition is essential. Once the microscopic meaning of entropy is understood, then it is easy to understand why entropy remains constant in a reversible process and increases in an irreversible process. All phenomena in the universe are either reversible or irreversible (there are no other possibilities), therefore the total entropy of the universe is always increasing, or at best remains constant. It never decreases. From the microscopic point of view, a thermodynamic system is composed of a very large number of particles (atoms or molecules). In this chapter we use classical mechanics to describe the microscopic states of the system.[1] The microscopic variables of the system consist of the position vector $\boldsymbol{r}_i$ and the momentum vector $\boldsymbol{p}_i$ of each particle. Therefore there are six variables for each particle,[2] $6N$ variables in total, where N is the number of particles in the system. There is usually a condition for these variables to meet, i.e., the total energy is a constant (for an isolated system). The total number of independent number of variables is therefore $6N - 1$, a very large number. On the other hand, from the thermodynamic point of view, the state variables of a system may be the temperature T, pressure

[1] A quantum mechanical description of the microscopic states of a thermodynamic system will be given in Chapter 10.

[2] For simplicity, we neglect the internal degrees of freedom for each particle.

P, and volume V, but only two of these are independent. Thus there is a large gap between the microscopic and thermodynamic description of a system. *The connection between these two descriptions is established by the microscopic definition of entropy.* Based on this definition, statistical thermodynamics (or statistical mechanics) is developed and, in many respects, is the theoretical basis for macroscopic thermodynamics, which is basically empirical. In this chapter we give a brief introduction of the microscopic definition of entropy and some of its consequences. Statistical thermodynamics will be introduced and discussed in more detail in the latter chapters of this book.

3.2 Macroscopic State vs. Microscopic State

We consider a thermodynamic system which consists of N particles. From the thermodynamic point of view, if the magnitudes of two of the state variables are given, then the macroscopic *equilibrium* state is specified. Moreover the state will not change with time. In order to obtain a simple connection with its microscopic state (for brevity, *microstate*) we consider an isolated system, and choose the internal energy U and volume V as the independent state variables. When U and V are given, then the macroscopic state (for brevity, *macrostate*) of the system is specified. However in order to specify the microstate of the system, it is necessary to specify $6N$ variables, which includes the three components of the position vectors $\boldsymbol{r}_i$ and the three components of the momentum vectors $\boldsymbol{p}_i$ for each particle. Moreover, the microstate will change continuously with time, because at least the position vectors $\boldsymbol{r}_i$ are functions of time. Apparently, for a given macrostate, there are a very large number of microstates (because N is a very large number) which are consistent with this macrostate. It is also apparent that not all microstates are consistent with the given macrostate. Hence we introduce the concept of a microscopically accessible state. **A microscopically accessible state is a microstate which is consistent with the given macrostate of the system.** We use mathematical equations to give a more thorough explanation. We consider a system whose macrostate is specified by the internal energy U and the volume V. In terms of the the microscopic variables, the total energy of the system E is given by

$$E = \sum_i \frac{\boldsymbol{p}_i^2}{2m} + \frac{1}{2} \sum_{i \neq j} u(\boldsymbol{r}_i, \boldsymbol{r}_j), \tag{3.1}$$

where m is the mass of each particle and $u(\boldsymbol{r}_i, \boldsymbol{r}_j)$ is the interaction energy between the ith and jth particles. For simplicity, we assume that the volume V is cubic with side length L ($V = L^3$). A microscopically accessible state is a set of the $6N$ variables $(\boldsymbol{r}_1, \boldsymbol{p}_1, \cdots, \boldsymbol{r}_N, \boldsymbol{p}_N)$ which satisfy the following inequalities:

$$U \leq E(\boldsymbol{r}_1, \boldsymbol{p}_1, \cdots, \boldsymbol{r}_N, \boldsymbol{p}_N) \leq U + \Delta U, \tag{3.2}$$

$$0 \leq x_i, y_i, z_i \leq L, \tag{3.3}$$

where x_i, y_i, z_i are components of $\boldsymbol{r}_i$. We note that U is a macroscopic quantity, and E is a sum of microscopic quantities, therefore it is reasonable to let U have a small range of uncertainty ΔU ($\Delta U/U \to 0$) in inequality (3.2).[3] A microscopic state which satisfies inequalities (3.2)–(3.3) is said to be a microscopically accessible state.

3.3 Entropy and the Multiplicity of a Macrostate

We consider a thermodynamic system of N particles whose macrostate is specified by the internal energy U and the volume V. When a set of the values of (U, V) are given, then the macrostate is completely specified. However for a given set of (U, V) there are very large number of microscopically accessible states which are consistent with the macrostate (U, V). We use the notation $\Omega(U, V)$ to denote the number of all the microscopically accessible states, which is known as the **multiplicity** of the macrostate.[4] This is a quantity related to the microscopic states. The microscopic total energy of the system E must be equal to the internal energy of the macrostate U; therefore in the notation $\Omega(U, V)$ we use U to denote the total energy instead of E. The connection between macroscopic thermodynamics and microscopic statistical mechanics is based on the important **Boltzmann[5] relation**,

$$S(U, V) = k \ln \Omega(U, V). \tag{3.4}$$

On the left hand side, $S(U, V)$ is the macroscopic entropy, and on the right hand side $\Omega(U, V)$ is the multiplicity of the macrostate (i.e., the

[3]If the condition $\Delta U/U \to 0$ is satisfied, then the magnitude of ΔU will not affect the result of the following analysis.

[4]Apparently $\Omega(U, V)$ is also a function of N, but for simplicity of notation, N is not explicitly shown.

[5]Ludwig Boltzmann, Austrian physicist (1844–1906).

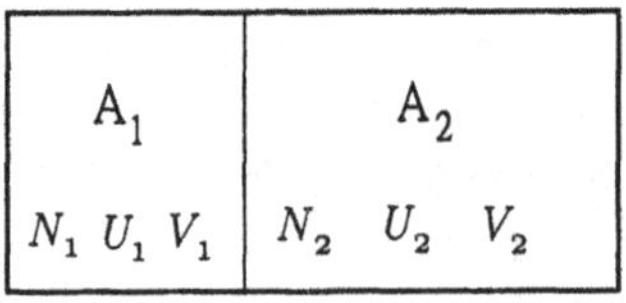

A_1	A_2
N_1 U_1 V_1	N_2 U_2 V_2

Figure 3.1: An isolated system $A^{(0)}$ is composed of A_1 and A_2, which are in thermal contact with each other.

total number of microscopically accessible states). The proportional constant k is called the *Boltzmann constant.*[6] The establishment of the Boltzmann relation is based on many theoretical and experimental tests and confirmations. We may gain some insight into the relation from the following discussions.

1. We mentioned at the end of Chapter 2, that entropy may be considered as a measure of the *randomness* of the system. Microscopically, a larger value of $\Omega(U, V)$ implies that there are *more ways* to distribute the particles of the system. We may say that the system is in a *more disordered* state. Therefore $\Omega(U, V)$ is related to randomness of the system. However S is not directly proportional to Ω, but is proportional to the logarithm of Ω. The reason why S is proportional to $\ln \Omega$ rather than directly to Ω, may be understood from Points 2 and 3 discussed below.

2. The Boltzmann relation satisfies the third law of thermodynamics.

   This means that when the absolute temperature T approaches zero, the entropy S also approaches zero. This is because as $T \to 0$, the system will be in the ground state, $\Omega = 1$ for a non-degenerate ground state, therefore $S = k \ln 1 = 0$. This is the third law of thermodynamics, which will be discussed in Chapter 6.

3. The Boltzmann relation makes S an *extensive* state variable.

   Consider two thermodynamic systems, A_1 and A_2, whose number of particles, internal energy, and volume are, respectively, N_1, U_1, V_1 and N_2, U_2, V_2, as shown in Fig. 3.1. We let A_1 and A_2 be in thermal contact, but the combined system of A_1 and A_2 is an isolated system $A^{(0)}$. A_1 and A_2 can exchange energy but cannot exchange volume and particles. Therefore, N_1, V_1, N_2 and V_2 are

[6] The numerical value of k will be derived in Sec. 3.6.

constants, but U_1 and U_2 are not. However the sum of the internal energies is a constant, i.e.,

$$U_1 + U_2 = U^{(0)} = \text{constant}, \tag{3.5}$$

where $U^{(0)}$ is the total internal energy of $A^{(0)}$. In Eq. (3.5), we have neglected the *mutual interaction energy* between A_1 and A_2, because the number of particles in A_1 and A_2 which have mutual interaction energy are negligibly small in comparison with the number of particles in A_1 and A_2. Now we want to prove that the entropy $S^{(0)}$ of $A^{(0)}$ is the sum of S_1 and S_2, which are, respectively, the entropies of A_1 and A_2, i.e.,

$$S^{(0)}(U^{(0)}, V^{(0)}) = S_1(U_1, V_1) + S_2(U_2, V_2). \tag{3.6}$$

According to the Boltzmann relation, the microscopic definitions of the above three entropies are

$$S^{(0)}(U^{(0)}, V^{(0)}) = k \ln \Omega^{(0)}(U^{(0)}, V^{(0)}), \tag{3.7}$$

$$S_i(U_i, V_i) = k \ln \Omega_i(U_i, V_i), \quad i = 1, 2. \tag{3.8}$$

The validity of Eq. (3.6) is proved in the following:

We take ε as a very small energy unit ($\varepsilon \ll U_1, U_2$) and define $\mathcal{N} \equiv U^{(0)}/\varepsilon$, for simplicity of discussion, we let $\mathcal{N}$ be an integer. Since $U^{(0)} = U_1 + U_2$, we may consider A_1 and A_2 as weakly interacting systems, therefore[7]

$$\Omega^{(0)}(U^{(0)}, V^{(0)}) = \sum_{n=0}^{\mathcal{N}} \Omega_1(n\varepsilon)\,\Omega_2(U^{(0)} - n\varepsilon), \tag{3.9}$$

and Eq. (3.7) can be rewritten as

$$S^{(0)}(U^{(0)}, V^{(0)}) = k \ln \sum_{n=0}^{\mathcal{N}} \Omega_1(n\varepsilon)\,\Omega_2(U^{(0)} - n\varepsilon). \tag{3.10}$$

Suppose that the largest term on the right hand side of Eq. (3.9) is the term with $n = n_0$, i.e.,

$$\Omega_1(n_0\varepsilon)\,\Omega_2(U^{(0)} - n_0\varepsilon) \;>\; \Omega_1(n\varepsilon)\,\Omega_2(U^{(0)} - n\varepsilon), \quad \text{for all } n \neq n_0. \tag{3.11}$$

[7]We suppose that the ground state energy of the systems are zero.

Therefore we may define $\overline{U}_1 \equiv n_0\varepsilon$ and $\overline{U}_2 \equiv U^{(0)} - n_0\varepsilon$ as the most probable energies for A_1 and A_2, respectively. The reason is that on the right hand side of Eq. (3.9), the term with $n = n_0$ contributes the most. We may rewrite the right hand side of Eq. (3.9) as

$$\sum_{n=0}^{\mathscr{N}} \Omega_1(n\varepsilon)\,\Omega_2(U^{(0)} - n\varepsilon) = c\,\Omega_1(n_0\varepsilon)\,\Omega_2(U^{(0)} - n_0\varepsilon), \qquad (3.12)$$

where c is a constant (independent of energy) with an order of magnitude N^{α} $(\alpha > 0)$. This can be seen from the fact that c is a number much smaller than $\mathscr{N}$, and the order of magnitude of $\mathscr{N}$ is N^b $(b > 1)$.[8] Equation (3.10) then becomes (omitting the variable V)

$$\begin{aligned} S^{(0)}(U^{(0)}) &= k \ln \sum_{n=0}^{\mathscr{N}} \Omega_1(n\varepsilon)\,\Omega_2(U^{(0)} - n\varepsilon) \\ &= k \ln \left[c\,\Omega_1(\overline{U}_1)\,\Omega_2(\overline{U}_2)\right] \\ &= k \ln c + k \ln \Omega_1(\overline{U}_1) + k \ln \Omega_2(\overline{U}_2) \\ &= S_1(\overline{U}_1) + S_2(\overline{U}_2) \quad \text{(macroscopic } N_i\text{)}. \qquad (3.13) \end{aligned}$$

In the last equality we have let $\ln c$ be zero, and therefore we get an *equal sign* rather than an approximate sign. This is due to the fact that, for a macroscopic system, we have $\ln c \sim \ln N$, but $\ln \Omega_i \sim N_i$ $(i = 1, 2)$,[9] therefore for a very large N_i $(N_i = 10^n$, $n > 10)$, $\ln c / \ln \Omega_i \sim n \ln 10/10^n$, which may be considered as zero. From Eq. (3.13) we know that only when $U_1 = \overline{U}_1$ and $U_2 = \overline{U}_2$, is $S^{(0)}$ equal to the sum of S_1 and S_2. We will prove in the next section that when $U_1 = \overline{U}_1$ and $U_2 = \overline{U}_2$, A_1 and A_2 are in thermal equilibrium, in which case the entropy of $A^{(0)}$ is the sum of the entropies of A_1 and A_2. This proves that entropy is an extensive variable, such that if the size of the system is doubled, the entropy will also be doubled.

In the above discussion we have used the condition that the number of particles N is very large (i.e., macroscopically large) in several places,

[8] We choose ε to be an energy unit which is much smaller than the average energy per particle $(U^{(0)}/N)$. Therefore there exists a positive number a such that $\varepsilon \sim N^{-a}(U^{(0)}/N)$ holds, and $\mathscr{N} = U^{(0)}/\varepsilon \sim N^b$ $(b = a + 1 > 1)$.

[9] We may take an ideal gas as an example to show that $\ln \Omega_i \sim N_i$, this will be given in Sec. 3.6.

such that we may use an equal sign rather than an approximate sign, because the errors introduced are negligibly small. Hence the application of microscopic statistical theory to macroscopic systems can be considered as very precise. However if the number of particles is not large enough to reach the macroscopic scale, the errors or fluctuations may be non-negligible.

3.4 The Most Probable Energy vs. Thermal Equilibrium

We define the quantity $W(U^{(0)}, U_1) = \Omega_1(U_1)\,\Omega_2(U^{(0)} - U_1)$, which has the physical meaning that when the internal energy of the subsystem A_1 has the value U_1, then the multiplicity of $A^{(0)}$ (the combined system of A_1 and A_2) is $W(U^{(0)}, U_1)$. Since the allowed range of U_1 is from 0 to $U^{(0)}$, the multiplicity of $A^{(0)}$ is

$$\Omega^{(0)}(U^{(0)}) = \sum_{U_1=0}^{U^{(0)}} W(U^{(0)}, U_1). \tag{3.14}$$

The condition for $\overline{U}_1$ to be the *most probable energy* is that when $U_1 = \overline{U}_1$, $W(U^{(0)}, U_1)$ has its maximum value, i.e.,

$$\delta W(U^{(0)}, U_1) = W\,\delta \ln W(U^{(0)}, U_1) = 0, \quad \text{when } U_1 = \overline{U}_1. \tag{3.15}$$

If we use U_2 to denote the energy of A_2, i.e., $U^{(0)} - U_1 = U_2$, then $W(U^{(0)}, U_1) = \Omega_1(U_1)\,\Omega_2(U_2)$, and the above variation becomes

$$\begin{aligned}\delta \ln W(U^{(0)}, U_1) &= \frac{\partial \ln \Omega_1(U_1)}{\partial U_1}\delta U_1 + \frac{\partial \ln \Omega_2(U_2)}{\partial U_1}\delta U_1 \\ &= \left[\frac{\partial \ln \Omega_1(U_1)}{\partial U_1} - \frac{\partial \ln \Omega_2(U_2)}{\partial U_2}\right]\delta U_1 \\ &= 0, \quad \text{when } U_1 = \overline{U}_1 \text{ and } U_2 = \overline{U}_2,\end{aligned} \tag{3.16}$$

where $\overline{U}_2 = U^{(0)} - \overline{U}_1$. Since δU_1 is an arbitrary nonzero quantity, and V_1 and V_2 are kept constant in the differentiations, the above equation can be written as

$$\left(\frac{\partial \ln \Omega_1(U_1)}{\partial U_1}\right)_{V_1}\Bigg|_{U_1=\overline{U}_1} = \left(\frac{\partial \ln \Omega_2(U_2)}{\partial U_2}\right)_{V_2}\Bigg|_{U_2=\overline{U}_2}. \tag{3.17}$$

From $S = k \ln \Omega$ and $dU = TdS - PdV$, the above condition becomes

$$\frac{1}{kT_1} = \frac{1}{kT_2}, \quad \text{or } T_1 = T_2, \tag{3.18}$$

this is the condition that A_1 and A_2 are in thermal equilibrium. Therefore the most probable energy $\overline{U}_1$ ($\overline{U}_2$) is the internal energy for A_1 (A_2) when it is in thermal equilibrium. From the above analysis, we know that when two systems are in thermal contact, they exchange energy constantly, and the respective internal energies are not constant in time (although their sum is a constant). There are energy fluctuations in A_1 and A_2. Yet when we measure them experimentally, most of the time (most probably) we will get the values $\overline{U}_1$ and $\overline{U}_2$, respectively, for the internal energy. Therefore we call these energies the most probable energies; they are also the energies that make A_1 and A_2 to be in thermal equilibrium.

3.5 Entropy and the Second Law of Thermodynamics

In the preceding section, we have defined $W(U_1) = \Omega_1(U_1)\,\Omega_2(U^{(0)} - U_1)$ which is the multiplicity for $A^{(0)}$ when the internal energy of A_1 has the value U_1. Since both A_1 and A_2 are macroscopic systems, $\Omega_1(U_1)$ increases rapidly as U_1 increases, while $\Omega_2(U^{(0)} - U_1)$ decreases rapidly as U_1 increases. This makes the product of Ω_1 and Ω_2 ($= W(U_1)$) have a sharp maximum as U_1 varies (cf. Fig. 3.2). The maximum occurs at $U_1 = \overline{U}_1$, i.e., $W(\overline{U}_1) = W_{\text{max}}$, and the width[10] of the maximum peak $\Delta\overline{U}_1$ is very small in comparison with $\overline{U}_1$. For macroscopic systems, it can be shown[11] that $\Delta\overline{U}_1/\overline{U}_1 \sim N_1{}^{-1/2}$ where N_1 is the number of particles in A_1. Therefore $\Delta\overline{U}_1/\overline{U}_1 \to 0$ for a macroscopic number of N_1.

The behavior of $W(U_1)$ vs. U_1, as shown in Fig. 3.2, is essential for the understanding of the principle of non-decrease of entropy. If initially the wall between A_1 and A_2 is an insulating wall (cf. Fig. 3.1), and the energy of A_1 is $U_1 = U_1^0 \neq \overline{U}_1$, then the value of W corresponding to U_1^0 will be outside of the maximum peak of W, i.e., U_1^0 is not included in

[10]The half width is usually defined as $W(\overline{U}_1 \pm \frac{1}{2}\Delta\overline{U}_1)/W_{\text{max}} = e^{-1}$.

[11]The proof for the case that $N_2 \gg N_1$ (i.e., A_2 is a heat reservoir for A_1) will be given in Chapter 10.

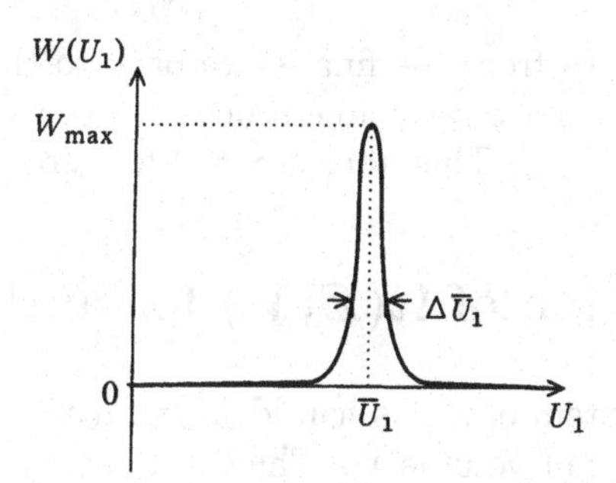

Figure 3.2: The diagram for $W(U_1)$ *vs.* U_1, when both N_1 and N_2 are macroscopically large numbers.

the range $\Delta\overline{U}_1$. Now let the wall between A_1 and A_2 become thermally conducting, i.e., A_1 and A_2 are in thermal contact and can exchange energy with each other. Apparently the initial states of A_1 and A_2 are not in thermal equilibrium, because $U_1^0 \neq \overline{U}_1$, which implies that the initial temperatures of A_1 and A_2 are different, i.e., $T_1^0 \neq T_2^0$. Therefore, after thermal contact, A_1 and A_2 will experience a period of non-equilibrium irreversible processes, until finally $T_1 = T_2$ and thermal equilibrium is reached. The final state is the state where $U_1 = \overline{U}_1$, and $W(U_1) = W_{max}$. It is apparent that the change in entropy in this process is greater than zero,

$$\Delta S = S_{final} - S_{initial} = k(\ln W_{max} - \ln W(U_1^0)) > 0. \tag{3.19}$$

This is consistent with the second law.

It is also easy to understand why this is an irreversible process microscopically. The reason is that the values of $W(U_1^0)$ and $W(\overline{U}_1)$ are, respectively, proportional to the probabilities that the system A_1 will be found in its initial and final states. Due to the fact that $W(U_1^0)/W(\overline{U}_1) \to 0$ when A_1 is a macroscopic system, the probability that the system will return from the final state to the initial state is extremely small (but not really equal to zero). In principle, there is still a *non-zero* probability that the system will *return* to the initial state, but one may have to wait for years, and when the system does return to its initial state, it may stay in the initial state for only a very short and undetectable period of time. Therefore, in practice, we say that this is an irreversible process. According to the above analysis, we may say that there is no inconsistency between macroscopic irreversibility and microscopic reversibility. When the final state and the initial state have the same probability W, then the system can return from the final state back to the initial state with ease. This is the case for a reversible process; in this case $S_{\text{final}} = S_{\text{initial}}$, because $W_{\text{final}} = W_{\text{initial}}$. Therefore $\Delta S = 0$ for a reversible process.

3.6 Calculation of $\Omega(E, V)$ for an Ideal Gas

We consider the system of a classical ideal gas with N particles, moving freely in a container of volume V. The total energy of the system is in the range from U to $U + \Delta U$ (we allow U to have a small uncertainty ΔU, such that $\Delta U/U \to 0$). The Hamiltonian H of the system is

$$H(q,p) = \sum_{i=1}^{N} \frac{\mathbf{p}_i^2}{2m}, \tag{3.20}$$

where m is the mass of each particle, $\mathbf{q}_i = (q_{ix}, q_{iy}, q_{iz})$ and $\mathbf{p}_i = (p_{ix}, p_{iy}, p_{iz})$ are, respectively, the position and the momentum vector of the ith particle. We use the notation (q,p) to denote a set of $6N$ variables $(\mathbf{q}_1, \cdots, \mathbf{q}_N, \mathbf{p}_1, \cdots, \mathbf{p}_N)$. Each set of $6N$ variables (q,p) denotes a microstate. Usually we call the space spanned by these $6N$ variables the $6N$-dimensional *phase space*. We note that $H(q,p)$ for the ideal gas is independent of the spatial variables $\mathbf{q}_i$. The multiplicity Ω is equal to the volume in the phase space, which is *consistent with the macrostate* of the system, divided by the volume occupied by a microstate.[12] Therefore [13]

$$\Omega'_N(U,V) = \frac{1}{\omega_{6N}} \int' d^3q_1 \cdots d^3q_N d^3p_1 \cdots d^3p_1 \cdots d^3p_N, \tag{3.21}$$

where ω_{6N} is an infinitesimal $6N$-dimensional phase space volume, which represents a unit of volume occupied by a microstate. A primed integral sign $\int'$ means that the range of integration must be consistent with the macrostate. This means that (q,p) should satisfy the following conditions:

$$U \le H(q,p) \le U + \Delta U, \quad 0 \le q_{ix}, q_{iy}, q_{iz} \le L, \tag{3.22}$$

[12]Theoretically, a point (q,p) in the phase space represents a microstate. In practical calculations, however, we need a small volume ω_{6N} in the phase space to represent a microstate. Of course, it is ideal to have ω_{6N} as small as possible. Due to the uncertainty principle of quantum mechanics, however $\delta q_{ix}\delta p_{ix} \ge h$, the smallest possible value is $\omega_{6N} = h^{3N}$. Since entropy is proportional to $\ln \Omega_N(U,V)$, different values of ω_{6N} will give only an unimportant additive constant in the entropy.

[13]For the moment we use Ω', instead of Ω, to denote the multiplicity. The reason for this will be explained in Sec. 3.8.

where $L^3 = V$ is the volume of the system. The lower limit in the integral in Eq. (3.21) is not 0 (the smallest possible value of $H(q,p)$ is chosen to be 0), therefore the integration is not an easy one. We define Σ'_N to be the following integral,

$$\begin{aligned}\Sigma'_N(U,V) &= \frac{1}{\omega_{6N}} \int_{0\le H(q,p)\le U} d^3q_1\cdots d^3q_N d^3p_1\cdots d^3p_N \\ &= \frac{V^N}{\omega_{6N}} \int_{0\le H(q,p)\le U} d^3p_1\cdots d^3p_N \\ &= \frac{V^N}{\omega_{6N}} \int_{0\le \sum_i (p_i^2/2m)\le U} d^3p_1\cdots d^3p_N \\ &= C_{3N}\frac{V^N}{\omega_{6N}}(2mU)^{3N/2}. \end{aligned} \tag{3.23}$$

In the above equation, the integration of $d^3q_1\cdots d^3q_N$ is V^N (since H is independent of the spatial variables). The integration of $d^3p_1\cdots d^3p_N$ will give the volume of a $3N$-dimensional sphere $\mathcal{V}_{3N} = C_{3N}(2mU)^{3N/2}$, where $(2mU)^{1/2}$ is the radius of the sphere (i.e., $\sum_i p_i^2 = 2mU$), and C_{3N} is a constant independent of the radius. Therefore,

$$\begin{aligned}\Omega'_N(U,V) &= \Sigma'_N(U+\Delta U, V) - \Sigma'_N(U,V) \quad \left(\frac{\Delta U}{U}\to 0\right) \\ &= \frac{\partial \Sigma'_N}{\partial U}\Delta U = \left(\frac{3N}{2}\right)\left(\frac{\Delta U}{U}\right)\Sigma'_N(U,V). \end{aligned} \tag{3.24}$$

From this result we get $\Omega'_N(U,V) \sim U^{3N/2-1}$, which gives us the condition $\ln \Omega'_N \sim N$ ($1 \ll 3N/2$ and can be neglected), which we used in Sec. 3.3. From the Boltzmann relation (3.4), we obtain the entropy

$$\begin{aligned}S'_N(U,V) &= k\ln \Sigma'_N(U,V) + k\ln\left(\frac{3N\Delta U}{2U}\right) \\ &\approx k\ln\Sigma'_N(U,V), \end{aligned} \tag{3.25}$$

where the last approximate sign can be replaced by an equal sign, because $\ln(3N\Delta U/2U)/\ln \Sigma'_N \to 0$ (we may take $3N\Delta U/2U \sim \Delta U/(U/N)$ as a number of order N, and $\ln N/N \to 0$). In Eq. (3.23) we take[14] $\omega_{6N} = h^{3N}$. From mathematics (see Appendix D) we know that when N is a very large number, $\ln C_{3N} \approx (3N/2)\ln\pi - (3N/2)\ln(3N/2) + 3N/2$.

[14]See footnote 12 in p. 82.

Therefore from Eq. (3.25) the entropy $S'(U,V)$ can be written as (dropping the subscript N)

$$S'(U,V) = Nk \ln \left[V \left(\frac{4\pi mU}{3h^2 N} \right)^{3/2} \right] + \frac{3}{2} Nk. \tag{3.26}$$

This is the entropy we have obtained from the multiplicity $\Omega'(U,V)$ and the Boltzmann relation. From $dU = TdS - PdV$ and the above equation for entropy, we obtain

$$\left(\frac{\partial S'}{\partial U} \right)_V = \frac{1}{T} = \frac{3}{2} Nk \frac{1}{U}, \tag{3.27}$$

$$\left(\frac{\partial S'}{\partial V} \right)_U = \frac{P}{T} = Nk \frac{1}{V}. \tag{3.28}$$

The above two equations give us

$$U = \frac{3}{2} NkT = \frac{3}{2} nRT, \tag{3.29}$$

$$PV = NkT = nRT. \tag{3.30}$$

Equation (3.29) tells us that the internal energy U of an ideal gas is independent of the volume V. Moreover, from this relation we can get the constant volume specific heat c_v per kilomole to be equal to[15] $3R/2$. Equation (3.30) is the equation of state for an ideal gas, $PV = NkT$, which is derived from microscopic considerations. Comparing this with the empirical equation, $PV = nRT$, we get the Boltzmann constant

$$k = \frac{n}{N} R = \frac{R}{N_A} = \frac{8.31 \times 10^3}{6.02 \times 10^{26}} = 1.38 \times 10^{-23}\, \mathrm{J\,K^{-1}}. \tag{3.31}$$

In macroscopic thermodynamics, both the equation of state and the specific heat of an ideal gas, can only be obtained by experimental means. There is no way to derive them from thermodynamic relations. However the fact that the internal energy U of an ideal gas is independent of the volume V (i.e., U is a function of T only) can be derived from thermodynamic relations. We will derive $U = U(T)$ thermodynamically in the next section.

[15]This is the constant volume specific heat for a monatomic ideal gas. The Hamiltonian in Eq. (3.20) indicates that each particle has only translational kinetic energy, hence the gas is monatomic.

3.7 Proof of $U = U(T)$ Thermodynamically

We start from the relation $dU = TdS - PdV$ to prove that the internal energy U of an ideal gas is independent of V. From $dS = \frac{1}{T}dU + \frac{P}{T}dV$ we have

$$\left(\frac{\partial S}{\partial V}\right)_T = \frac{1}{T}\left(\frac{\partial U}{\partial V}\right)_T + \frac{P}{T},$$
$$\left(\frac{\partial S}{\partial T}\right)_V = \frac{1}{T}\left(\frac{\partial U}{\partial T}\right)_V. \tag{3.32}$$

Since dS is an exact differential, we have $\partial^2 S/\partial T\partial V = \partial^2 S/\partial V\partial T$, and Eq. (3.32) gives

$$\frac{\partial}{\partial T}\left[\frac{1}{T}\left(\frac{\partial U}{\partial V}\right)_T + \frac{P}{T}\right]_V = \frac{\partial}{\partial V}\left[\frac{1}{T}\left(\frac{\partial U}{\partial T}\right)_V\right]_T. \tag{3.33}$$

Differentiate the above equation and simplify the terms to get

$$\left(\frac{\partial U}{\partial V}\right)_T = T^2\left[\frac{\partial}{\partial T}\left(\frac{P}{T}\right)\right]_V. \tag{3.34}$$

For an ideal gas $P/T = nR/V$, the partial differentiation of the right hand side of Eq. (3.34) is zero, therefore

$$\left(\frac{\partial U}{\partial V}\right)_T = 0 \text{ (ideal gas)}. \tag{3.35}$$

The above equation tells us that under the condition of constant temperature, the internal energy of an ideal gas is independent of the volume, i.e., U is a function of T only. Mathematically we may write $U = U(T)$. From $dU = đQ - PdV$, the constant volume heat capacity C_V of an ideal gas is

$$C_V = \left(\frac{đQ}{dT}\right)_V = \left(\frac{dU}{dT}\right)_V = \frac{dU}{dT}, \tag{3.36}$$

that is

$$dU = C_V dT \text{ (ideal gas)}. \tag{3.37}$$

If C_V is independent of T, we can integrate the above equation to get

$$U = C_V T = nc_v T \text{ (ideal gas)}. \tag{3.38}$$

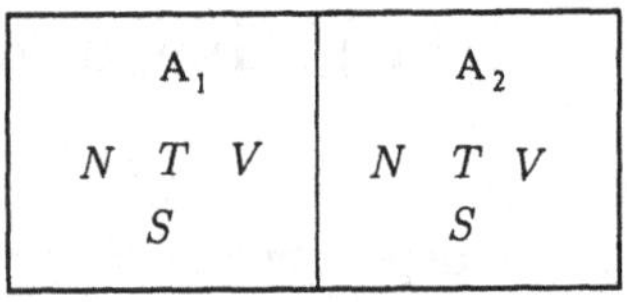

A_1	A_2
$N\ T\ V$ S	$N\ T\ V$ S

Figure 3.3: The Gibbs paradox.

Here $c_v = C_V/n$ is the constant volume specific heat (n is the number of kilomoles), and we may take $U = 0$ when $T = 0$. The value of c_v, however, cannot be derived from macroscopic thermodynamic relations. This is one of the examples that the microscopic statistical theory is more powerful than macroscopic thermodynamics.

3.8 The Gibbs Paradox

If we substitute $U = \frac{3}{2}NkT$ into Eq. (3.26), then the expression of the entropy for an ideal gas becomes

$$S'(T,V) = Nk\left(\ln V + \frac{3}{2}\ln T + s_0\right), \tag{3.39}$$

$$s_0 = \frac{3}{2}\ln\left(\frac{2\pi mk}{h^2}\right) + \frac{3}{2}.$$

Equation (3.39) is an expression for the entropy of an ideal gas which is derived microscopically. Although this expression can lead to the correct equation of state and the internal energy $U = U(T)$ for an ideal gas, yet it does contain a paradox, known as the **Gibbs[16] paradox**. In the following we will explain what this paradox is.

We consider two identical ideal gas systems A_1 and A_2, which have the same number of particles N, temperature T, and volume V. The systems are in thermal contact, and the combined system A_1+A_2 is an isolated system, as shown in Fig. 3.3. The entropies for A_1 and A_2 are equal to each other, which can be expressed by Eq. (3.39). In order to emphasize that S' depends on the number of particles, we rewrite the expression of Eq. (3.39) as

$$S'_N(T,V) = Nk\left(\ln V + \frac{3}{2}\ln T + s_0\right). \tag{3.40}$$

[16]Josiah W. Gibbs, American physicist (1839–1903).

Now we remove the partition between A_1 and A_2, and let the molecules of A_1 and A_2 freely mix. Since the initial temperatures and pressures are equal for A_1 and A_2, after mixing the temperature and pressure will remain unchanged. The number of particles will become $2N$, and the volume $2V$. The entropy after mixing is therefore

$$S'_{2N}(T, 2V) = 2Nk\left(\ln 2V + \frac{3}{2}\ln T + s_0\right). \tag{3.41}$$

The change of entropy after the mixing is

$$\Delta S' = S'_{2N}(T, 2V) - 2S'_N(T, V) = 2Nk \ln 2 > 0. \tag{3.42}$$

The entropy change $\Delta S' > 0$ implies that it is an irreversible process, but in fact this is a reversible process, because we can put back the partition and the system returns to its original situation. The systems A_1 and A_2 and their surroundings are all the same as before mixing. Note that we do not need any extra effort to accomplish this, if we assume the mass of the partition is negligibly small. The only difference that we know about is that A_1 and A_2 exchange some of their molecules, but this can not be detected experimentally, because the molecules are *indistinguishable*. Thus both A_1 and A_2 return to their initial states, and the process is reversible. We should have $\Delta S = 0$, but the expression for entropy Eq. (3.40) leads to a contradictory result $\Delta S' > 0$; this is known as the **Gibbs paradox**. The origin of the Gibbs paradox clearly comes from the fact that the same kind of molecules are *indistinguishable*, but we have assumed that they are *distinguishable*, as their orbits can be traced in classical mechanics. However this is incorrect, because all the particles are identical, there is no way to tell the difference between any two of them.[17] Therefore the exchange of any two particles would not give a new microstate. In the calculation of the multiplicity, Eq. (3.21) does not take this factor into account, which results in the Gibbs paradox. Gibbs suggested that the *correct multiplicity* should be of the form

$$\Omega_N(U, V) = \frac{1}{N!}\frac{1}{\omega_{6N}}\int' d^3q_1 \cdots d^3q_N d^3p_1 \cdots d^3p_1 \cdots d^3p_N. \tag{3.43}$$

This means that the number $\Omega'_N(U, V)$ given by Eq. (3.21) is wrong by a factor of $N!$, which comes from the exchange among the N particles. We need to divide the original $\Omega'_N(U, V)$ by $N!$ to get the correct multiplicity,

[17]For a classical gas, however, we still assume that the orbit of each molecules can be traced.

Eq. (3.43). The correct entropy is then $S'_N(U,V)$, given in Eq. (3.40), minus $k\ln N!$. By using the Stirling[18] approximation (see Appendix C)

$$\ln N! = N\ln N - N \quad (\text{for large } N), \tag{3.44}$$

the *correct entropy* is

$$S_N(T,V) = Nk\left(\ln\frac{V}{N} + \frac{3}{2}\ln T + s'_0\right), \quad s'_0 = s_0 + 1. \tag{3.45}$$

It is easy to prove that there is no Gibbs paradox for the entropy $S_N(T,V)$ given in Eq. (3.45), because when N changes to $2N$, V also changes to $2V$, and T remains unchanged; Eq. (3.45) yields $S_{2N}(T,2V) = 2S_N(T,V)$, therefore $\Delta S = S_{2N}(T,2V) - 2S_N(T,V) = 0$. It is a reversible process and there is no Gibbs paradox. Moreover, the entropy given by Eq. (3.45) will produce the same Eqs. (3.29)–(3.30). Therefore Eq. (3.45) gives the correct microscopic formula for the entropy of an ideal gas.[19]

Before closing this section, we remark that in Fig. 3.3, if A_1 and A_2 are ideal gases with *different kinds of molecules*, then removing the partition and then putting it back again is an irreversible process. After the partition is removed, the molecules of A_1 and A_2 will occupy the entire volume of A_1 and A_2, i.e., the volume is $2V$. In this situation, the total entropy of the system is the sum of the entropies of A_1 and A_2, whose molecules occupy the same volume of size $2V$, and the number of molecules and temperature remain unchanged. Therefore from Eq. (3.45) we have

$$\Delta S = 2Nk\ln 2 > 0. \tag{3.46}$$

This is the amount of the entropy change after two different kinds of molecules are mixed, and is called the *entropy of mixing*.

3.9 Thermal Entropy and Configuration Entropy

The entropy discussed above is the entropy due to the distribution of energy, and is called the **thermal entropy**. This means that the entropy is associated with thermal energy. There is another type of entropy which is associated with the configuration of the system. This type of entropy is called **configuration entropy**. The energy of the system in different

[18] James Stirling, British mathematician (1696–1770).

[19] The properties of $\ln\Omega'_N$ for large N we mentioned in Sec. 3.6 still hold for $\ln\Omega_N$.

configurations may be different, but may sometimes also have the same value. For the latter case we say the system is degenerate. There are two frequently encountered configuration entropies, the ***spin entropy*** due to the spins of the particles, and the ***substitutional entropy*** occurring in alloys. We give brief explanations for these two entropies.

1. Spin entropy:

   We use the simplest case of spin-$\frac{1}{2}$ particles[20] as an example. Suppose the system consists of N *localized*[21] particles, each with a spin equal to 1/2. The spin of each particle has two possible directions, i.e., spin up or spin down. If the interactions between the spins are negligible and there is no external magnetic field, then the energy of the system is independent of the spin direction of each particle. In this case, it is apparent that the spin entropy is $S = k \ln 2^N$. This is because the probability of spin up or down is equal for each particle, as it does not depend on the energy of the system. Therefore the multiplicity is 2^N, and the spin entropy is a constant. Consequently spin entropy is important only when energy depends on the spin directions of the particles. In this case there are different probabilities for a particle to have spin up or spin down.

   We consider the case where there is an external magnetic field B. The magnetic potential energy due to B is $-\boldsymbol{\mu} \cdot \boldsymbol{B}$, where $\boldsymbol{\mu}$ is the magnetic moment associated with the spin.[22] The magnetic potential energy for each particle can then have two possible values, $+\mu B$ and $-\mu B$. Now suppose that the number of particles with energy $+\mu B$ and $-\mu B$ are N_+ and N_-, respectively, we have

$$N_+ \mu B + N_-(-\mu B) = U, \tag{3.47}$$

$$\Omega = \frac{N!}{N_+! N_-!}, \quad N_+ + N_- = N, \tag{3.48}$$

[20]This means that each particle has a spin equal to $\hbar/2$.

[21]Usually this refers to atoms or molecules in a solid. The position of each atom (molecule) is fixed in the sense that it may have a small vibration around a *fixed position*, whose vibrational motion is usually not considered in the spin entropy. If the particles are atoms or molecules in a gas, they can exchange their positions easily and frequently; then the situation is much more complicated. In this case one needs to consider what statistics the particles have. We will discuss the statistics of non-localized particles in Chapter 10.

[22]The magnetic moment is proportional to the angular momentum (here the spin) for a charged particle. See Sec. 11.9 for the details.

where U is the ***total magnetic energy*** of the system (kinetic energy is not included). For a macroscopic system all of N, N_+ and N_- are very large numbers, so we may use the Stirling approximation (3.44) to calculate the entropy $S = k \ln \Omega$. The calculation of the entropy and the thermodynamic properties one can obtain from the entropy is left as an exercise for the reader (cf. Problem 3.6).

2. Substitutional entropy:

   We use a binary alloy as an example. The alloy is composed of N_A A atoms and N_B B atoms. In general the nearest neighbor pair interaction energies of AA, BB, and AB will be different. Therefore the multiplicity Ω, and the entropy S, will be a function of the energy of the system U. The calculation of S as a function of U (or temperature T) and the other thermodynamic properties of the system is an interesting, but not easy, problem in statistical mechanics. We will not go any further here. The simplest case is when all the pair interaction energies are equal; then the total energy of the system is independent of the geometrical arrangement of the A and B atoms. If there are a total of N lattice sites, and $N = N_A + N_B$, then the Ω due to the geometrical arrangement is given by Eq. (3.48):

$$\Omega = \frac{N!}{N_A!\, N_B!}, \quad N_A + N_B = N. \tag{3.49}$$

   The degrees of degeneracy is therefore Ω.

   There is another type of substitutional entropy. A crystal may have only one kind of atoms, but there are two different kinds of sites, ***regular sites*** (R-sites) and ***interstitial sites*** (I-sites). An I-site is not a R-site, but it is in between several R-sites, and hence the name interstitial. In the ground state, all atoms are positioned at R-sites, and the system has minimum energy. However, when the temperature increases, it is possible that some of the atoms may be displaced from a R-site to an I-site. The displaced atoms cost energy, as well as increase the entropy of the system. When an atom is excited to an I-site, there will be an empty R-site, which can be anywhere in the R-sites. The excited atom can be in any of the I-sites, therefore the multiplicity will be increase by a factor of $(N_R!)(N_I!)$, where N_R and N_I are, respectively, the total number of R-sites and I-sites. This is one example of substitutional entropy.

3.10 Problems

3.1. A box is separated by a partition which divides its volume in the ratio 3:1. The larger portion of the box contains 1000 molecules of Ne gas; the smaller, 100 molecules of He gas. A hole is punched in the partition and the molecules can go through the hole to the other portion of the box. A final thermal equilibrium for the molecules is then reached.

(a) Find the mean number of molecules of each type on either side of the partition.

(b) What is the probability of finding 1000 molecules of Ne gas in the larger portion and 100 molecules of He gas in the smaller (i.e., the same distribution as in the initial system)?

3.2. Consider an ideal gas with N molecules which are confined to move in a two-dimensional space of area A. Calculate $\Sigma'_N(U, A)$ for this gas, which is the two-dimensional version of Eq. (3.23). From this calculate the entropy $S_N(U, A)$ for the two-dimensional ideal gas.

3.3. Consider a container which is divided into two parts, 1 and 2, by a removable wall. Part 1, having volume $2V$, is filled with $2n$ kilomoles of gas A; and part 2, having volume V, is filled with n kilomoles of gas B. Both gases have the same temperature T, and can be treated as ideal gases. Now the removable wall is removed.

(a) Calculate the entropy change *by using* Eq. (2.30) *and* Eq. (3.38) for: (i) A and B are identical gases; and (ii) A and B are different gases.

(b) Show that (i) is reversible; and (ii) is irreversible.

(c) Check the entropy changes for (i) and (ii) by using Eq. (3.45), the microscopic formula for the entropy S of an ideal gas.

3.4. Consider a container which is divided into two parts, 1 and 2, by a removable wall. Part 1 is filled with N_1 molecules of gas A; and part 2 is filled with N_2 molecules of gas B. Both gases have the same temperature T and pressure P, and can be treated as ideal gases. Now the removable wall is removed. By using the microscopic formula for the ideal gas entropy Eq. (3.45), show that the *entropy of mixing* is given by

$$\Delta S_{\text{mixing}} = -Nk[x \ln x + (1 - x) \ln(1 - x)],$$

where $N = N_1 + N_2$, $x = N_2/N$. Show that ΔS_{mixing} is always greater than 0 when $0 < x < 1$.

3.5. The energy level of a localized oscillator is given by $\varepsilon = (n + 1/2)\hbar\nu$, with $n = 0, 1, 2, 3, \cdots$. Consider a system consisting of N such weakly interacting oscillators which has the total energy $U = \frac{1}{2}N\hbar\nu + M\hbar\nu$, where M is an integer.

(a) Find the multiplicity $\Omega_N(M)$.
(b) Find the temperature T of the system as a function of N and M (or N and U), under the conditions $N \gg 1$ and $M \gg 1$.
(c) Find U as a function of T, i.e., $U = U(T)$.

3.6. Consider a system of N weakly interacting particles, each fixed in position with only two energy states, 0 or ε ($\varepsilon > 0$), for each particle. The total energy of the system is U. Assume that N is a very large number.

(a) Make use of the relation $S = k \ln \Omega$ to calculate the entropy of the system in terms of N, N_0, and N_1, where N_0 and N_1 are the number of particles with energy 0 and ε, respectively.
(b) Express S in terms of N, ε, and U.
(c) Use the relation $dU = TdS - PdV$ to find N_0 and N_1 as a function of N, T, and ε. Note that ε may be a function of V.

3.7. Consider a crystal which consists of two *equal* number of sublattice sites: R-sites and I-sites (R=regular, I=interstitial). In the ground state, each of the R-sites is occupied by an atom, and all the I-sites are empty. An atom can be excited from a R-site to an I-site with an excitation energy $\varepsilon > 0$. Suppose the total number of atoms is N, and at some temperature T there are n atoms in the I-sites. Neglect the vibrational motion of the atoms.

(a) What is the entropy at temperature T? Express your answer in terms of N and n. Simplify your result by assuming that $N >> 1$, $n >> 1$, and $N - n >> 1$.
(b) Find the equilibrium n as a function of T, N, and ε.

3.8. Assuming that a silver-gold alloy is a random mixture of gold and silver atoms, calculate the increase in entropy when 10 g of gold are mixed with 20 g of silver to form an ideal homogeneous alloy. The gram atomic weights of gold and silver are 198 and 107.9, respectively.

Chapter 4

Consequences of the First and Second Laws

In this chapter we discuss some of the results which can be derived from the first and second laws.

4.1 Calculations of the Entropy Change

The first topic we want to discuss is how to use certain formulas, derived from the first and second laws, to calculate the entropy change for a system which changes from state 1 to state 2. In Chapter 2 we have derived some formulas for calculating the entropy change ΔS for a system in simple cases. In this chapter we will discuss in more detail the calculation method for more general cases. In general, the system under consideration may not be an isolated system, therefore the total change in entropy is the sum of the entropy changes of the system and the surroundings. Thus

$$\Delta S_{\text{total}} = \Delta S + \Delta S_{\text{surroundings}}, \tag{4.1}$$

where ΔS is the entropy change for the system under consideration. Since S is a state variable, the change of S depends only on the initial state 1 and the final state 2, independent of the path.[1] Therefore we

[1] Although ΔS for the system under consideration is independent of the path, $\Delta S_{\text{surroundings}}$ will depend on the path. Therefore ΔS_{total} will also depend on the path.

can always choose a reversible process to calculate ΔS, no matter what the real process is. Of course, the *chosen process* must bring the system from state 1 to state 2 just as the *real process* does. We can then use Eq. (2.59) to calculate the entropy change,

$$dS = \frac{đQ^{(r)}}{T} = \frac{dU}{T} + \frac{P}{T}\,dV, \tag{4.2}$$

where the superscript (r) in $đQ^{(r)}$ indicates that it is the heat absorbed by the system in the *reversible process*. Integrating the above equation from the initial state 1 to the final state 2, we get

$$\Delta S = S_2 - S_1 = \int_1^2 \frac{đQ^{(r)}}{T} = \int_1^2 \frac{dU}{T} + \int_1^2 \frac{P}{T} dV. \tag{4.3}$$

This is the main equation we will use as the starting point for the calculation of ΔS. Before doing any real calculations, we emphasize again that the second law requires that $\Delta S_{\text{total}} \geq 0$ for an *isolated system*. Therefore for a system which is not isolated, $\Delta S < 0$ does not violate the second law, moreover ΔS is independent of the path. In the following we discuss several frequently used reversible processes and derive the formula for calculating ΔS in each case.

1. A reversible adiabatic process:

   In this case $đQ^{(r)} = 0$, therefore

   $$\Delta S = \int_1^2 \frac{đQ^{(r)}}{T} = 0. \tag{4.4}$$

   This means that the entropy change for a system is zero when it undergoes a *reversible adiabatic process*. This is known as an *isentropic process*, i.e., constant entropy process. We emphasize here that for an *irreversible adiabatic process*, even though there is no heat absorbed or rejected by the system, yet the entropy change is, in general, not zero. An example is the adiabatic free expansion for an ideal gas. The process is adiabatic but irreversible, therefore the entropy change ΔS is not equal to zero. We will come back to this problem in Sec. 4.2 and will there explain how to calculate the entropy change ΔS for the irreversible adiabatic free expansion of an ideal gas.

2. A reversible isothermal process:

   In this process T =constant, therefore

$$\Delta S = \int_1^2 \frac{đQ^{(r)}}{T} = \frac{\Delta Q^{(r)}}{T}, \tag{4.5}$$

   where $\Delta Q^{(r)}$ is the total amount of heat absorbed by the system during the isothermal process. The most common examples of this process are the phase transitions. For a given temperature and pressure, the entropies for a given substance are different in different phases. This means that the multiplicities are different for different phases, and this leads to the existence of latent heat in phase transitions. From Eq. (4.5) the latent heat per kilomole is $l = T\Delta s$, where $\Delta s = s_2 - s_1$ is the entropy difference per kilomole between the two phases. The entropy for the phase stable at higher temperature is greater than that of the phase stable at lower temperature.

3. A reversible isochoric process:

   In this process V =constant,$dV = 0$,and Eq. (4.3) can be written as

$$\Delta S = \int_1^2 \frac{đQ^{(r)}}{T} = \int_{T_1}^{T_2} C_V \frac{dT}{T}, \tag{4.6}$$

   where $C_V = (đQ^{(r)}/dT)_V$ is the constant volume heat capacity. Suppose C_V can be approximated as a constant (independent of T) in the temperature range from T_1 to T_2, then C_V can be taken out of the integration sign, and Eq. (4.6) becomes

$$\Delta S = C_V \int_{T_1}^{T_2} \frac{dT}{T} = C_V \ln \frac{T_2}{T_1} \quad (C_V = \text{constant}). \tag{4.7}$$

4. A reversible isobaric process:

   In this case P =constant and Eq. (4.3) becomes

$$\Delta S = \int_1^2 \frac{đQ^{(r)}}{T} = \int_{T_1}^{T_2} C_P \frac{dT}{T}, \tag{4.8}$$

   where $C_P = (đQ^{(r)}/dT)_P$ is the constant pressure heat capacity. Suppose C_P can be approximated as a constant (independent of T) in the temperature range from T_1 to T_2, then C_P can be taken out of the integration sign, and, by Eq. (4.8), we have

$$\Delta S = C_P \int_{T_1}^{T_2} \frac{dT}{T} = C_P \ln \frac{T_2}{T_1} \quad (C_P = \text{constant}). \tag{4.9}$$

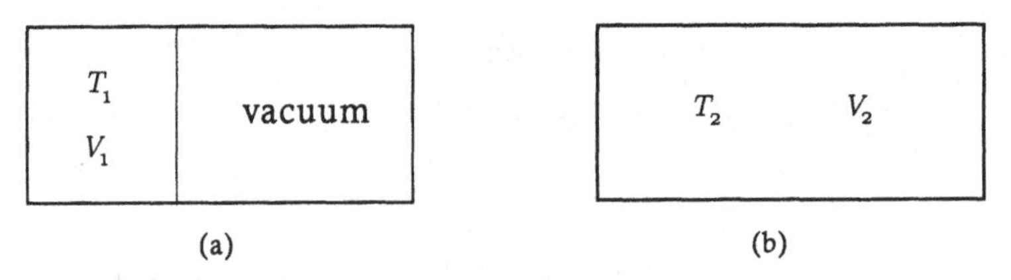

Figure 4.1: An adiabatic free expansion of an ideal gas: (a) before expansion, and (b) after expansion.

4.2 Numerical Examples

In this section we give two examples which illustrate how to calculate the entropy change ΔS.

1. The adiabatic free expansion of an ideal gas:

As shown in Fig. 4.1(a), a container with thermally insulated walls is divided into two parts by a partition. The left part of the container is filled with an ideal gas at temperature T_1 and volume V_1, and the right part is a vacuum. Now remove the partition, letting the gas molecules freely expand to the right part. In this process the gas does no work. In the final equilibrium state the gas has temperature T_2 and volume V_2, as shown in Fig. 4.1(b). During the expansion, the number of molecules of the gas remains unchanged. The question is: How to calculate the entropy change of the gas $\Delta S = S(T_2, V_2) - S(T_1, V_1)$ for the expansion? We note that the walls of the container are thermally insulated, therefore, during the expansion, there is no heat absorbed or rejected by the gas. However this is an irreversible process,[2] therefore $\Delta S > 0$ by the second law, even though $\Delta Q = 0$.

In order to calculate ΔS, we have to choose a *reversible process* which will bring the system from the initial state (T_1, V_1) to the final state (T_2, V_2), where T_1, V_1, and V_2 can be considered as known quantities. However T_2 is unknown, and it has to be found before we can calculate the entropy change. In the free expansion process no work is done and no heat is absorbed by the system, therefore by the first law the internal energy U of the system remains unchanged. From Eq. (3.35) we know that the internal energy U of an ideal gas is a function of T only, independent of V. Therefore the internal energy U and the temperature T of the gas

[2]The gas cannot be returned to its initial state by simply replacing the partition.

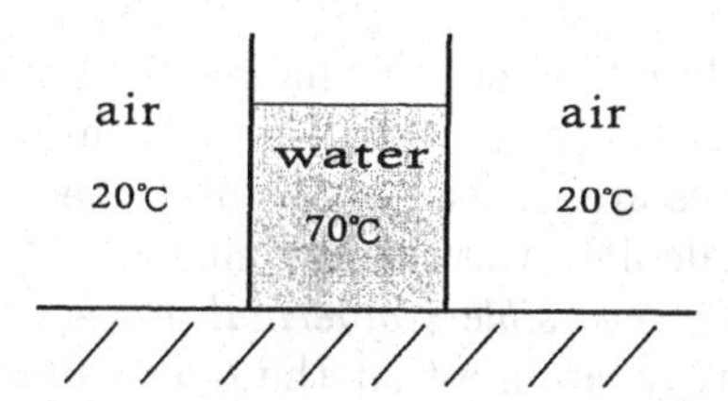

Figure 4.2: A glass of 70°C water is placed in open 20°C air. The final temperature of the water is equal to the air temperature, 20°C.

remain unchanged after the expansion,[3] i.e., $T_2 = T_1$. This result tells us that the temperatures of the initial and final states are equal. Thus, for the purpose of calculating ΔS, *we may choose a reversible isothermal process to replace the irreversible free expansion.* Equation (4.5) can then be used to calculate the entropy change ΔS. Since this is an isothermal process, we have $\Delta U = \Delta Q - \Delta W = 0$, thus $\Delta Q = \Delta W$. Therefore from Eq. (4.5)

$$\begin{aligned} \Delta S &= \frac{\Delta Q}{T_1} = \frac{\Delta W}{T_1} = \frac{1}{T_1}\int_{V_1}^{V_2} PdV \\ &= nR\int_{V_1}^{V_2}\frac{dV}{V} = nR\ln\frac{V_2}{V_1}. \end{aligned} \tag{4.10}$$

Since $V_2 > V_1$, $\Delta S > 0$, which is consistent with the requirement of the second law. Note that $\Delta S_{\text{surroundings}} = 0$ in the real process, because the system under consideration is an isolated system. However in the reversible isothermal process we have chosen $\Delta S_{\text{surroundings}} = -\Delta Q/T_1 = -\Delta S$, therefore $\Delta S_{\text{total}} = 0$ in this case. Thus although ΔS is independent of the process, ΔS_{total} does depend on the process.

2. Thermal contact between two bodies with different temperatures:

As shown in Fig. 4.2, a glass of 70°C water is put in open 20°C air. After a certain period of time the final temperature of the water is equal to the air temperature, 20°C. What is the total entropy change (if the entropy of the glass can be neglected)?

The solution is as follows:

[3] We assume that there is no heat exchange between the gas and the container during the expansion process.

We need to calculate the entropy changes for both the water and the air. Since the system (water plus air) is in open air, therefore the initial and final pressures are the same. We may therefore use a *reversible isobaric process* to calculate the entropy change ΔS for the water. For the air, we may use a *reversible isothermal* process to calculate the entropy change.[4] We may assume that the C_P of water is approximately a constant between 20°C and 70°C, then from Eq. (4.9) we have

$$\begin{aligned}\Delta S_{\text{water}} &= C_P^{\text{water}} \int_{T_1}^{T_2} \frac{dT}{T} = C_P^{\text{water}} \ln \frac{T_2}{T_1} \\ &= -nc_P^{\text{water}} \ln \frac{343}{293}. \end{aligned} \tag{4.11}$$

This is the entropy change for the water, where n is the number of kilomoles for the water molecules in the glass. T_1 and T_2 in the above equation should be in the *absolute temperature scale* rather than in the Celsius temperature scale. i.e., $T_1 = 343$ K and $T_2 = 293$ K. The entropy change for the air can be obtained from Eq. (4.5)

$$\begin{aligned}\Delta S_{\text{air}} &= \frac{\Delta Q}{T_{\text{air}}} = \frac{nc_P^{\text{water}}(70-20)}{T_{\text{air}}} \\ &= nc_P^{\text{water}} \frac{50}{293}, \end{aligned} \tag{4.12}$$

where ΔQ is the total amount of heat transferred from the water to the air. The total entropy change (including water and air) is

$$\begin{aligned}\Delta S_{\text{total}} &= \Delta S_{\text{air}} + \Delta S_{\text{water}} \\ &= nc_P^{\text{water}} \left(\frac{50}{293} - \ln \frac{343}{293} \right). \end{aligned} \tag{4.13}$$

This is the answer we obtain. We will not give the exact numerical number of the answer, but just point out one point which is worth mentioning. We note that this is an irreversible process, because we cannot re-heat the 20°C water back to 70°C *without extra effort or cost.* According to the second law, the quantity ΔS_{total} in Eq. (4.13) must be greater

[4]For the air, this is also an isobaric process. However if we use Eq. (4.9) to calculate the entropy change for the air, we will not get a correct result. This is because C_P^{air} is very large (open air is a very large system) and it is difficult to calculate (or even estimate) its magnitude, and the initial and final temperatures can be considered as equal. That is understandable because open air is a huge heat reservoir, so its temperature will not change by absorbing heat from a glass of water.

than zero. Of course we may calculate the numerical value of ΔS_{total} in Eq. (4.13) to prove that it is a positive quantity. But the conclusion that ΔS_{total} must be greater than zero is ***a general result***, not just for water with temperature 70°C. In fact, if the initial temperatures of the water and the air are different, then the process is always an irreversible one. Therefore we always have[5] $\Delta S_{\text{total}} > 0$. We will prove this general result in the following.

We define a function $f(x)$, and use the properties of $f(x)$ to prove that if the initial temperatures of the water and the air are different, then $\Delta S_{\text{total}} > 0$ always holds. We define the quantity inside the parenthesis on the right hand side of Eq. (4.13) as $f(x)$, i.e.,

$$f(x) = \frac{x}{a} - \ln\frac{a+x}{a}. \tag{4.14}$$

Obviously, in Eq. (4.13), $a = 293$ and $x = 50$. Thus a represents the air temperature and x the temperature difference between water and air. Differentiate $f(x)$ with respect to x,

$$\frac{df}{dx} = \frac{1}{a} - \frac{1}{a+x}, \tag{4.15}$$

thus $df/dx > 0$ for $x > 0$ and $df/dx < 0$ for $-a < x < 0$. Since $f(0) = 0$, no matter whether $x > 0$ or $-a < x < 0$, we always get the result $f(x) > 0$. We therefore arrive at the conclusion: we always have $\Delta S_{\text{total}} > 0$, no matter whether the water is initially hotter than the air ($x > 0$) or the air is hotter than the water ($x < 0$), because $C_P > 0$ for all materials.[6] This satisfies the requirement for an irreversible process.

4.3 Entropy Change for an Ideal Gas

An ideal gas is a relatively simple system. Under different conditions, many of the state variables of the system can be calculated analytically and expressed in relatively simple expressions. Therefore in many topics in thermodynamics, an ideal gas serves as an excellent system to illustrate the physics or mathematics involved. In this section we will

[5]Only if the water and the air have the same initial temperature, is the process a reversible one, and $\Delta S_{\text{total}} = 0$ in this case.

[6]We will prove in Chapter 5 that the heat capacity (C_V and C_P) of any material should be greater than zero, otherwise the material will be unstable and cannot exist.

calculate the entropy change for an ideal gas under different conditions. The formulas we obtained in this section will be useful for latter chapters of the book, where an ideal gas will again be used as an example.

We consider one kilomole of an ideal gas. All the extensive variables can be divided by the number of kilomoles to become intensive variables. We use lower case letters to denote these intensive variables, thus u is the internal energy per kilomole (i.e., the specific internal energy), v the specific volume, s the specific entropy, etc.

1. T and v as independent variables:

For an ideal gas the internal energy u is independent of the volume v. Therefore as stated in Eq. (3.37) $du = c_v dT$; Eq. (4.2) becomes

$$ds = \frac{du}{T} + \frac{P}{T}dv = c_v\frac{dT}{T} + R\frac{dv}{v}, \tag{4.16}$$

where $Pv = RT$ for an ideal gas has been used. Therefore, when an ideal gas changes from the state (T_1, v_1) to (T_2, v_2), we may integrate the above equation to get the entropy change

$$\Delta s = c_v \ln\frac{T_2}{T_1} + R\ln\frac{v_2}{v_1}, \tag{4.17}$$

where we have used the fact that c_v is a constant independent of T (at least for a monatomic gas). Except for the case where the third law is involved, we are interested only in the relative values of entropy, rather than its absolute value. The above equation can therefore be rewritten as

$$s(T, v) = c_v \ln T + R\ln v + s_0, \tag{4.18}$$

where s_0 is a constant. Therefore, when the specific volume v is kept constant, with increasing (decreasing) temperature T, the specific entropy s will increase (decrease). The same is true when T is kept constant: for increasing (decreasing) v, the entropy s will increase (decrease).

2. T and P as independent variables:

In this case it is more convenient to use the enthalpy h defined in Eq. (1.16), $h = u + Pv$, instead of using the internal energy u. The reason is that the natural variables of h are T and P. From $du = dh - Pdv - vdP$, Eq. (4.2) becomes

$$ds = \frac{dh}{T} - \frac{v}{T}dP = c_P\frac{dT}{T} - R\frac{dP}{P}, \tag{4.19}$$

where we have used the relation $Pv = RT$ and $dh = c_p dT$ for an ideal gas.[7] The change in entropy, when the system changes from the state (T_1, P_1) to (T_2, P_2), can be obtained by integrating the above equation to yield

$$\Delta s = c_p \ln \frac{T_2}{T_1} - R \ln \frac{P_2}{P_1}, \tag{4.20}$$

where c_p is assumed to be a constant. The above equation can be rewritten as

$$s(T, P) = c_p \ln T - R \ln P + s_0, \tag{4.21}$$

where s_0 is a constant. When the pressure P is kept constant, increasing (decreasing) the temperature T will increase (decrease) the entropy s. On the other hand, when T is kept constant, for *increasing* (decreasing) P, the entropy s will *decrease* (increase). Thus s and P change in different directions.

3. v and P as independent variables:

We may use Eq. (4.20) to derive the equation we want. As $R = c_p - c_v$, Eq. (4.20) becomes

$$\begin{aligned} \Delta s &= c_p \ln \frac{T_2}{T_1} - (c_p - c_v) \ln \frac{P_2}{P_1} \\ &= c_p \ln \left(\frac{T_2}{T_1} \frac{P_1}{P_2} \right) + c_v \ln \frac{P_2}{P_1} \\ &= c_p \ln \frac{v_2}{v_1} + c_v \ln \frac{P_2}{P_1}. \end{aligned} \tag{4.22}$$

The above equation can be rewritten as

$$s(v, P) = c_p \ln v + c_v \ln P + s_0, \tag{4.23}$$

where s_0 is a constant. Therefore when P is kept constant, increasing (decreasing) v will increase (decrease) the entropy s. Also when v is kept constant, increasing (decreasing) P, the entropy s will increase (decrease).

[7] For an ideal gas, $u = u(T)$ and $Pv = RT$ hence $h = u + Pv = h(T)$, which implies that h depends only on T, independent of P. Therefore $dh = c_p dT$.

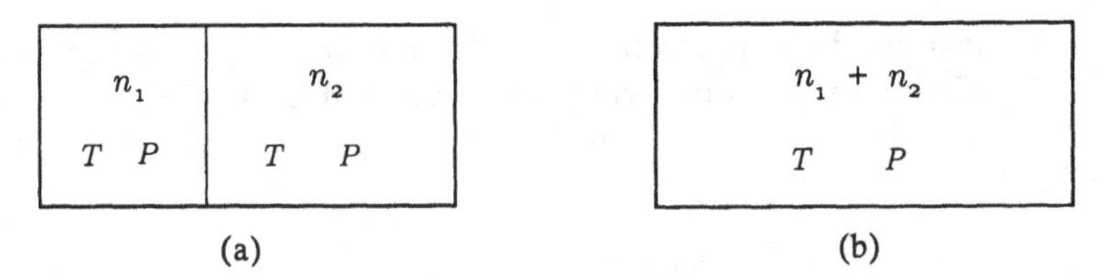

Figure 4.3: The entropy of mixing for two ideal gases of different kinds: (a) before mixing, and (b) after mixing.

4.4 The Entropy of Mixing

In Chapter 3 we have used a formula, which was derived microscopically, to compute the entropy change when two isolated ideal gases of different kinds mix together to become a single homogeneous system. This is an irreversible process and the entropy will increase. The amount of the increase in entropy is called the entropy of mixing. In this section we will use the macroscopic thermodynamic equations (derived by using the first and second laws) to compute the entropy of mixing. As shown in Fig. 4.3(a), a container is divided into two parts by a partition. One part of the container contains n_1 kilomoles of an ideal gas of type 1, the other part contains n_2 kilomoles of an ideal gas of type 2. Both gases have the same temperature T and pressure P, and can be considered as isolated systems initially. Now the partition is removed, and the gas molecules mix homogeneously. We want to calculate the total entropy change of the mixing process.

To calculate the total entropy change, we have to calculate the entropies of the final state and the initial state. We consider the initial state first. The initial state is two isolated ideal gases, therefore the total entropy is the sum of the entropies of the two ideal gases. Since T and P are known, from Eq. (4.21) we obtain the total entropy of the initial state as

$$\begin{aligned} S_{\text{initial}} &= n_1 s_1(T,P) + n_2 s_2(T,P) \\ &= (n_1 c_{P_1} + n_2 c_{P_2}) \ln T \\ &\quad -(n_1+n_2)R \ln P + n_1 s_{01} + n_2 s_{02}, \end{aligned} \tag{4.24}$$

where c_{P_1} and c_{P_2} are the constant specific heat for gas 1 and gas 2, respectively. Now we consider the final state. It is a homogeneous mixture of two different ideal gases. Since before mixing the two gases have the same T and P, after mixing the temperature T will remain unchanged.

However the pressure for each kind of gas will be reduced, because the volume available for the molecules is enlarged. It is easy to deduce from the ideal gas law $PV = nRT$ that the pressure for gas 1 (P_1) and for gas 2 (P_2) are given by

$$P_1 = \frac{n_1}{n_1 + n_2} P, \quad P_2 = \frac{n_2}{n_1 + n_2} P. \tag{4.25}$$

Here P_1 and P_2 are called the *partial* pressures. The total pressure of the mixture is the sum of P_1 and P_2, i.e., $P_{\text{total}} = P_1 + P_2 = P$, which is equal to the pressure before mixing. We may consider the final state as two independent ideal gases with the same temperature T, but with pressures P_1 and P_2, respectively. They are still ideal gases, therefore Eq. (4.21) can be used to calculate the entropy. However care must be taken as these two gases have different pressures. The total entropy of the final state is therefore

$$\begin{aligned} S_{\text{final}} &= n_1 s_1(T, P_1) + n_2 s_2(T, P_2) \\ &= (n_1 c_{P_1} + n_2 c_{P_2}) \ln T - n_1 R \ln P_1 - n_2 R \ln P_2 \\ &\quad + n_1 s_{01} + n_2 s_{02} \\ &= S_{\text{initial}} - n_1 R \ln \frac{n_1}{n_1 + n_2} - n_2 R \ln \frac{n_2}{n_1 + n_2}. \end{aligned} \tag{4.26}$$

Hence the entropy of mixing is

$$\begin{aligned} \Delta S_{\text{mixing}} &= S_{\text{final}} - S_{\text{initial}} \\ &= R \left(n_1 \ln \frac{n_1 + n_2}{n_1} + n_2 \ln \frac{n_1 + n_2}{n_2} \right). \end{aligned} \tag{4.27}$$

Equation (4.27) tells us that ΔS_{mixing} is always greater than 0, because $\ln x > 0$ for $x > 1$. This is consistent with the requirement of the second law, as the mixing process is an irreversible process. We note that when $n_1 = n_2 = n$, then from Eq. (4.27) we obtain $\Delta S_{\text{mixing}} = 2nR \ln 2$, which is exactly equal to the result obtained by using the microscopic Eq. (3.46).

Next we should check whether the problem of the Gibbs paradox occurs in the macroscopic equations. The answer is no, because if the two gases molecules are identical, then after removing the partition, the container contains only one kind of molecule, and it is meaningless to define any partial pressure. The total pressure P of the final state is exactly the same as the initial pressure P, and the number of kilomoles is $n_1 + n_2$. Therefore $\Delta S_{\text{mixing}} = 0$, and there is no paradox.

4.5 The Tds Equations

In this section we start from the relation $Tds = du + Pdv$ to derive three useful Tds equations which use (T, v), (T, P) and (v, P) as independent variables. We first write down the three equations, after which the derivations will be given.

$$Tds = c_v\, dT + T\left(\frac{\partial P}{\partial T}\right)_v dv = c_v\, dT + \frac{T\beta}{\kappa} dv, \tag{4.28}$$

$$Tds = c_P dT - T\left(\frac{\partial v}{\partial T}\right)_P dP = c_P dT - Tv\beta\, dP, \tag{4.29}$$

$$\begin{aligned} Tds &= c_P\left(\frac{\partial T}{\partial v}\right)_P dv + c_v\left(\frac{\partial T}{\partial P}\right)_v dP \\ &= \frac{c_P}{\beta v} dv + \frac{c_v\kappa}{\beta} dP, \end{aligned} \tag{4.30}$$

where $\beta = v^{-1}(\partial v/\partial T)_P$ is the expansivity and $\kappa = -v^{-1}(\partial v/\partial P)_T$ the isothermal compressibility. We have introduced these two quantities in Sec. 1.13. Before we proceed to derive the three Tds equations, we first derive the *cyclic relation* (also known as the *chain relation*) in calculus.

Consider three variables x, y, and z which satisfy a relation of the form $f(x, y, z) = 0$. If the function f is a differentiable function, then x, y, and z satisfy the following cyclic relation

$$\left(\frac{\partial x}{\partial y}\right)_z \left(\frac{\partial y}{\partial z}\right)_x \left(\frac{\partial z}{\partial x}\right)_y = -1. \tag{4.31}$$

The proof is as follows.

We may consider x as a function of y and z, i.e., $x = x(y, z)$, then

$$dx = \left(\frac{\partial x}{\partial y}\right)_z dy + \left(\frac{\partial x}{\partial z}\right)_y dz. \tag{4.32}$$

Next we may consider y as a function of z and x, i.e., $y = y(x, z)$, then

$$dy = \left(\frac{\partial y}{\partial x}\right)_z dx + \left(\frac{\partial y}{\partial z}\right)_x dz. \tag{4.33}$$

Substituting Eq. (4.33) into Eq. (4.32), we obtain

$$dx = \left(\frac{\partial x}{\partial y}\right)_z \left(\frac{\partial y}{\partial x}\right)_z dx + \left[\left(\frac{\partial x}{\partial y}\right)_z \left(\frac{\partial y}{\partial z}\right)_x + \left(\frac{\partial x}{\partial z}\right)_y\right] dz. \tag{4.34}$$

In the above equation we choose $dx \neq 0$ and $dz = 0$, then

$$\left(\frac{\partial y}{\partial x}\right)_z = \frac{1}{\left(\frac{\partial x}{\partial y}\right)_z}. \tag{4.35}$$

This relation is known as the ***reciprocal relation.*** Now, if we choose $dx = 0$ and $dz \neq 0$ in Eq. (4.34) and use Eq. (4.35), we get the *cyclic relation*,

$$\left(\frac{\partial x}{\partial y}\right)_z \left(\frac{\partial y}{\partial z}\right)_x \left(\frac{\partial z}{\partial x}\right)_y = -1.$$

Now we are ready to derive the three Tds equations given above. The derivations are given in the following.

1. Take (T, v) as independent variables, i.e., $u = u(T, v)$, and $ds = T^{-1}(du + Pdv)$, therefore

$$du = \left(\frac{\partial u}{\partial T}\right)_v dT + \left(\frac{\partial u}{\partial v}\right)_T dv, \tag{4.36}$$

$$ds = \frac{1}{T}\left(\frac{\partial u}{\partial T}\right)_v dT + \frac{1}{T}\left[\left(\frac{\partial u}{\partial v}\right)_T + P\right] dv. \tag{4.37}$$

Since ds is an exact differential, we have

$$\frac{\partial}{\partial v}\left[\frac{1}{T}\left(\frac{\partial u}{\partial T}\right)_v\right]_T = \frac{\partial}{\partial T}\left[\frac{1}{T}\left(\frac{\partial u}{\partial v}\right)_T + \frac{P}{T}\right]_v, \tag{4.38}$$

after simplification we have

$$\left(\frac{\partial u}{\partial v}\right)_T = T\left(\frac{\partial P}{\partial T}\right)_v - P. \tag{4.39}$$

Substituting the above equation into Eq. (4.37), and using the definition of constant volume specific heat $c_v = (đq/dT)_v = (\partial u/\partial T)_v$, we obtain the first Tds equation (4.28);

$$Tds = c_v\, dT + T\left(\frac{\partial P}{\partial T}\right)_v dv = c_v\, dT + \frac{T\beta}{\kappa} dv.$$

In the last equality we have used the cyclic relation between T, v, P, i.e.,

$$\begin{aligned}\left(\frac{\partial P}{\partial T}\right)_v &= -\left(\frac{\partial T}{\partial v}\right)_P^{-1}\left(\frac{\partial v}{\partial P}\right)_T^{-1} \\ &= -\left(\frac{\partial v}{\partial T}\right)_P\left(\frac{\partial v}{\partial P}\right)_T^{-1} = \frac{\beta}{\kappa}.\end{aligned} \tag{4.40}$$

2. Take (T, P) as the independent variables. In this case expressing ds in terms of dh ($h = u + Pv$) is more convenient than using du, since the natural variables for h are T and P,

$$dh = \left(\frac{\partial h}{\partial T}\right)_P dT + \left(\frac{\partial h}{\partial P}\right)_T dP. \tag{4.41}$$

Replacing du by dh,

$$\begin{aligned} ds &= \frac{1}{T}(du + Pdv) = \frac{1}{T}(dh - vdP) \\ &= \frac{1}{T}\left(\frac{\partial h}{\partial T}\right)_P dT + \frac{1}{T}\left[\left(\frac{\partial h}{\partial P}\right)_T - v\right] dP. \end{aligned} \tag{4.42}$$

By using the condition that ds is an exact differential, we obtain

$$\left(\frac{\partial h}{\partial P}\right)_T = -T\left(\frac{\partial v}{\partial T}\right)_P + v = -Tv\beta + v. \tag{4.43}$$

Therefore, from Eqs. (4.42) and (4.43), we get the second Tds equation (4.29),

$$Tds = c_P dT - T\left(\frac{\partial v}{\partial T}\right)_P dP = c_P dT - Tv\beta\, dP,$$

where $c_P = (\partial h/\partial T)_P$ is the constant pressure specific heat.

3. Take (v, P) as the independent variables. At first we use (T, v) as independent variables, and then express dT in terms of dv and dP.

$$\begin{aligned}
Tds &= du + Pdv = \left(\frac{\partial u}{\partial T}\right)_v dT + \left(\frac{\partial u}{\partial v}\right)_T dv + Pdv \\
&= \left(\frac{\partial u}{\partial T}\right)_v \left[\left(\frac{\partial T}{\partial P}\right)_v dP + \left(\frac{\partial T}{\partial v}\right)_P dv\right] + \left[\left(\frac{\partial u}{\partial v}\right)_T + P\right] dv \\
&= \left(\frac{\partial u}{\partial T}\right)_v \left(\frac{\partial T}{\partial P}\right)_v dP \\
&\quad + \left[\left(\frac{\partial u}{\partial T}\right)_v \left(\frac{\partial T}{\partial v}\right)_P + \left(\frac{\partial u}{\partial v}\right)_T + P\right] dv. \qquad (4.44)
\end{aligned}$$

Therefore we have

$$\left(\frac{\partial s}{\partial P}\right)_v = \frac{1}{T}\left(\frac{\partial u}{\partial T}\right)_v \left(\frac{\partial T}{\partial P}\right)_v = \frac{c_v}{T}\left(\frac{\partial T}{\partial P}\right)_v = \frac{c_v \kappa}{T\beta}, \qquad (4.45)$$

$$\begin{aligned}
\left(\frac{\partial s}{\partial v}\right)_P &= \frac{1}{T}\left[\left(\frac{\partial u}{\partial T}\right)_v \left(\frac{\partial T}{\partial v}\right)_P + \left(\frac{\partial u}{\partial v}\right)_T + P\right] \\
&= \frac{c_P}{T}\left(\frac{\partial T}{\partial v}\right)_P = \frac{c_P}{Tv\beta}. \qquad (4.46)
\end{aligned}$$

In obtaining the second equality of Eq. (4.46) we have used Eq. (4.39),

$$\left(\frac{\partial u}{\partial v}\right)_T = T\left(\frac{\partial P}{\partial T}\right)_v - P,$$

thus the sum of the second and third terms in the parenthesis of the first equality of Eq. (4.46) is $T(\partial P/\partial T)_v$. Moreover we made use of $(\partial u/\partial T)_v = c_v$, and Eq. (1.20),

$$\begin{aligned}
c_P - c_v &= \left(\frac{\partial v}{\partial T}\right)_P \left[P + \left(\frac{\partial u}{\partial v}\right)_T\right] \\
&= T\left(\frac{\partial P}{\partial T}\right)_v \left(\frac{\partial v}{\partial T}\right)_P. \qquad (4.47)
\end{aligned}$$

The first term in the parenthesis of the first equality of Eq. (4.46) can be written as

$$\begin{aligned}
\left(\frac{\partial u}{\partial T}\right)_v\left(\frac{\partial T}{\partial v}\right)_P &= c_v\left(\frac{\partial T}{\partial v}\right)_P \\
&= \left[c_P - T\left(\frac{\partial P}{\partial T}\right)_v\left(\frac{\partial v}{\partial T}\right)_P\right]\left(\frac{\partial T}{\partial v}\right)_P \\
&= c_P\left(\frac{\partial T}{\partial v}\right)_P - T\left(\frac{\partial P}{\partial T}\right)_v\left(\frac{\partial v}{\partial T}\right)_P\left(\frac{\partial T}{\partial v}\right)_P \\
&= c_P\left(\frac{\partial T}{\partial v}\right)_P - T\left(\frac{\partial P}{\partial T}\right)_v .
\end{aligned}$$

Therefore we get the second equality of Eq. (4.46). The third equality comes from the definition of the expansivity β. From Eqs. (4.45)–(4.46) we obtain the third Tds equation,

$$Tds = c_P\left(\frac{\partial T}{\partial v}\right)_P dv + c_v\left(\frac{\partial T}{\partial P}\right)_v dP = \frac{c_P}{\beta v}dv + \frac{c_v\kappa}{\beta}dP.$$

Here we have used the definition of β to replace $(\partial T/\partial v)_P$ and Eq. (4.40) to replace $(\partial T/\partial P)_v$.

4.6 Applications

1. Calculation of the amount of heat absorbed by a system undergoing various processes:

   Tds is the amount of heat $đq$ absorbed by a system during an infinitesimal process, therefore the three Tds equations (4.28)–(4.30) are useful for calculating the heat absorbed by a system undergoing various processes. In principle any one of the three equations can be used for the calculations, but in practice, for a given process, the degree of difficulty of the calculations may be quite different for different choices of the three equations. Therefore for a given process, it is desirable to choose a *suitable equation* for the calculation. For instance, suppose we want to calculate the heat absorbed by a system in a reversible isochoric process while the temperature of the system changes from T_1 to T_2. Apparently it is much simpler to use the first Tds equation (4.28), because for an isochoric process $dv = 0$, and the Tds equation contains a term proportional to dT

only. However for the same isochoric process, if we want to calculate the heat absorbed by the system when the pressure changes from P_1 to P_2, then the third Tds equation (4.30) is apparently a better choice than the first equation. We use an ideal gas as an example. Suppose we want to calculate the heat absorbed by n kilomoles of an ideal gas which undergoes a reversible isothermal process, when the pressure changes from P_1 to P_2. Apparently the second Tds equation (4.29) is the best choice. Let $dT = 0$; Eq. (4.29) becomes

$$\begin{aligned}\Delta Q &= n\Delta q = -n\int_{P_1}^{P_2} Tv\beta\, dP = -n\int_{P_1}^{P_2} v\, dP \\ &= -nRT\int_{P_1}^{P_2}\frac{dP}{P} = -nRT\ln\frac{P_2}{P_1},\end{aligned} \tag{4.48}$$

where we have used the known results $\beta = v^{-1}(\partial v/\partial T)_P = 1/T$ and $v = RT/P$ for an ideal gas. Equation (4.48) tells us that, in an isothermal process, a gas absorbs heat, i.e., $\Delta Q > 0$, for an expansion ($P_2 < P_1$). However it rejects heat, i.e., $\Delta Q < 0$, for a compression ($P_2 > P_1$).

2. Derivation of the expression for $c_p - c_v$ for a general system:
In Sec. 1.9 we used the definition of the heat capacities, C_P and C_V, and the enthalpy H to derive an expression for $C_P - C_V$ for a general system. Now we will use the Tds equations to derive an expression for $c_p - c_v$. In deriving the third Tds equation, we used the expression of $C_P - C_V$ derived in Sec. 1.9, therefore in the present derivation of $c_p - c_v$, we should not use the third equation. We use only the first and second Tds equations in this derivation. From Eqs. (4.28)–(4.29) we have

$$Tds = c_v\, dT + \frac{T\beta}{\kappa}dv = c_p dT - Tv\beta\, dP. \tag{4.49}$$

Rearranging the terms and using $T = T(v, P)$, we obtain

$$\begin{aligned}dT &= \frac{T\beta}{\kappa(c_p - c_v)}dv + \frac{Tv\beta}{c_p - c_v}dP \\ &= \left(\frac{\partial T}{\partial v}\right)_P dv + \left(\frac{\partial T}{\partial P}\right)_v dP.\end{aligned} \tag{4.50}$$

Comparing the coefficients of dv and dP for both sides of the second equality,

$$\left(\frac{\partial T}{\partial v}\right)_P = \frac{T\beta}{\kappa(c_P - c_v)}, \tag{4.51}$$

$$\left(\frac{\partial T}{\partial P}\right)_v = \frac{Tv\beta}{c_P - c_v}. \tag{4.52}$$

Now we can solve Eq. (4.51) for $c_P - c_v$ to obtain

$$c_P - c_v = \frac{T\beta}{\kappa}\left(\frac{\partial v}{\partial T}\right)_P = \frac{Tv\beta^2}{\kappa}. \tag{4.53}$$

One can solve Eq. (4.52) for $c_P - c_v$ to get the same result. This will be left as an exercise for the reader. Equation (4.53) may be applied to an ideal gas, using $\beta = 1/T$ and $\kappa = 1/P$, to obtain

$$c_P - c_v = Tv\frac{1}{T^2}P = \frac{Pv}{T} = R. \tag{4.54}$$

This is the same result as we obtained in Sec. 1.9, i.e., Eq. (1.26). Equation (4.53) can be used to estimate the experimental value of c_v for liquids and solids. Experimentally it is easier to measure c_P than c_v, because in the laboratory, for liquids or solids, it is easier to keep the pressure constant than to keep the volume constant. Therefore the experimental value of c_v is usually estimated from c_P by using Eq. (4.53). One still, however, needs the experimental values of v, β and κ to estimate c_v. All these quantities are usually functions of the temperature (assuming the pressure is 1 atm), and therefore one needs data for different temperatures.

3. Relation of the isothermal compressibility κ and the adiabatic compressibility κ_s:

   We defined two quantities called the *isothermal compressibility* κ and the *adiabatic compressibility* κ_s in Sec. 1.13 by Eqs. (1.42) and (1.44), respectively,

$$\kappa = -\frac{1}{v}\left(\frac{\partial v}{\partial P}\right)_T, \quad \kappa_s = -\frac{1}{v}\left(\frac{\partial v}{\partial P}\right)_s.$$

   They represent the same physical quantity, the rate of change of the volume when the pressure changes, but under different conditions. We expect them to be related to each other. Indeed, the ratio

κ/κ_s has a simple relation which can be found from one of the Tds equations.

We use the third Tds equation (4.30). By letting $ds = 0$ (for an adiabatic process), we have

$$\frac{c_P}{\beta v}\, dv_s + \frac{c_v \kappa}{\beta} dP_s = 0, \tag{4.55}$$

where the subscripts s in dv_s and dP_s denote that the process is adiabatic. Simplifying the equation, we obtain

$$-\frac{1}{v}\left(\frac{\partial v}{\partial P}\right)_s = \kappa \frac{c_v}{c_P} = \frac{\kappa}{\gamma}, \quad \gamma \equiv \frac{c_P}{c_v}. \tag{4.56}$$

The left hand side of Eq. (4.56) is just the definition of κ_s, therefore

$$\kappa_s = \frac{\kappa}{\gamma}. \tag{4.57}$$

From this relation we can also obtain the difference $\kappa - \kappa_s$ (see Problem 4.20). Since $\gamma \geq 1$, it follows that $\kappa_s \leq \kappa$. The reason for this is that when a system is compressed adiabatically ($dP > 0$, S=constant), the temperature will either increase or decrease, depending on whether the expansivity $\beta > 0$ or $\beta < 0$.[8] However in either case the volume of the system will expand due to the temperature change, and therefore the volume change ($|dv|$) is smaller in an adiabatic process than in an isothermal process, i.e., $|dv_s| < |dv_T|$ (note that $dv < 0$ for $dP > 0$).[9]

We have seen that κ appears in several thermodynamic relations, such as $c_P - c_v$, the Tds equations etc. However some physical phenomena are related to κ_s rather than κ. For example, the flow or vibration of air molecules are closer to an adiabatic process than to an isothermal process. The thermal conductivity of air is very small, consequently the heat transfer by air may be neglected, therefore most of the phenomena associated with air may be approximated as adiabatic processes. An example is the speed of sound v_s, which is given by

$$v_s = \sqrt{\frac{1}{\rho \kappa_s}} = \sqrt{\frac{\gamma}{\rho \kappa}}, \tag{4.58}$$

[8]This statement can be proved by the thermodynamic relations derived in Chapter 5 (see Problem 5.5).

[9]This is due to the fact that κ, and therefore κ_s, must be greater than 0 for any substance. See Sec. 5.9 for the proof.

where ρ is the mass density of air. If air is approximated as a diatomic ideal gas, then $\gamma \approx 7/5 = 1.40$, $\kappa = 1/P \approx 10^{-5}$ Pa^{-1}, and $\rho \approx 1.2$ kg m^{-3}. Substituting these values into Eq. (4.58) we get $v_s \approx 340$ m s^{-1}, which is very close to the experimental value $v_s = 343$ m s^{-1}.

4.7 Problems

4.1. Consider a body, whose constant pressure heat capacity C_P is 10 J K^{-1}, and is initially at the temperature $T_1 = 200$ K. The body is then brought into thermal contact with a heat reservoir at temperature T_2. Calculate the changes in entropy for the body, the reservoir, and the universe if (a) $T_2 = 400$ K, (b) $T_2 = 600$ K, and (c) $T_2 = 100$ K.

4.2. The initial state of one mole of a monatomic ideal gas is $P = 5$ atm and $T = 300$ K. The system undergoes a reversible process which reduces the pressure to 3 atm. Calculate the entropy change of the gas in each of the following processes:

(a) If the process is isothermal.

(b) If the process is adiabatic.

(c) If the process is isochoric.

4.3. A monatomic ideal gas undergoes a reversible process such that the specific volume changes from v_1 to v_2. Calculate the change in specific entropy Δs in each of the following processes:

(a) If the process is isobaric.

(b) If the process is isothermal.

(c) If the process is adiabatic.

4.4. One kg of water is heated by an electric resistor from 20°C to 90°C at constant pressure ($P = 1$ atm). Assume that the constant heat capacity of water is 1 kcal kg^{-1} (4180 J kg^{-1}) and is independent of temperature.

(a) Calculate the change in internal energy of the water.

(b) Calculate the entropy change of the water.

(c) Calculate the maximum mechanical work achievable by using this water as a heat reservoir to run an engine whose lower temperature reservoir is at 20°C.

4.5. Consider one mole of a monatomic ideal gas which undergoes a reversible expansion.

(a) The expansion is conducted at constant pressure, during which the entropy of the gas increases by 14.41 J K^{-1}, and the gas absorbs 6236 J of heat. Calculate the initial and final temperatures of the gas.

(b) The expansion is conducted at a constant temperature, during which it doubles its volume, performs 1729 J of work, and increases its entropy by 5.763 J K^{-1}. Calculate the temperature at which the expansion was conducted.

4.6. Consider the increase in entropy when an ideal gas is heated from T_1 to T_2 at (a) constant volume, and (b) constant pressure. Show that the entropy increase in case (b) is γ times that in case (a), where $\gamma = C_P/C_V$.

4.7. The constant volume specific heat c_v of an ideal gas is given by $c_v = a + bT$, where a and b are constants. Show that the change in specific entropy in going from state (T_1, v_1) to state (T_2, v_2) is

$$\Delta s = a \ln\left(\frac{T_2}{T_1}\right) + b(T_2 - T_1) + R \ln\left(\frac{v_2}{v_1}\right).$$

4.8. A mixture of n_1 moles of gas A and n_2 moles of gas B are contained in a container with a movable piston. Both gases A and B can be considered as ideal gases. Suppose that the constant volume specific heat of A and B are c_{v1} and c_{v2}, respectively, and $c_{v1} \neq c_{v2}$ (e.g. A is monatomic and B is diatomic). Show that in a reversible adiabatic process for the mixture, the following relations hold: (a) $TV^{\gamma-1}$ =constant, and (b) PV^{γ} =constant, where

$$\gamma = \frac{n_1 c_{P1} + n_2 c_{P2}}{n_1 c_{v1} + n_2 c_{v2}}.$$

4.9. A mass m of a liquid at a temperature T_1 is mixed with an equal mass of the same liquid at a temperature T_2. The system is thermally insulated. Assume that the constant pressure specific heat

c_P of the liquid is independent of temperature. Show that the entropy change of the universe is

$$2mc_P \ln \frac{(T_1 + T_2)/2}{\sqrt{T_1 T_2}}.$$

Show that when $T_1 \neq T_2$ this is an irreversible process.

4.10. Show that $\left(\frac{\partial c_v}{\partial v}\right)_T = T\left(\frac{\partial^2 P}{\partial T^2}\right)_v$.

4.11. Show that $\left(\frac{\partial c_P}{\partial P}\right)_T = -T\left(\frac{\partial^2 v}{\partial T^2}\right)_P = -Tv\left[\beta^2 + \left(\frac{\partial \beta}{\partial T}\right)_P\right]$.

4.12. Experimentally it has been found for a given gas that the product of the pressure P and the specific volume v is a function of the temperature T only, and that the internal energy u is also a function of T only. What can one say about the equation of state of this gas?

4.13. The equation of state of a substance is given by $(P + b)v = RT$. What information can be deduced about the entropy, the internal energy, and the enthalpy of the substance? What other measurement(s) must be made to determine all of the properties of the substance?

4.14. A substance has the properties that $\left(\frac{\partial u}{\partial v}\right)_T = 0$ and $\left(\frac{\partial h}{\partial P}\right)_T = 0$.

(a) Show that the equation of state must be of the form $T = aPV$, where a is a constant.

(b) What additional information is necessary to specify the entropy of the system?

4.15. (a) Show that for reversible changes in temperature at constant volume, $c_v = T(\partial s/\partial T)_v$.

(b) Assume that $c_v = aT + bT^3$ for a metal at low temperatures, where a and b are constants. Calculate the variation of the specific entropy s with temperature.

4.16. Assume that c_P for an ideal gas is given by $c_P = a + bT$, where a and b are constants.

(a) What is the expression for c_v for this gas?

(b) What is the expression for the specific entropy for this gas instead of Eq. (4.21)?

(c) What is the expression for the specific enthalpy for this gas?

4.17. Consider a van der Waals gas.

(a) Show that c_v is a function of T only.

(b) Show that the specific internal energy is

$$u = \int c_v \, dT - \frac{a}{v} + \text{constant.}$$

(c) Show that the specific enthalpy is

$$h = \int c_v \, dT - \frac{2a}{v} + \frac{RTv}{v-b} + \text{constant.}$$

(d) Show that the specific entropy is

$$s = \int \frac{c_v}{T} \, dT + R \ln (v - b) + \text{constant.}$$

(e) If c_v is considered as a constant, show that in a reversible adiabatic process $T(v-b)^{R/c_v} = \text{constant}$.

4.18. Show that for a van der Waals gas

$$c_p - c_v = R \frac{1}{1 - \dfrac{2a(v-b)^2}{RTv^3}}.$$

4.19. Consider one kilomole of a van der Waals gas which undergoes a reversible isothermal process. During the process the specific volume is changed from v_1 to v_2, while the temperature is kept constant at T. Show that the total amount of work done by the gas in this process is

$$\Delta w = RT \ln \left(\frac{v_2 - b}{v_1 - b} \right) + a \left(\frac{1}{v_2} - \frac{1}{v_1} \right).$$

4.20. Show that the difference between the isothermal and the adiabatic compressibilities is

$$\kappa - \kappa_s = \frac{T\beta^2 v}{c_p}.$$

4.21. The pressure on a block of copper at a temperature of 0°C is increased isothermally and reversibly from 1 atm to 1000 atm. Assume that the expansion coefficient β, the isothermal compressibility κ, and the mass density ρ are constants and respectively equal to 5×10^{-5} K^{-1}, 8×10^{-12} Pa^{-1}, and 9×10^{3} kg m^{-3}.

(a) Calculate the work done on the copper per kilogram.

(b) Calculate the heat rejected .

(c) How do you account for the fact that the heat rejected is greater than the work done?

4.22. (a) The temperature of a block of copper is increased from T_1 to T_2 without any appreciable change in its volume v_1. Show that the change in its specific entropy is

$$\Delta s = c_P \ln\left(\frac{T_2}{T_1}\right) - \frac{v_1 \beta^2}{\kappa}(T_2 - T_1).$$

(b) Calculate Δs in units of J kg^{-1} K^{-1}, if $T_1 = 300$ K and $T_2 = 320$ K. For copper we may take $c_P = 390$ J kg^{-1} K^{-1}, $\beta = 5 \times 10^{-5}$ K^{-1}, $\kappa = 8 \times 10^{-12}$ Pa^{-1}, and the mass density 9×10^{3} kg m^{-3}.

4.23. A substance has a constant pressure heat capacity per gram c_s in the solid state and that in the liquid state is c_l. Both of the heat capacities can be considered as temperature independent quantities. The latent heat per gram of this substance is q_0 at the temperature T_0. Calculate the entropy difference between the supercooled liquid state and the solid state of M grams of this material at the same temperature T_1 ($T_1 < T_0$). Assume that the specific heat of the supercooled liquid is also equal to c_l. The pressure is assumed at $P = 1$ atm.

4.24. For a solid whose equation of state is given by Eq. (1.47), and for which c_v and c_P are independent of T in the neighborhood of T_0 and P_0, show that the specific internal energy u and specific enthalpy h are given by

$$u = c_v(T - T_0) + \left[\left(2\beta T_0 + \frac{v}{v_0} - 1\right)\frac{1}{2\kappa} - P_0\right](v - v_0) + u_0,$$

$$h = c_P(T - T_0) + v_0\left[1 - \beta T_0 - \frac{\kappa}{2}(P - P_0)\right](P - P_0) + h_0.$$

Chapter 5

Thermodynamic Potentials

5.1 Introduction

One of the major applications of thermodynamics is to study the conditions necessary for a system to attain thermodynamic equilibrium under different external situations. The second law tells us that when an isolated system undergoes any spontaneous process the entropy of the system either increases or remains unchanged. The former is an irreversible process, while the latter is a reversible process. Therefore when *an isolated system* attains thermodynamic equilibrium the system must be *in a state of maximum entropy*, otherwise it will continue to change to states with larger entropy. When the system reaches the state with maximum entropy then the state of the system will stop changing, and thermodynamic equilibrium is attained. This is known as the *principle of maximum entropy.* However this principle applies only to isolated systems, therefore it is not so useful. This is because we are always interested in systems which can make contact with their surroundings, so that we can make observations and/or measurements of the system, etc. Therefore we have to generalize the principle of maximum entropy to be applicable to non-isolated systems. This is what we will do in this chapter. We will introduce a couple of thermodynamic potentials which are useful in the study of the conditions necessary for a system to attain thermodynamic equilibrium under various external situations.

In thermodynamics we define four *thermodynamic potentials*, all of which are state variables of the system under consideration. In addition to units (which are those of energy), the thermodynamic potentials, in some sense, play a role rather similar to that of potentials in mechanics. We recall the role played by a potential in mechanics. Suppose we place a particle in a potential field. If the particle is placed at the potential minimum, then the particle will stay at rest (there is no force exerted on it) so it remains at the potential minimum, because this is the place where the particle is at a stable equilibrium. If the particle is placed at a place other than the minimum of the potential, then the particle will experience a force and will start to move, and the motion will not stop, unless there is a dissipative friction force. A friction force will cause a particle to lose energy in the course of its motion. Then, after a sufficiently long time, the particle will lose all of its kinetic energy and come to rest at the potential minimum, where it is in a state of stable equilibrium. The above description in *mechanics* may be applied to *thermodynamics* if we make the following replacements: "particle" is replaced by "thermodynamic system", "potential" is replaced by "thermodynamic potential", and "dissipative friction force" is replaced by "irreversible processes". Then we have the following description in thermodynamics:

If a thermodynamic system is in a state with minimum thermodynamic potential, then the system is in a stable equilibrium state, and the state of the system will not change with time. If a thermodynamic system is not in a state with minimum thermodynamic potential, then the system is not stable, and its state will change with time. It will undergo irreversible processes and an equilibrium state will be reached when the system achieves the state with minimum thermodynamic potential.

Note that under different external situations, different thermodynamic potentials should be used. In this chapter in the subsequent sections, we will introduce the two most important and frequently used thermodynamic potentials: the *Helmholtz free energy* and the *Gibbs free energy.* We will also discuss open systems, for which the number of particles of the system may vary. For these systems we introduce a third thermodynamic potential, which is known as the *chemical potential.* Finally, at the end of this chapter we will study the conditions for a thermodynamic equilibrium to be stable. These are important conditions, they are such that a substance which violates these conditions cannot exist in nature, because it is not stable thermodynamically.

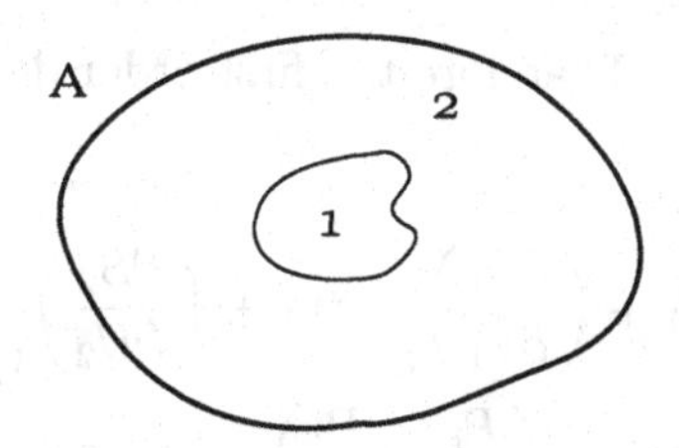

Figure 5.1: An isolated system A is divided into two arbitrary subsystems, 1 and 2.

5.2 Thermodynamic Equilibrium for an Isolated System

First, we use the ***principle of maximum entropy*** to study the conditions of thermodynamic equilibrium for an isolated system. We consider an isolated system, A, with fixed internal energy, U_{tot}, and volume, V_{tot}. In order to study the conditions for the thermodynamic equilibrium of A, we divide it into two arbitrary subsystems 1 and 2, as shown in Fig. 5.1. The internal energy and volume of subsystems 1 and 2 are, respectively, (U_1, V_1) and (U_2, V_2). The interaction energy between subsystems 1 and 2 (i.e., the energy shared by 1 and 2) is very small[1] in comparison with U_1 and U_2; therefore the total internal energy U_{tot} of A is the sum of the internal energies of 1 and 2. The total volume V_{tot} of A is also the sum of the volumes of 1 and 2, i.e.,

$$U_1 + U_2 = U_{\text{tot}} = \text{constant}, \quad V_1 + V_2 = V_{\text{tot}} = \text{constant}. \tag{5.1}$$

Consequently the total entropy S_{tot} of A is the sum of the entropies of 1 and 2,

$$S_{\text{tot}} = S_1(U_1, V_1) + S_2(U_2, V_2). \tag{5.2}$$

We divide A into two subsystems to enable us to vary U_1 (or U_2) and V_1 (or V_2), while both U_{tot} and V_{tot} have fixed values. When A is in a state of thermodynamic equilibrium, the variations of U_1 and V_1 must make

[1]Only particles near the surface which separates 1 and 2 are involved in the interaction energy. The number of particles near the surface are very small in comparison with the total number of particles in 1 or 2.

S_{tot} a maximum value. Therefore the first order derivatives of S_{tot} must be zero:

$$\begin{aligned} dS_{\text{tot}} &= \left(\frac{\partial S_1}{\partial U_1}\right)_{V_1} dU_1 + \left(\frac{\partial S_1}{\partial V_1}\right)_{U_1} dV_1 + \left(\frac{\partial S_2}{\partial U_2}\right)_{V_2} dU_2 + \left(\frac{\partial S_2}{\partial V_2}\right)_{U_2} dV_2 \\ &= \left(\frac{1}{T_1} - \frac{1}{T_2}\right) dU_1 + \left(\frac{P_1}{T_1} - \frac{P_2}{T_2}\right) dV_1 = 0, \end{aligned} \tag{5.3}$$

where we have used the conditions $dU_2 = -dU_1$, $dV_2 = -dV_1$, and the thermodynamic relations

$$\left(\frac{\partial S}{\partial U}\right)_V = \frac{1}{T} \text{ and } \left(\frac{\partial S}{\partial V}\right)_U = \frac{P}{T}. \tag{5.4}$$

Since dU_1 and dV_1 can vary independently, the respective coefficients of dU_1 and dV_1 in Eq. (5.3) must vanish independently. We then get the conditions for thermodynamic equilibrium:

$$T_1 = T_2, \quad \text{and} \quad P_1 = P_2. \tag{5.5}$$

This means that the temperatures and pressures of subsystems 1 and 2 must be equal to each other. The division of 1 and 2 is arbitrary, therefore the conditions that an isolated system A is in a state of thermodynamic equilibrium are that the ***temperature and pressure must be uniform in the entire system.*** However the vanishing of the first order derivatives of S_{tot} does not guarantee that S_{tot} is a maximum. It is possible that it is a minimum. One still needs the conditions that the second order derivatives of S_{tot} are less than zero to guarantee that S_{tot} is a maximum. These requirements will impose additional conditions for thermodynamic equilibrium. There is much more involved mathematically in the requirements on the second order derivatives, so we defer their discussion to a latter section in this chapter.

The principle of maximum entropy is applicable only to isolated systems, so its applications are rather limited. It is therefore desirable to extend the principle to non-isolated systems, then the conditions of thermodynamic equilibrium can be established under various external situations. We will do this in the subsequent sections. The key procedure for this purpose is to introduce a couple of thermodynamic potentials.

5.3 Thermodynamic Potentials

Before we introduce these thermodynamic potentials, we introduce the **Legendre**[2] **transformation** in mathematics. Consider a function with two independent variables $f(x, y)$ with a differential

$$df = Xdx + Ydy. \tag{5.6}$$

By definition x and X are a *canonically conjugate pair*, and so are y and Y. In this expression, x and y are the independent variables. The purpose of the Legendre transformation is to introduce a new function whose independent variables are (x, Y) instead of (x, y), i.e., y is replaced by its conjugate pair partner Y. This can be done by defining a new function g which is f minus yY, i.e., $g = f - yY$, then

$$dg = df - d(yY) = df - ydY - Ydy = Xdx - ydY. \tag{5.7}$$

Therefore $g = g(x, Y)$, as we wanted. This is known as the Legendre transformation. By using this transformation, in addition to g defined above, we can define two additional functions, one with (X, y), and the other with (X, Y) as the independent variables. That is to say, the two variables in the same conjugate pair can be interchanged, i.e., x may be replaced by X and y may be replaced by Y, and vice versa.[3] Thermodynamic potentials are defined in this way.

In thermodynamics, the first state variable defined with energy as its units is the *internal energy* U, which can be considered as the first thermodynamic potential. We will start with U to define new thermodynamic potentials. From Eq. (2.57), which is a combination of the first and second laws,

$$dU = TdS - PdV,$$

we have $U = U(S, V)$. Therefore (S, V) are the natural independent variables for U, and T, S and $-P$, V are the two canonically conjugate pairs. We use U as the starting function and make a Legendre transformation to get the state variable *enthalpy* H, i.e., Eq. (1.16):

$$\begin{aligned} H &= U - (-P)V = U + PV, \\ dH &= dU + PdV + VdP = TdS + VdP, \end{aligned} \tag{5.8}$$

[2]Adrien M. Legendre, French mathematician (1752–1833).

[3]We are unable to define a function with the independent variables (x, X) or (y, Y). Therefore for a function with two independent variables, we have a combination of four pairs of independent variables, thus we can define at most four different functions. In general a thermodynamic system has two independent variables, therefore we can define four different thermodynamic potentials.

which we had already defined in Chapter 1. From Eq. (5.8) the natural independent variables for H are S and P, i.e., $H = H(S, P)$. Enthalpy can also be considered as one of the thermodynamic potentials. The natural independent variables for both U and H contain the state variable S, which is not an ideal variable to be controlled externally, especially in phase equilibrium.[4] Therefore the usefulness of U and H are less obvious, although they play the role of potentials as we have mentioned above. However both U and H are useful state variables in thermodynamics; in particular we can define the constant volume and constant pressure heat capacities in terms of U and H, in a rather convenient way, as in Eq. (1.19).

The most important and widely used thermodynamic potentials are the **Helmholtz**[5] **free energy** (also known as the **Helmholtz function**) F and the **Gibbs free energy** (also known as the **Gibbs function**) G. Both F and G can be defined as Legendre transformations of U. The independent variables for the Helmholtz free energy F are (T, V), which is defined by a Legendre transformation of U replacing the variable S by T:

$$F = U - TS, \tag{5.9}$$

$$dF = dU - TdS - SdT = -SdT - PdV. \tag{5.10}$$

From the expression of dF we have $F = F(T, V)$. The independent variables for the Gibbs free energy G are (T, P), which can be defined by a Legendre transformation of F, replacing the variable V by P:

$$G = F - (-P)V = F + PV, \tag{5.11}$$

$$dG = dF + PdV + VdP = -SdT + VdP. \tag{5.12}$$

Therefore $G = G(T, P)$. We see that the independent variables (T, V) (for F) or (T, P) (for G) are ideal variables which can be controlled in the laboratory. Therefore both F and G are the *thermodynamic potentials* most frequently used in the study of the conditions for thermodynamic equilibrium, especially in phase transitions. We will discuss the properties of F and G in the next two sections.

---

[4]The thermodynamic potentials are particularly useful in the study of phase equilibrium; see Secs. 5.4–5.5.

[5]Herman L. F. Helmholtz, German physicist (1821–1894).

5.4 The Helmholtz Free Energy

We consider a system which is in thermal contact with a heat reservoir at temperature T. The volume of the system changes from V_1 to V_2. This is an isothermal process, which can either be reversible or irreversible (e.g., when there is dissipative heat produced). From the second law of thermodynamics,

$$\Delta S_{\text{tot}} = \Delta S + \Delta S_{\text{reservoir}} \geq 0, \tag{5.13}$$

where ΔS and $\Delta S_{\text{reservoir}}$ are the entropy change of the system and the heat reservoir. The equal sign applies to reversible processes only. Suppose during this process, the heat released by the reservoir is ΔQ, therefore the entropy change for the reservoir is $\Delta S_{\text{reservoir}} = -\Delta Q/T$, as this is an isothermal process. Substituting this result into Eq. (5.13) we have[6]

$$\Delta S - \frac{\Delta Q}{T} \geq 0, \quad \text{or} \quad \Delta Q \leq T\Delta S. \tag{5.14}$$

From the first law $\Delta U = \Delta Q - \Delta W$, $\Delta W = -\Delta U + \Delta Q \leq -\Delta U + T\Delta S$, and hence

$$\Delta W \leq -\Delta F = F(T, V_1) - F(T, V_2), \tag{5.15}$$

where T is the same for both the initial and final states. Therefore *in an isothermal process, the change of the Helmholtz free energy (take the absolute value) is the maximum amount of work the system can do.* This implies that the function F *stores* some energy, which can be freed to do work during an isothermal process. The function F is therefore known as the *Helmholtz free energy.* If we consider an isochoric process, $\Delta V = 0$ and $\Delta W = 0$, then we have

$$\Delta F \leq 0 \ (T, V \text{ constant}). \tag{5.16}$$

Therefore in an *irreversible* isothermal ($\Delta T = 0$) and isochoric ($\Delta V = 0$) process, the Helmholtz free energy F of the final state will be smaller than the F of the initial state; while in a reversible process the Helmholtz free energy remains constant. Therefore we have the conclusion: **In an isothermal and isochoric process, the system will attain an equilibrium state only when the system is in a state such that the Helmholtz free energy F has a minimum value.** Temperature and

[6]For the system the reversible heat is $\Delta Q^{(r)} = T\Delta S = \Delta Q + \delta q$, $\delta q > 0$ is the dissipative heat. Therefore $T\Delta S > \Delta Q$. See also Eq. (2.62).

volume are two state variables which can be controlled with relative ease in a laboratory (especially for gas systems), therefore the above criterion for thermodynamic equilibrium is rather useful. However for a system having only two independent variables (most of the thermodynamic systems belong to this category), if T and V are fixed the state of the system is also fixed. The above criterion seems useless. Nevertheless if a system can have two or more different states for the same T and V, then the state with the smallest value of F is the stable state, and the states with larger F are unstable. We say that these different states (with the same T and V) belong to different phases of the system. Therefore the condition that F must be a minimum for the system to be in an equilibrium state is useful in the study of phase equilibrium.

5.5 The Gibbs Free Energy

We consider a system which is in thermal contact with a heat reservoir at temperature T, where the external pressure on the system is P. The system then undergoes an isothermal and isobaric process, with T and P kept constant. In this process the heat reservoir releases heat ΔQ to the system, there is a volume change of the system ΔV, and work ΔW is done by the system. From Eq. (5.14), $\Delta Q \leq T\Delta S$, and the first law of thermodynamics, $\Delta U = \Delta Q - \Delta W$, $\Delta W = P\Delta V$, we have

$$\Delta U + P\Delta V - T\Delta S \leq 0. \tag{5.17}$$

Since $G = U + PV - TS$, when T and P are kept constant, Eq. (5.17) becomes

$$\Delta G \leq 0 \ \ (T, P \text{ constant}). \tag{5.18}$$

Note that the equal sign holds for reversible processes only. The above inequality implies that in an isothermal and isobaric process, the Gibbs free energy G always decreases for irreversible processes, while it remains unchanged for reversible processes. The above statement can be summarized in the following conclusion: **In an isothermal and isobaric process, the system will attain an equilibrium state only when the system is in a state such that the Gibbs free energy G has a minimum value.** Temperature T and pressure P are two variables which can be controlled with relative ease in a laboratory (or even outside laboratories), therefore the minimum G condition is very useful in studying the thermodynamic equilibrium phenomena. For liquids and solids the condition of constant pressure is much easier to achieve than the condition of constant volume. Therefore for these systems G is more useful than F, experimentally. However theoretically it is easier to calculate[7] F than G, thus both G and F are equally useful. As in the case of the Helmholtz free energy F, the Gibbs free energy G is useful in the study of phase equilibrium. Thus if a system can have two or more different states (phases) for a given T and P, then the state with the smallest value of G is the stable phase and the states with higher G are unstable. The application of the principle that at constant T and P, G must be minimum for the phenomena of phase equilibrium will be given in Chapter 7.

If, in addition to the PdV work, the system has non-mechanical work W_{NM} (electric, magnetic etc.), then $\Delta W = P\Delta V + W_{\mathrm{NM}}$, and

$$W_{\mathrm{NM}} \le -\Delta G \ (T, P \text{ constant}). \tag{5.19}$$

Therefore in an isothermal and isobaric process, the change in the Gibbs free energy G of a system (take the absolute value) is the maximum amount of non-mechanical work the system can do. This implies that the function G *stores* some energy which can be freed to do work during an isothermal and isobaric process. The function G is therefore also known as the **Gibbs free energy**.

5.6 The Maxwell Relations

In the previous sections, we started from the expression for dU and obtained the two pairs of conjugate variables (T, S) and (P, V). By using the Legendre transformations we defined four thermodynamic potentials U, H, F and G. All of them are state variables, and therefore, their differentials are all exact differentials. By using the condition for an exact differential, we can derive four useful thermodynamic relations, known as the *Maxwell*[8] *relations*. From $U = U(S, V)$, $H = H(S, P)$, $F = F(T, V)$ and $G = G(T, P)$, we have

$$dU = TdS - PdV = \left(\frac{\partial U}{\partial S}\right)_V dS + \left(\frac{\partial U}{\partial V}\right)_S dV, \tag{5.20}$$

[7] The microscopic energy levels for a system depend on the volume, therefore keeping the volume constant makes the theoretical calculations easier. The relation between the microscopic energy levels and F will be given in Chapter 10.

[8] James C. Maxwell, British physicist (1831–1879).

$$dH = TdS + VdP = \left(\frac{\partial H}{\partial S}\right)_P dS + \left(\frac{\partial H}{\partial P}\right)_S dP, \tag{5.21}$$

$$dF = -SdT - PdV = \left(\frac{\partial F}{\partial T}\right)_V dT + \left(\frac{\partial F}{\partial V}\right)_T dV, \tag{5.22}$$

$$dG = -SdT + VdP = \left(\frac{\partial G}{\partial T}\right)_P dT + \left(\frac{\partial G}{\partial P}\right)_T dP. \tag{5.23}$$

We start with Eq. (5.20), and get

$$\left(\frac{\partial U}{\partial S}\right)_V = T \quad \text{and} \quad \left(\frac{\partial U}{\partial V}\right)_S = -P. \tag{5.24}$$

From the condition for an exact differential, Eq. (1.7), we obtain

$$\left(\frac{\partial T}{\partial V}\right)_S = -\left(\frac{\partial P}{\partial S}\right)_V. \tag{5.25}$$

This is the first Maxwell relation. The second, the third, and the fourth relations can be obtained by the same method. From Eqs. (5.21)–(5.23) we find

$$\left(\frac{\partial T}{\partial P}\right)_S = \left(\frac{\partial V}{\partial S}\right)_P, \tag{5.26}$$

$$\left(\frac{\partial S}{\partial V}\right)_T = \left(\frac{\partial P}{\partial T}\right)_V, \tag{5.27}$$

$$\left(\frac{\partial S}{\partial P}\right)_T = -\left(\frac{\partial V}{\partial T}\right)_P. \tag{5.28}$$

Equations (5.25)–(5.28) are called the **Maxwell relations**. They are very useful in thermodynamics. In thermodynamics one may frequently encounter some partial differentiations whose physical meaning is not clear. In many cases the partial differentiations may be transformed, in part by using the Maxwell relations, into quantities which have a clear physical meaning, or can even be measured experimentally.

It is helpful if one can memorize these four relations. There is a simple trick for memorizing them, and the memory may last a long time. From the equation of dU, we know that (T, S) and (P, V) are two conjugate pairs. We may consider any one of the Maxwell relations as an equation of the form

$$\frac{A}{B} = \frac{C}{D},$$

where A (B, C, D) represents one of the variables T, S, P and V (omitting the ∂ signs, i.e., T representing ∂T, etc.). The key point is that the cross products across the equal sign, i.e., AD and BC, must be TS or PV (the order of the variables may be reversed, i.e., it may be ST or VP.) Of course when AD is TS, then BC must be PV, and vice versa. Two more things need to be remembered: (1) The variable in the denominator on one side of the equation is the variable which is held constant on the other side. (2) When T and P are *not on the same side* of the equation, then there is a *minus sign* in the equation. (Note that T and P are the most *natural variables* in the laboratory.) The reader may check this trick to see whether it works.

In the derivations of thermodynamic formulas or relations, the Maxwell relations are quite often helpful. We have already used the Maxwell relations to derive some thermodynamic relations, such as in Sec. 3.7 and in Sec. 4.5. In those sections, the Maxwell relations had not yet been derived; we used the condition of an exact differential to derive the relations, which takes more steps and time. Now that the Maxwell relations have been derived, we can use them directly, without using the condition for an exact differential.

5.7 Open Systems and the Chemical Potential

In this section we will introduce another thermodynamic potential, known as the *chemical potential.* We consider an open system, which can exchange energy and mass with its surroundings. This means that the number of particles (molecules) of the system is not a constant. Usually we use the number of kilomoles n to denote the number of particles; the temperature T and pressure P are the other independent variables. Therefore if the system contains only one type of particle, the Gibbs free energy can be written as $G = G(T, P, n)$, which includes n as a independent variable. If there are more than one type of particle in the system,[9] denoting n_i as the number of kilomoles for the ith type particle, the Gibbs free energy is then

$$G = G(T, P, n_1, n_2, \cdots). \tag{5.29}$$

[9]The same type of particle in different phases is considered as two types of particles. For example if a system contains both the molecules of water and ice, then the system is considered to have two types of molecules.

Differentiation gives

$$\begin{aligned} dG &= \left(\frac{\partial G}{\partial T}\right)_{P,n_j} dT + \left(\frac{\partial G}{\partial P}\right)_{T,n_j} dP + \sum_i \left(\frac{\partial G}{\partial n_i}\right)_{T,P,n_{j\neq i}} dn_i \\ &= -SdT + VdP + \sum_i \mu_i \, dn_i, \end{aligned} \tag{5.30}$$

where we have used the relations $(\partial G/\partial T)_{P,n_j} = -S$, $(\partial G/\partial P)_{T,n_j} = V$, and have defined the **chemical potential** μ_i for the ith type of particle:

$$\mu_i(T, P, n_1, n_2, \cdots) = \left(\frac{\partial G}{\partial n_i}\right)_{T,P,n_{j\neq i}}. \tag{5.31}$$

If we use different sets of independent variables, then the chemical potential μ_i can be derived from the other thermodynamic potentials,

$$\mu_i(S, V, n_1, n_2, \cdots) = \left(\frac{\partial U}{\partial n_i}\right)_{S,V,n_{j\neq i}}, \tag{5.32}$$

$$\mu_i(S, P, n_1, n_2, \cdots) = \left(\frac{\partial H}{\partial n_i}\right)_{S,P,n_{j\neq i}}, \tag{5.33}$$

$$\mu_i(T, V, n_1, n_2, \cdots) = \left(\frac{\partial F}{\partial n_i}\right)_{T,V,n_{j\neq i}}. \tag{5.34}$$

Note that all these μ_i's have the same physical meaning, the only difference is that they depend on different independent variables. Therefore, for an open system, we have the following expansions of the differentials,

$$dU = TdS - PdV + \sum_i \mu_i \, dn_i, \tag{5.35}$$

$$dH = TdS + VdP + \sum_i \mu_i \, dn_i, \tag{5.36}$$

$$dF = -SdT - PdV + \sum_i \mu_i \, dn_i. \tag{5.37}$$

If we write the first law, Eq. (1.3), as

$$dU = đQ - đW = TdS - đW,$$

and compare it with Eq. (5.35), we find

$$đW = PdV - \sum_i \mu_i \, dn_i. \tag{5.38}$$

Therefore $\sum_i \mu_i \, dn_i$ can be considered as the *chemical work done on* the system. If we add one kilomole of the ith type of particle to the system, the internal energy of the system will be increased by μ_i, the chemical potential of the ith type of particle. This is a configuration work,[10] where μ_i is the generalized force and dn_i is the generalized displacement, therefore μ_i and n_i form a canonically conjugate pair.

5.8 The Gibbs-Duhem Relation

If a system is composed of several types of particles then, in addition to U and V, the number of kilomoles for each type of particle, n_1, n_2, $\cdots$, will also be the independent variables for the entropy of the system S, i.e., $S = S(U, V, n_1, n_2, \cdots)$. We note that S itself and all the independent variables in the parenthesis, U, V, and all n_i, are extensive variables. Therefore when U, V, and all the n_i's are all increased by a factor of α, S will also be increased by the same factor α, i.e.,

$$\alpha S(U, V, n_1, n_2, \cdots) = S(\alpha U, \alpha V, \alpha n_1, \alpha n_2, \cdots). \tag{5.39}$$

Now let $\alpha = 1 + \varepsilon$, with ε a very small quantity, then S on the right hand side of the above equation can be expanded as a Taylor series. To first order derivatives, we have

$$\begin{aligned}(1+\varepsilon)S = \; & S + \left(\frac{\partial S}{\partial U}\right)_{V,n_j} \varepsilon U \\ & + \left(\frac{\partial S}{\partial V}\right)_{U,n_j} \varepsilon V + \sum_i \left(\frac{\partial S}{\partial n_i}\right)_{U,V,n_{j\neq i}} \varepsilon\, n_i. \end{aligned} \tag{5.40}$$

Simplifying the equation, we get

$$S = \left(\frac{\partial S}{\partial U}\right)_{V,n_j} U + \left(\frac{\partial S}{\partial V}\right)_{U,n_j} V + \sum_i \left(\frac{\partial S}{\partial n_i}\right)_{U,V,n_{j\neq i}} n_i. \tag{5.41}$$

Making use of Eq. (5.35), one obtains

$$\left(\frac{\partial S}{\partial U}\right)_{V,n_j} = \frac{1}{T}, \tag{5.42}$$

[10]See Sec. 1.7 for the definition of a configuration work.

$$\left(\frac{\partial S}{\partial V}\right)_{U,n_j} = \frac{P}{T}, \tag{5.43}$$

$$\left(\frac{\partial S}{\partial n_i}\right)_{U,V,n_{j\neq i}} = -\frac{\mu_i}{T}. \tag{5.44}$$

Substituting Eqs. (5.42)–(5.44) into Eq. (5.41), we have

$$S = \frac{1}{T}U + \frac{P}{T}V - \sum_i \frac{\mu_i}{T} n_i$$

or

$$U = TS - PV + \sum_i \mu_i n_i. \tag{5.45}$$

Differentiating the above equation, we get

$$dU = TdS + SdT - PdV - VdP + \sum_i (\mu_i \, dn_i + n_i \, d\mu_i). \tag{5.46}$$

Comparing the above equation and Eq. (5.35), one obtains

$$SdT - VdP + \sum_i n_i \, d\mu_i = 0. \tag{5.47}$$

This is called the **Gibbs-Duhem[11] relation**. By substituting Eq. (5.45) into the definition of the Gibbs free energy G in Eq. (5.11), one gets

$$G = U - TS + PV = \sum_i \mu_i n_i. \tag{5.48}$$

From this equation we obtain a very important result. When a system is composed of only one type of particle, and there are n kilomoles, then

$$\mu = \mu(T, P) = \frac{G}{n} \equiv g(T, P). \tag{5.49}$$

This means that **the chemical potential of the system μ is exactly equal to the Gibbs free energy per kilomole** g. It is worth noting that this simple relation holds only for the Gibbs free energy G. For the other energy functions, U, H, or F, the simple relation $\mu = X/n$ ($X = U$, H, or F) *does not hold.* The main reason is that the independent variables for U, H, or F *contain at least one extensive variable*, such

[11]Pierre M. M. Duhem, French physicist (1861–1916).

as the volume V or entropy S. Therefore when n varies, the extensive variable (V or S) should also change in order to keep the properties of the system unchanged (i.e., the same as before n varied). Therefore the relation between n and these state variables is much more complicated, not a simple proportional relation. However, the independent variables for G are T and P, which are both *intensive variables*. Therefore when n varies, T and P can be kept constant while the properties of the system remain unchanged. Thus G is simply proportional to n, as in Eq. (5.49).

We can extend the relation of μ and g in Eq. (5.49) to a system composed of several types of particles. The specific Gibbs free energy g_i for the ith type of particle is defined as the increase of the total Gibbs free energy G if the i type of particle is increased by one kilomole, while T, P and all other n_j's ($j \neq i$) are kept constant, i.e.,

$$g_i \equiv \left(\frac{\partial G}{\partial n_i}\right)_{T,P,n_{j\neq i}} = \mu_i, \tag{5.50}$$

where the last equality is due to Eq. (5.31). Therefore $\mu_i = g_i$, and Eq. (5.48) can be rewritten as

$$G = G(T, P, n_1, n_2, \cdots) = \sum_i n_i \, g_i(T, P, n_1, n_2, \cdots). \tag{5.51}$$

It may be worth mentioning that $g_i = \mu_i$ can also be obtained by a direct differentiation of Eq. (5.48),

$$dG = \sum_i \mu_i \, dn_i + \sum_i n_i \, d\mu_i.$$

At constant T and P, the last sum in the above equality is zero due to the Gibbs-Duhem relation (5.47), i.e.,

$$\sum_i n_i \, d\mu_i = 0 \ \ (T, P \text{ constant}). \tag{5.52}$$

Therefore

$$dG = \sum_i \mu_i \, dn_i \ \ (T, P \text{ constant}),$$

which can also be obtained from Eq. (5.30) by setting $dT = dP = 0$. Therefore $g_i = \mu_i$ as in the case that the system contains only one type of particle.

5.9 Conditions of Stable Equilibrium

In Sec. 5.2 we used the principle of maximum entropy to derive the conditions for thermodynamic equilibrium for an isolated system. The conditions are that both the temperature T and pressure P must be homogeneous, i.e., everywhere in the system T and P must have the same value. However we had used only the condition that the first order derivatives of the total entropy S_{tot} must be zero. We still need the condition that the second order derivatives must be less than zero to make the entropy a maximum. This will impose additional conditions on the system. This is what we will do in this section.

Referring to Fig. 5.1, we consider an isolated system A which is divided into two arbitrary subsystems 1 and 2. The internal energy and volume for 1 and 2 are (U_1, V_1) and (U_2, V_2). The total entropy S_{tot} of A is (Eq. (5.2))

$$S_{\text{tot}} = S_1(U_1, V_1) + S_2(U_2, V_2).$$

From the condition that the first order derivatives of the above equation are zero, we obtain the equilibrium conditions $T_1 = T_2$ and $P_1 = P_2$. We denote the equilibrium state as state 0. Now we go one step further in expanding S_{tot} out to the second order derivatives to find the conditions that the equilibrium state 0 is a stable one. We expand S_{tot} around the state 0 to second order in the derivatives. The first order derivatives at state 0 are zero, and we keep only the second order derivative terms,

$$\begin{aligned} dS_{\text{tot}} &= \frac{1}{2}\sum_{i=1}^{2}\left[d\left(\frac{\partial S_i}{\partial U_i}\right)_{V_i}^0 dU_i + d\left(\frac{\partial S_i}{\partial V_i}\right)_{U_i}^0 dV_i\right] \\ &= \frac{1}{2}\sum_{i=1}^{2}\left[d\left(\frac{1}{T_i}\right) dU_i + d\left(\frac{P_i}{T_i}\right) dV_i\right]^0 \\ &= -\frac{1}{2T}\sum_{i=1}^{2}[dT_i\, dS_i - dP_i\, dV_i]^0, \end{aligned} \tag{5.53}$$

where the superscript 0 means that the value of the derivatives are evaluated at state 0. The right hand side of the first equality contains the second order derivative terms where the two d's are defined as

$$d\left(\frac{\partial S_i}{\partial U_i}\right)_{V_i}^0 = \left[\frac{\partial}{\partial U_i}\left(\frac{\partial S_i}{\partial U_i}\right)_{V_i}\right]_{V_i}^0 dU_i + \left[\frac{\partial}{\partial V_i}\left(\frac{\partial S_i}{\partial U_i}\right)_{V_i}\right]_{U_i}^0 dV_i,$$

$$d\left(\frac{\partial S_i}{\partial V_i}\right)_{U_i}^0 = \left[\frac{\partial}{\partial U_i}\left(\frac{\partial S_i}{\partial V_i}\right)_{U_i}\right]_{V_i}^0 dU_i + \left[\frac{\partial}{\partial V_i}\left(\frac{\partial S_i}{\partial V_i}\right)_{U_i}\right]_{U_i}^0 dV_i.$$

In the second equality of Eq. (5.53) we make use of Eq. (2.60), i.e., $(\partial S_i/\partial U_i)_{V_i} = 1/T_i$ and $(\partial S_i/\partial V_i)_{U_i} = P_i/T_i$. The third equality is obtained by the relations $T = T_1 = T_2$ and $dS_i = dU_i/T_i - P_i\, dV_i/T_i$.

On the right hand side of the last equality of Eq. (5.53), each subsystem has four state variables S, T, P and V, but only two of them are independent. First we take T and P as the independent variables, then expand dS_i and dV_i in terms of dT_i and dP_i to obtain

$$dS_i = \left(\frac{\partial S_i}{\partial T_i}\right)_{P_i}^0 dT_i + \left(\frac{\partial S_i}{\partial P_i}\right)_{T_i}^0 dP_i, \tag{5.54}$$

$$dV_i = \left(\frac{\partial V_i}{\partial T_i}\right)_{P_i}^0 dT_i + \left(\frac{\partial V_i}{\partial P_i}\right)_{T_i}^0 dP_i. \tag{5.55}$$

Substituting the above two equations into the right hand side of the last equality of Eq. (5.53) one obtains

$$\begin{aligned} dS_{\text{tot}} &= -\frac{1}{2T}\sum_{i=1}^{2}\left[\left(\frac{\partial S_i}{\partial T_i}\right)_{P_i}^0 (dT_i)^2 + \left(\frac{\partial S_i}{\partial P_i}\right)_{T_i}^0 dT_i\, dP_i \right. \\ &\quad \left. - \left(\frac{\partial V_i}{\partial T_i}\right)_{P_i}^0 dT_i\, dP_i - \left(\frac{\partial V_i}{\partial P_i}\right)_{T_i}^0 (dP_i)^2\right] \\ &= -\frac{1}{2T}\sum_{i=1}^{2}\left[\left(\frac{\partial S_i}{\partial T_i}\right)_{P_i}^0 (dT_i)^2 - 2\left(\frac{\partial V_i}{\partial T_i}\right)_{P_i}^0 dT_i\, dP_i \right. \\ &\quad \left. - \left(\frac{\partial V_i}{\partial P_i}\right)_{T_i}^0 (dP_i)^2\right], \end{aligned} \tag{5.56}$$

where in the last equality we have used the Maxwell relation (5.28), $(\partial S_i/\partial P_i)_{T_i} = -(\partial V_i/\partial T_i)_{P_i}$. It is still not easy to establish the condition $dS_{\text{tot}} \leq 0$ from Eq. (5.56). We note that

$$\left(\frac{\partial S}{\partial T}\right)_P = \frac{C_P}{T} \quad \text{and} \quad \left(\frac{\partial S}{\partial T}\right)_V = \frac{C_V}{T},$$

then from the equation for $C_P - C_V$ (4.53),

$$C_P - C_V = \frac{TV\beta^2}{\kappa},$$

we have

$$\left(\frac{\partial S}{\partial T}\right)_P = \left(\frac{\partial S}{\partial T}\right)_V - \left(\frac{\partial P}{\partial V}\right)_T \left(\frac{\partial V}{\partial T}\right)_P^2 . \tag{5.57}$$

Here we have used the definitions of the expansivity β and the isothermal compressibility κ,

$$\beta = \frac{1}{V}\left(\frac{\partial V}{\partial T}\right)_P , \quad \kappa = -\frac{1}{V}\left(\frac{\partial V}{\partial P}\right)_T .$$

Substituting Eq. (5.57) into the last equality of Eq. (5.56), we obtain

$$\begin{aligned} dS_{\text{tot}} &= -\frac{1}{2T}\sum_{i=1}^{2}\left[\left(\frac{\partial S_i}{\partial T_i}\right)_{V_i}(dT_i)^2 - \left(\frac{\partial P_i}{\partial V_i}\right)_{T_i}\left(\frac{\partial V_i}{\partial T_i}\right)_{P_i}^2 (dT_i)^2 \right. \\ &\quad \left. - 2\left(\frac{\partial V_i}{\partial T_i}\right)_{P_i} dT_i\, dP_i - \left(\frac{\partial V_i}{\partial P_i}\right)_{T_i}(dP_i)^2\right]^0 \\ &= -\frac{1}{2T}\sum_{i=1}^{2}\left\{\left(\frac{\partial S_i}{\partial T_i}\right)_{V_i}(dT_i)^2 - \left(\frac{\partial P_i}{\partial V_i}\right)_{T_i}\left[\left(\frac{\partial V_i}{\partial T_i}\right)_{P_i}^2 (dT_i)^2 \right.\right. \\ &\quad \left.\left. + 2\left(\frac{\partial V_i}{\partial T_i}\right)_{P_i}\left(\frac{\partial V_i}{\partial P_i}\right)_{T_i} dT_i\, dP_i + \left(\frac{\partial V_i}{\partial P_i}\right)_{T_i}^2 (dP_i)^2\right]\right\}^0 \\ &= -\frac{1}{2T}\sum_{i=1}^{2}\left[\left(\frac{\partial S_i}{\partial T_i}\right)_{V_i}(dT_i)^2 - \left(\frac{\partial P_i}{\partial V_i}\right)_{T_i}(dV_i)^2\right]^0 . \end{aligned} \tag{5.58}$$

The second equality is just a rearrangement of the terms. The last equality is obtained by noting that the terms inside the parenthesis [· · ·] in the preceding equality are the square of dV_i in the expansion of Eq. (5.55),

$$\begin{aligned} (dV_i)^2 &= \left(\frac{\partial V_i}{\partial T_i}\right)_{P_i}^2 (dT_i)^2 + 2\left(\frac{\partial V_i}{\partial T_i}\right)_{P_i}\left(\frac{\partial V_i}{\partial P_i}\right)_{T_i} dT_i\, dP_i \\ &\quad + \left(\frac{\partial V_i}{\partial P_i}\right)_{T_i}^2 (dP_i)^2 . \end{aligned} \tag{5.59}$$

Now we are ready to obtain the conditions that the second order derivatives of S_{tot} are less than zero; this can be obtained from Eq. (5.58). The differentials dT_i and dV_j can vary independently (regardless of whether $i = j$ or $i \neq j$), therefore we easily get the conditions for S_{tot} to be a maximum,

$$\left(\frac{\partial S_i}{\partial T_i}\right)_{V_i}^0 > 0, \quad \text{and} \quad \left(\frac{\partial P_i}{\partial V_i}\right)_{T_i}^0 < 0. \tag{5.60}$$

The conditions of Eq. (5.60) can be applied to any thermodynamic system, because the subscript i (denoting a subsystem), can be any small part of a large isolated system. Therefore the conditions that a thermodynamic system is in a stable state (S_{tot} is a maximum) are (we drop the superscript 0 in the following discussion, because we are always dealing with equilibrium states)

$$C_V = T\left(\frac{\partial S}{\partial T}\right)_V > 0, \tag{5.61}$$

and

$$\kappa = -\frac{1}{V}\left(\frac{\partial V}{\partial P}\right)_T > 0. \tag{5.62}$$

In the above inequalities C_V is the constant volume heat capacity, and κ is the isothermal compressibility. Therefore from the principle of maximum entropy, we obtain four conditions for an isolated thermodynamic system to be in a *stable equilibrium state*: **the temperature and pressure must be everywhere the same in the entire system, and both C_V and κ of the system must be positive.** It should be noted that, if either C_V or κ were less than zero for a thermodynamic system, then the system would be unstable and it could not exist at all.

In mechanics, we know that a particle is in equilibrium when the total force on it is zero. The equilibrium may be stable or unstable. When the position of the particle is slightly displaced from the equilibrium position, the force on it will no longer be zero. If there is a restoring force which can bring the particle back to its equilibrium position, then it is a stable equilibrium. If there are no such restoring forces, then the equilibrium is unstable. This analysis may also be applied to thermodynamic equilibrium. We get an important result: *in thermodynamics, only stable equilibrium can exist in nature.* Unstable equilibrium can not exist because *thermodynamic fluctuations* will cause the system to deviate from the equilibrium state and the system can never return back to equilibrium, if the equilibrium is an unstable one. The conditions of thermodynamic equilibrium (T and P are uniform in the entire system) are obtained by requiring that the first order derivatives of S_{tot} are zero. Moreover the equilibrium must be a stable one, this requires $C_V > 0$ and

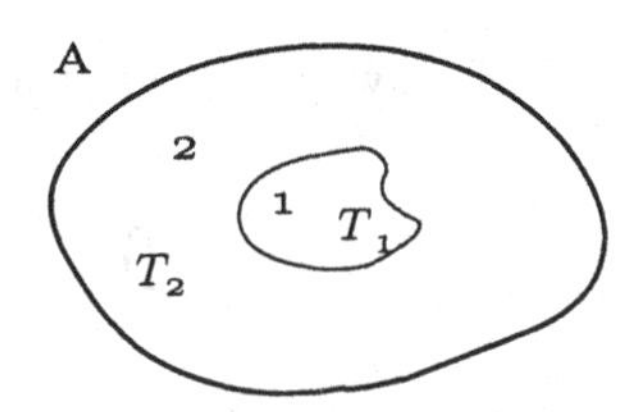

Figure 5.2: Due to thermal fluctuations, the temperature T_1 of the subsystem 1 may be greater than T_2 of 2 for a short moment, then heat will flow from 1 to 2. For stable equilibrium to be maintained, the heat flow must make T_1 decrease and T_2 increase, to re-establish the equilibrium condition $T_1 = T_2$.

$\kappa > 0$ which are obtained by requiring that the second order derivatives of S_{tot} are negative.

Now we will explain why $C_V > 0$ is necessary for the system to have a restoring force when there is a temperature fluctuation. As shown in Fig. 5.2, we divide an isolated system A into two arbitrary subsystems 1 and 2. First we assume that the volumes of 1 and 2 (V_1 and V_2) are kept constant. The condition of thermodynamic equilibrium is $T_1 = T_2$ (the condition $P_1 = P_2$ is not necessary because V_1 and V_2 are constants). The wall between 1 and 2 however is thermally conducting, therefore 1 and 2 can exchange energy. In fact the energy exchanging process is a continuous one; it never stops. Therefore it is certain that the situation $T_1 \neq T_2$ will occur frequently, and this is known as a thermal fluctuation. Suppose at some moment $T_1 > T_2$, then by the second law, heat will flow from the hotter body (subsystem 1) to the colder body (subsystem 2). If this heat flow makes T_1 decrease and T_2 increase, then the equilibrium condition $T_1 = T_2$ will be restored. Therefore this is a stable equilibrium, and $C_V = (đQ/dT)_V > 0$ is the condition for stable equilibrium. The condition $C_V > 0$ means that the temperature of a body will increase (decrease) when the body absorbs (rejects) heat. This condition implies that in the universe any material with $C_V < 0$ cannot exist, because the material would not be stable against *temperature fluctuations*.

Next we consider the problem of *volume fluctuations*. For simplicity we assume that the subsystems 1 and 2 are in thermal equilibrium $T_1 = T_2$, and both T_1 and T_2 are fixed (and 1 and 2 exchange no mass). We let V_1 and V_2 vary but $V_1 + V_2 = V_0$ is a constant. Thermodynamic equilibrium requires $P_1 = P_2 = P_0$, and the volumes are at their equilibrium values $V_1 = V_1^0$, $V_2 = V_2^0$, and $V_1^0 + V_2^0 = V_0$. Suppose at some moment volume $V_1 > V_1^0$ and $V_2 < V_2^0$, due to a pressure fluctuation. In this situation, stable equilibrium requires that $P_1 < P_0$ and $P_2 > P_0$, thus $P_2 > P_1$. When $P_2 > P_1$ is satisfied, then, from mechanics, we know that V_1 will decrease and V_2 will increase. This makes the system return to the equilibrium state $V_1 = V_1^0$ and $V_2 = V_2^0$ (i.e., $P_1 = P_2 = P_0$). Therefore the condition for stable equilibrium against volume (or pressure) fluctuations is $\kappa = -V^{-1}(\partial V/\partial P)_T > 0$. This inequality means that when the pressure on a body increases, the volume of the body should decrease, and vice versa.

5.10 Problems

5.1. Calculate the change in entropy, Helmholtz free energy and Gibbs free energy when a mole of ideal gas is compressed from 1 atm to 100 atm at 27°C.

5.2. Calculate the values of ΔU, ΔH, ΔF, and ΔG for the following processes. Also show that in (c) an absolute value of the entropy is required.

(a) The process in Problem 3.3, (i) if A and B are the same molecules, and (ii) if A and B are different molecules.

(b) An adiabatic free expansion of one kilomole of an ideal gas at temperature T and pressure P to double its volume.

(c) A reversible adiabatic expansion of one kilomole of an ideal gas from (T_1, P_1) to (T_2, P_2).

5.3. For water at 27°C and the pressure P between 0 and 1000 atm, the volume per mole v can be written as

$$v = a + bP + cP^2,$$

where a, b, and c are constants. Also

$$\left(\frac{\partial v}{\partial T}\right)_P = A + B\,P,$$

where A and B are constants. Find the work necessary at 27°C to compress 1 mole of water from 1 atm to 1000 atm and also find the increase in its internal energy. The numerical values of the constants are: $a = 18.0$ cm^3 mole^{-1}, $b = -7.15 \times 10^{-4}$ cm^3 mole^{-1} atm^{-1}, $c = 4.60 \times 10^{-8}$ cm^3 mole^{-1} atm^{-2}, $A = 4.5 \times 10^{-3}$ cm^3 mole^{-1} deg^{-1}, and $B = 1.4 \times 10^{-6}$ cm^3 mole^{-1} atm^{-1} deg^{-1}.

5.4. A solid is experimentally found to satisfy the relation

$$\left(\frac{\partial V}{\partial T}\right)_P = a + bP + cP^2,$$

at temperature T_0 and in the pressure range $P_A \leq P \leq P_B$, where a, b, and c are functions of T independent of P. How much will the entropy increase when the solid is compressed from a pressure P_A to P_B at the constant temperature T_0?

5.5. Show that the temperature of a system, which is under a reversible adiabatic compression, may either increase or decrease depending on whether the expansivity of the system is greater or less than 0.

5.6. The energy equation of a substance is found to be $u = aT^2v$, where a is a constant. What information can be deduced about the entropy of the substance? What other measurement(s) must be made to determine the entropy?

5.7. Show that for an ideal gas the specific Helmholtz free energy f and the specific Gibbs free energy g are given by

(a) $f = c_v T - c_v T \ln T - RT \ln v - s_0 T + \text{constant}$.

(b) $g = c_p T - c_p T \ln T + RT \ln P - s_0 T + \text{constant}$.

5.8. Show that if the Helmholtz free energy F is known as a function of T and V, then

$$H = F - T\left(\frac{\partial F}{\partial T}\right)_V - V\left(\frac{\partial F}{\partial V}\right)_T,$$

$$G = F - V\left(\frac{\partial F}{\partial V}\right)_T,$$

where H is the enthalpy and G is the Gibbs free energy.

5.9. The specific Gibbs free energy of a gas is given by

$$g = RT \ln\left(\frac{P}{P_0}\right) - AP,$$

where A is a function of T only, and P_0 is a constant pressure.

Find the expressions for the following quantities:

(a) The equation of state.
(b) The specific entropy.
(c) The specific Helmholtz free energy.
(d) The constant pressure specific heat capacity c_p.
(e) The constant volume specific heat capacity c_v.

5.10. The specific Gibbs free energy of a gas is given by

$$g = RT \ln\left(\frac{v_0}{v}\right) + Bv,$$

where B is a function of T alone, and v_0 is a constant volume.

(a) Show explicitly that this form of the Gibbs free energy does not completely specify the properties of the gas.
(b) What further information is necessary so that the properties of the gas can be completely specified?

5.11. Show that $\left(\frac{\partial S}{\partial V}\right)_P = \frac{C_P}{\beta TV}$.

5.12. Show that $\left(\frac{\partial S}{\partial P}\right)_V = \frac{\kappa C_P}{\beta T} - \beta V$.

5.13. Show that $\left(\frac{\partial F}{\partial P}\right)_V = -\frac{\kappa S}{\beta}$.

5.14. Show that $\left(\frac{\partial F}{\partial V}\right)_P = -\frac{S}{\beta V} - P$.

5.15. Show that $\left(\frac{\partial H}{\partial S}\right)_V = T\left(1 + \frac{\beta V}{\kappa C_V}\right)$.

5.16. Show that $\left(\frac{\partial H}{\partial V}\right)_S = -\frac{C_P}{\kappa C_V}$.

5.17. Show that $T\left(\frac{\partial S}{\partial T}\right)_H = \frac{C_P}{1 - \beta T}$.

5.18. Show that $\left(\frac{\partial^2 G}{\partial P^2}\right)_T \left(\frac{\partial^2 F}{\partial V^2}\right)_T = -1$.

5.19. The equation of state of a certain gas is $(P + b)v = RT$, where b is a constant.

(a) Find $c_P - c_v$.

(b) Find the change in specific entropy in an isothermal process.

(c) Show that c_v is independent of v.

5.20. (a) Show that the Gibbs-Duhem relation leads to

$$\sum_i x_i \, d\mu_i = 0 \quad (T, P \text{ constant}),$$

where x_i is the mole fraction of the ith component.

(b) Show that for a two-component system

$$\left(\frac{\partial \mu_1}{\partial \ln x_1}\right)_{T,P} = \left(\frac{\partial \mu_2}{\partial \ln x_2}\right)_{T,P}.$$

5.21. The expression for the specific Helmholtz free energy for a certain system is

$$f = RT \ln\left(\frac{v_0}{v}\right) + aT^2 v,$$

where a and v_0 are constants. What is (or are) the condition(s) that the constant a must meet in order that the system has a reasonable specification of the properties for normal temperatures and pressures?

5.22. Consider n kilomoles of a van der Waals gas at temperature T and volume $V = nv$. Show that the total Helmholtz free energy of the gas is

$$F(n, T, V) = -nRT \ln(V - nb) - \frac{an^2}{V} + c(T),$$

where $c(T)$ is a constant which depends on T alone.
[*Hint*: It may be easier to find the Gibbs free energy G first, which can be found as a function of V at constant T and n.]

Chapter 6

Cooling and the Third Law

6.1 Introduction

Low temperature physics is an important research area. Many experiments are conducted under very low temperature conditions, in order to reduce the interference of the effects of thermal fluctuations. Moreover some important and interesting phenomena exist only when the temperature is low enough. The most well-known may be the phenomenon of superconductivity in some metals, discovered in the early twentieth century. Applications of low temperature techniques are important not only in physics but also in other branches of science and technology, such as chemistry, life science, medicine and engineering. In this chapter we will introduce some basic thermodynamic principles which may be used to produce low temperatures in the laboratory.

There exists a third law of thermodynamics which applies when the temperature is extremely low, approaching absolute zero. In this chapter we will discuss three different statements of the third law and explain whether they are equivalent to one another or not. We will also show that, because of the existence of the third law, there are some restrictions on the properties of all thermodynamic systems. Notably the constant volume (pressure) heat capacity of any substance must approach zero as the temperature approaches zero.

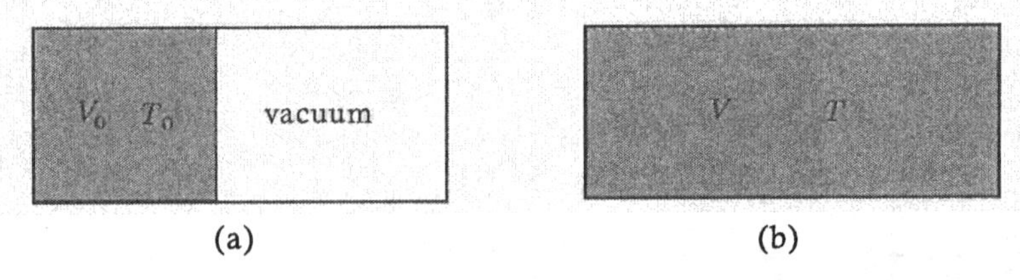

Figure 6.1: Adiabatic free expansion of a gas. The wall of the container is thermally insulating. (a) Before expansion. The left side of the container contains a gas of volume V_0 and temperature T_0. The right side of the container is vacuum. (b) The partition between the gas and vacuum is removed. The gas molecules expand freely to the whole container, with volume V and temperature T.

6.2 Gay-Lussac–Joule Experiment

In the mid nineteenth century Gay-Lussac[1] performed an experiment, now known as the adiabatic free expansion of a gas. He did the experiment not because he wanted to obtain a low temperature, but rather because he wanted to study whether the internal energy of a gas depends on the volume, in addition to the temperature. He found that, within experimental errors, the temperature of the gas remained unchanged after the adiabatic free expansion. Later Joule did the same experiment, and confirming the results of Gay-Lussac's experiment. Although both Gay-Lussac and Joule did the experiment by using a *real gas*,[2] which is expected to behave differently from an ideal gas, yet within experimental errors their result is the same as the ideal gas result. The most important source of the experimental errors may be due to the fact that it is not easy to really do the experiment under the *adiabatic condition*. The heat capacity of the gas is much smaller than that of the container. Therefore any temperature change of the gas will be undetectable because of the large heat capacity of the wall of the container.

Now we describe the adiabatic free expansion experiment and use the known thermodynamic relations to see how we can obtain the internal energy U as a function of T and V, i.e., $U = U(T, V)$. We consider a solid container with thermally insulating walls. Initially there is a partition which divides the container into two parts. The left portion of the container is filled with a gas with temperature T_0 and volume V_0, and

[1]Joseph L. Gay-Lussac, French chemistry (1778–1850).

[2]In nature there exists no ideal gas.

the right portion is a vacuum, as shown in Fig. 6.1(a). The partition is then removed, so the molecules of the gas can *freely expand* to the whole volume of the container. The final state of the gas has temperature T and volume V, as shown in Fig. 6.1(b). In the process the gas does no work (*free* expansion), and it is assumed that the wall is insulating, therefore no heat is absorbed or rejected by the gas. Apparently this is a constant U process, because $\Delta U = \Delta Q - \Delta W = 0$. It is more convenient to use the intensive variables u and v, where $u = U/n$ and $v = V/n$ (n is the number of kilomoles) are the specific internal energy and specific volume of the gas. From the cyclic relation (4.31) for T, v, and u,

$$\left(\frac{\partial T}{\partial v}\right)_u \left(\frac{\partial v}{\partial u}\right)_T \left(\frac{\partial u}{\partial T}\right)_v = -1,$$

and the reciprocal relation (4.35), we have

$$\eta \equiv \left(\frac{\partial T}{\partial v}\right)_u = -\frac{\left(\frac{\partial u}{\partial v}\right)_T}{\left(\frac{\partial u}{\partial T}\right)_v} = -\frac{1}{c_v}\left(\frac{\partial u}{\partial v}\right)_T, \tag{6.1}$$

where η is called the **Joule coefficient.** The Joule coefficient can be used to measure the relation between the specific internal energy u and the specific volume v when the system undergoes an isothermal process. The Gay-Lussac–Joule's experiment is a u =constant process, the result of T being unchanged implies $(\partial T/\partial v)_u$=0 and thus $\eta = 0$. From Eq. (6.1) one obtains the result that u is independent of v, i.e., $u = u(T)$ or $U = U(T)$. We have proved that $U = U(T)$ for an ideal gas in Sec. 3.7; therefore the result of the Gay-Lussac–Joule experiment is the ideal gas result. However this result is not reliable, because of unavoidable experimental errors. We use the van der Waals gas as an example to show that η is not zero for a *real gas.*

From the van der Waals equation of state Eq. (1.35),

$$P = \frac{RT}{v-b} - \frac{a}{v^2}, \tag{6.2}$$

$du = Tds - Pdv$, and the Maxwell relation (5.27), we have

$$\begin{aligned}\left(\frac{\partial u}{\partial v}\right)_T &= T\left(\frac{\partial s}{\partial v}\right)_T - P \\ &= T\left(\frac{\partial P}{\partial T}\right)_v - P = \frac{a}{v^2}.\end{aligned} \tag{6.3}$$

Therefore for a van der Waals gas

$$\eta = -\frac{1}{c_v}\left(\frac{\partial u}{\partial v}\right)_T = -\frac{a}{c_v v^2} = \left(\frac{\partial T}{\partial v}\right)_u , \tag{6.4}$$

$$\begin{aligned} T_2 - T_1 &= \int_{v_1}^{v_2} (-\frac{a}{c_v v^2}) dv \quad (u = \text{constant}) \\ &= \frac{a}{c_v}\left(\frac{1}{v_2} - \frac{1}{v_1}\right). \end{aligned} \tag{6.5}$$

For an ideal gas $a = 0$, therefore $\eta = 0$. For a van der Waals gas $a > 0$ and $v_2 > v_1$, therefore $T_2 < T_1$. This means that for an adiabatic free expansion of a van der Waals gas, whose behavior may be considered to be close to a *real gas*, the temperature decreases in the process. Experimentally, however, the gas may absorb heat from the container, and the decrease in temperature may not be measured.

6.3 Cooling by the Joule-Thomson Experiment

The Gay-Lussac–Joule experiment is a non-quasi-static irreversible process. Although the final temperature T of the gas may be lower than the initial temperature T_0, but because of the difficulty of insulating the gas from the wall of the container, the temperature difference $T - T_0$ may be negligibly small and cannot be measured. Later Joule and Thomson[3] designed a different experiment which can minimize the effect of the container and can lower the temperature of a gas under certain conditions. It is known as the **Joule-Thomson experiment** or the **Joule-Kelvin experiment**. This experiment is a **throttling process**, which is conducted under a *steady-state* condition, where the effect of the wall of the container can be reduced to an acceptable level. To say a system is in a steady-state means that the system is under the influence of an external force (or field), but the state variables of the system remain constant, independent of time, although they may be not uniform in space. The state variables involved in this case are the temperature and pressure. In the following we will explain the throttling process of the Joule-Thomson experiment.

As shown in Fig. 6.2, a cylindrical container is divided into two parts by a porous plug in the middle of the container. Initially all the gas

[3]William Thomson, Lord Kelvin, British physicist (1824–1907).

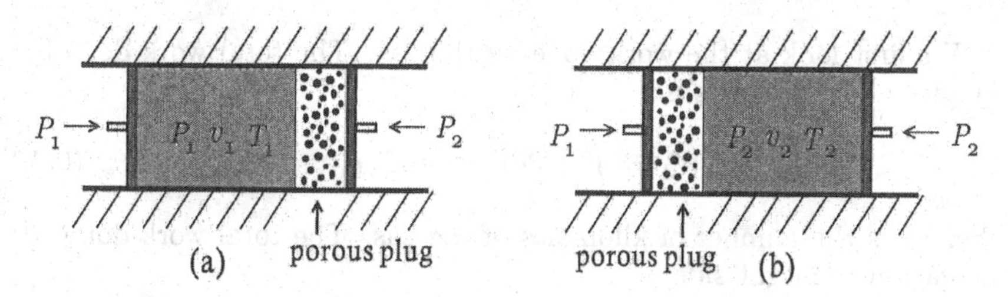

Figure 6.2: The throttling process in the Joule-Thomson experiment. (a) Initially all the gas molecules are on the left side, which is pushed by an external pressure from the left. The molecules can pass through the porous plug and enter the right side of the container. In the process the temperature and the pressure of the left side are kept constant at T_1 and P_1. (b) Finally all the gas molecules are on the right side of the container, where the temperature and the pressure are kept constant at T_2 and P_2.

molecules are on the left side, which is pushed by an external pressure P_1 from the left. The molecules can pass through the porous plug and enter the right side of the container. In the process the gas on the left side is in a steady-state (T_1, P_1, v_1), where T_1, P_1 and v_1 are, respectively, the temperature, the pressure and the specific volume. Gradually all the molecules on the left side will enter the right side of the container. During the process the right side is also kept in a steady-state (T_2, P_2, v_2). The porous plug makes it possible for the molecules to pass through the partition, and yet the pressures can be kept at different values on the two sides. The temperatures of the left and the right sides of the container are kept at T_1 and T_2 ($T_1 \neq T_2$), which make the steady-state possible. The wall and the gas are in thermal equilibrium (both on the left and on the right) so that there is no net exchange of thermal energy between the gas and the wall. In the experiment we may choose the initial temperature and pressure, T_1 and P_1, and also the final pressure P_2 (the pressure of the piston on the right side). We cannot choose the final temperature T_2, which is determined by thermodynamic relations. If we get the result $T_2 < T_1$, this implies that we get a cooling effect. Of course we need the condition that $P_1 > P_2$ in the beginning for the gas molecules to be able to pass through the porous plug to enter the right side. In the following we analyze the thermodynamic theory involved in the throttling process.

We first look at the work done by the gas. The total work done by the gas on the left side is

$$W_1 = \int_{nv_1}^{0} PdV = -nP_1v_1, \tag{6.6}$$

where n is the number of kilomoles of the gas. The total work done by the gas on the right side is

$$W_2 = \int_{0}^{nv_2} PdV = nP_2v_2. \tag{6.7}$$

Since $W_1 < 0$ and $W_2 > 0$, an external work is done on the gas on the left side; the gas on the right does work on its surroundings. The total net work done ***by the gas*** in the process is

$$\begin{aligned}\Delta W &= W_1 + W_2 = n(P_2v_2 - P_1v_1) \\ &= n(u_1 - u_2).\end{aligned} \tag{6.8}$$

The last equality holds because there is no net heat transferred between the gas and the container, thus $\Delta Q = Q_1 + Q_2 = 0$, and $\Delta W = -\Delta U = -(nu_2 - nu_1)$. From the last equality of the above equation, we get

$$u_1 + P_1v_1 = u_2 + P_2v_2, \quad \text{or} \quad h_1 = h_2, \tag{6.9}$$

where h is the specific ehthalpy of the gas, $h = H/n$. Therefore the Joule-Thomson experiment is ***a constant specific enthalpy process.***

In order to study whether a Joule-Thomson experiment can be used as a method of cooling, we define the **Joule-Thomson coefficient** (also known as the **Joule-Kelvin coefficient**) μ,

$$\mu \equiv \left(\frac{\partial T}{\partial P}\right)_h = -\frac{1}{c_P}\left(\frac{\partial h}{\partial P}\right)_T. \tag{6.10}$$

The last equality is derived by using the cyclic relation of the three state variables T, P and h, and the same procedure as was used in deriving Eq. (6.1) has been employed. For an ideal gas $h = u + Pv = u(T) + RT$, therefore h is a function of T only, thus $\mu = 0$. However for a real gas μ may be either greater or less than zero; it may also be zero.

We use the van der Waals gas as an example to study the properties of μ for a real gas. We plot the $h = h_i$=constant curves on a T-P diagram. Different values of h_i $(i = 1, 2, 3\cdots)$ will give different curves, as shown in Fig. 6.3. In some range of the values of h_i, the h =constant curves on

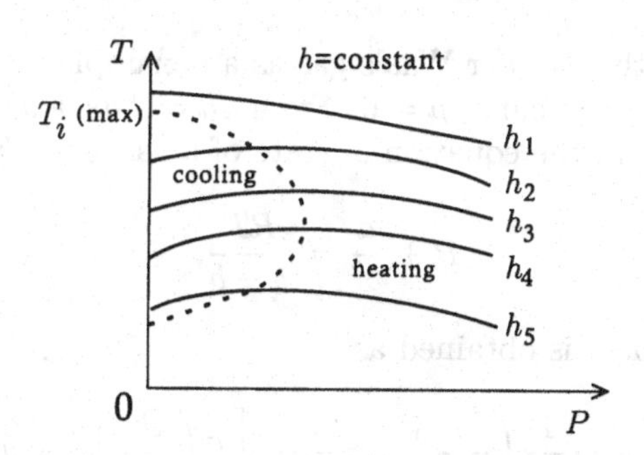

Figure 6.3: The inversion curve (dashed curve) is obtained from the curves of h=constant by connecting all the inversion points (P_i, T_i), which are not labeled.

the T-P diagram have a maximum. For a given h =constant curve, the location of the maximum on the T-P plot is denoted as (P_i, T_i), which is called the **inversion point**. At the inversion point $\mu = 0$, because at that point $(\partial T/\partial P)_h = 0$, which is the condition for a maximum. The physical significance of an inversion point can be easily understood from Fig. 6.3. On the left hand side of an inversion point (i.e., in the region of smaller P) $\mu > 0$, therefore when P increases, T will also increase, or when P decreases, T will decrease too. On the other hand, on the right hand side of an inversion point (i.e., in the region of the larger P) $\mu < 0$, therefore the changes of P and T are in the opposite directions (i.e., T decreases as P increases, or T increases as P decreases). The curve which passes through all the inversion points is called the **inversion curve**, which is the dashed curve shown in Fig. 6.3. We denote the region enclosed by the inversion curve and the T axis as the *cooling* region, which is the region that can be used to have a cooling effect. The reason is that if the initial state of a Joule-Thomson experiment is in this region, then the temperature of the final state T_2 will be lower than the temperature of the initial state T_1, because $P_2 < P_1$. Therefore a cooling effect is obtained. From the diagram, we know that if we want to use the Joule-Thomson experiment as a method for cooling, the initial state of the experiment must be in the cooling region. This includes a suitable choice of the gas, because different gases have different cooling regions. If the initial state is not in the cooling region, then the Joule-Thomson experiment will increase the temperature of the gas rather than decrease the temperature.

Now we use the van der Waals gas as an example to study the properties of the inversion curve $\mu = 0$. From $dh = Tds + vdP$, the Maxwell relation (5.28), and the equation of state of a van der Waals gas (1.35),

$$P + \frac{a}{v^2} = \frac{RT}{v-b},$$

the expression for μ is obtained as

$$\begin{aligned}\mu &= -\frac{1}{c_P}\left(\frac{\partial h}{\partial P}\right)_T = -\frac{1}{c_P}\left[T\left(\frac{\partial s}{\partial P}\right)_T + v\right] \\ &= -\frac{1}{c_P}\left[-T\left(\frac{\partial v}{\partial T}\right)_P + v\right] \\ &= -\frac{1}{c_p}\frac{RTv^3b - 2av(v-b)^2}{RTv^3 - 2a(v-b)^2}.\end{aligned} \tag{6.11}$$

By setting the last equality of the above equation equal to zero, i.e., $\mu = 0$, we obtain the **inversion temperature** T_i as a function of the specific volume v:

$$T_i = \frac{2a(v-b)^2}{Rbv^2}. \tag{6.12}$$

By using the equation of state (6.2) we can eliminate T_i, to obtain the relation between the **inversion pressure** P_i and the specific volume v as

$$P_i = \frac{a(2v-3b)}{bv^2}. \tag{6.13}$$

Eliminating the variable v in Eqs. (6.12)–(6.13), we obtain the inversion curve for the van der Waals gas.[4] The shape of the curve is close to that of the experimental results for many real gases, but numerically the fittings are poorer. From Eq. (6.12) we get the maximum inversion temperature $T_i(\text{max})$,

$$T_i(\text{max}) = \frac{2a}{Rb}, \tag{6.14}$$

which is the value of T_i when $v \to \infty$. The physical significance of $T_i(\text{max})$ is that *if we use the Joule-Thomson experiment to cool down a gas, then the initial temperature of the gas must be smaller than* $T_i(\text{max})$. Otherwise the initial state cannot be in the *cooling* region as shown in Fig. 6.3, and no cooling effect can be obtained. Most of the gases have

[4]Alternatively, we may consider v as a parameter in Eqs. (6.12)–(6.13) and the inversion curve for the van der Waals gas can be obtained by the graphical method.

about the same value for b (which is related to the size of a molecule), therefore $T_i(\max)$ is approximately proportional to the value of a. In Table 6.1 we tabulate the experimental values of $T_i(\max)$ and theoretical values of $2a/Rb$ (including a and b) for four different gases, CO_2, N_2, H_2, and ^4He. We see that the theoretical van der Waals values for $T_i(\max)$ are in reasonably good agreement with the experiments, especially for those gases with smaller values of $T_i(\max)$.

Table 6.1 Values of $T_i(\max)$: Theory *vs.* Experiment

Gas	$2a/Rb$ (Theory)	$T_i(\max)$ (Experiment)	a ($Jm^3kilomole^{-2}$)	b ($m^3kilomole^{-1}$)
CO_2	2040 K	~1500 K	366×10^3	.0429
N_2	855 K	~610 K	139×10^3	.0391
H_2	220 K	~200 K	24.6×10^3	.0266
^4He	35 K	~40 K	3.44×10^3	.0234

6.4 Cooling by Adiabatic Demagnetization

The method of Joule-Thomson can be used to cool gaseous substances only. The method can no longer be used when the gas is liquefied. Under a pressure of 1 atm, all gases will be liquefied if the temperature is low enough. Helium has the lowest liquefaction temperature among all gases. There are two isotopes of helium, ^4He (two protons and two neutrons in the nucleus) and ^3He (two protons and one neutron in the nucleus). The liquefaction temperatures for ^4He and ^3He are 4.2 K and 3.2 K, respectively. Liquid helium (for both ^4He and ^3He) remains in the liquid state even when the temperature is lowered to absolute zero at a pressure of 1 atm, while all other liquids are solidified at a temperature much higher than 4.2 K. Therefore liquid helium serves as an ideal low temperature heat reservoir in the laboratory. However, most of the helium atoms found in nature are ^4He, therefore liquid ^4He is the most commonly used low temperature heat reservoir. If one needs a temperature lower than 4.2 K, then by rapid vaporization of liquid ^4He the temperature may be pushed down to about 1 K.[5] If one needs a

[5] Rapid vaporization of liquid ^3He may give a temperature as low as 0.3 K.

temperature below 1 K, the most frequently used method of cooling is **adiabatic demagnetization**, which we will discuss in this section.

The method of adiabatic demagnetization is based on the thermodynamic properties of a paramagnetic salt. One of the characteristics of a paramagnetic salt is that, when there is no external magnetic field, the total magnetic moment of the system is zero. However when there is an external magnetic field, the system will have a net total magnetic moment which is *parallel* to the external magnetic field. The physical origin of the paramagnetism is that the system is composed of atoms, each (or some) of which has a permanent moment, and the interaction energy between the neighboring magnetic moments is much smaller than the thermal energy kT. Therefore each magnet can be considered as free (i.e., the direction is free), and, if there is no external magnetic field, the sum of the magnetic moments will be zero. However if there is an external magnetic field each magnet has a magnetic potential energy, which is at a minimum when the magnetic moment is parallel to the field. Therefore there is a net total magnetic moment for the system which is parallel to the magnetic field. The relation between the magnetic entropy of the system S and the temperature T is plotted in Fig. 6.4. The external magnetic field is denoted as B. From the diagram one can see that, for a given temperature T, the entropy of the system is smaller when $B \neq 0$ than for the case with $B = 0$, i.e., $S(T, 0) > S(T, B \neq 0)$. The reason is that when there is a magnetic field there is a preferred direction for each magnetic moment, therefore the multiplicity of the system becomes smaller.[6]

Now we explain why a reversible adiabatic demagnetization process can be used as a method of cooling. Referring to Fig. 6.4, the initial state of a paramagnetic salt is a, which is a state with temperature T_1 and zero magnetic field ($B = 0$). The system then undergoes the following two processes, $a \to b$ and $b \to c$, to reach the final state c, which is a state with temperature T_2 and zero magnetic field ($B = 0$). We analyze the thermodynamics involved in these two processes.

1. $a \to b$ is a reversible isothermal magnetization process. The state of the system changes from a to b, during which the temperature is kept constant at T_1, and the magnetic field B is increased from 0 to $B_1 (\neq 0)$. In this process the entropy of the system decreases, and the amount of change is $\Delta S = S(T_1, B_1) - S(T_1, 0) < 0$. Therefore

[6] The direction of the external field B is irrelevant, therefore one needs to consider the case of $B > 0$ only, because $S(T, B) = S(T, -B)$.

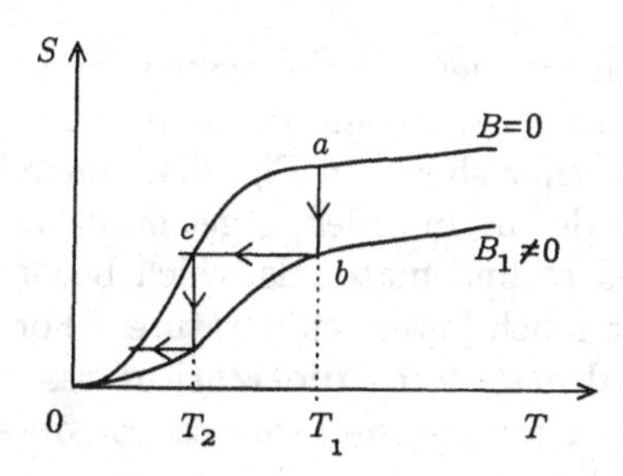

Figure 6.4: Cooling by adiabatic demagnetization. The initial state is a and the final state is c.

the system rejects heat ($\Delta Q = T_1 \Delta S < 0$) to the heat reservoir, in order to keep the temperature unchanged. It is known that when a magnetic field is applied to a paramagnetic system the energy of the system decreases,[7] and the decreased energy must be absorbed by the heat reservoir, otherwise the temperature of the system will increase.

2. $b \to c$ is a reversible adiabatic demagnetization process. In this process the entropy S is a constant, the magnetic field decreases from B_1 to 0, and the temperature changes from T_1 to T_2. It is easy to see from Fig. 6.4 that $T_2 < T_1$, and therefore there is a cooling effect. In Chapter 8 we will use thermodynamic relations to prove that for a paramagnetic system demagnetization ($dB < 0$) will lower the temperature ($dT < 0$) in an isentropic process (S=constant). Therefore there is a cooling effect.

Next we consider c as a new initial state and repeat processes 1 and 2 many times. In this way the temperature can be pushed down to about 10^{-3} K. At such low temperatures paramagnetic materials may become ferromagnetic or anti-ferromagnetic. This is because at these low temperatures the mutual interaction energy between two neighboring magnetic moments is no longer small in comparison with the thermal energy, and thus the direction of a magnetic moment will be influenced by the neighboring magnetic moments, even when there is no external magnetic field. Therefore the directions of each of the magnetic moments are no longer free, and the effect of the external magnetic field

[7]The magnetic potential energy for a magnetic moment $\boldsymbol{M}$ in a magnetic field $\boldsymbol{B}$ is $-\boldsymbol{M} \cdot \boldsymbol{B} = -MB < 0$ for a paramagnetic system.

on the total magnetic moment of the system diminishes. This means that the S-T diagram for the system in the presence of a magnetic field will be different from that shown in Fig. 6.4. Therefore the above theory of cooling breaks down. In order to get even lower temperatures by this method, we need to find materials which become ferromagnetic or anti-ferromagnetic at much lower temperatures. For this purpose **cooling by nuclear adiabatic demagnetization** was developed. By using this method, the lowest temperature that can be obtained is about 10^{-9} to 10^{-10} K. The materials used in this method possess *nuclear magnetic moments*, which come from the magnetic moment of the protons and neutrons inside the nucleus. The magnetic moments of a paramagnetic salt discussed above come from the orbital electrons of the atoms, which are much larger than the nuclear magnetic moments, because the mass of a proton (or a neutron) is almost 2000 times larger than that of an electron. Moreover, the distance between two neighboring nuclear magnetic moments is much larger than that between two neighboring atomic magnetic moments. These two factors make the nuclear magnetic moments may remain paramagnetic down to temperatures as low as 10^{-9} K.

6.5 Laser Cooling

Laser cooling is a technique which began its development in the 1970's, and has made tremendous progress in the latter years of the twentieth century. At present the lowest temperature obtained by this method is about 10^{-5} K. The temperature can be lowered further to about 10^{-7} K (or even 10^{-9} K) by the assistance of the method of evaporative cooling. The theoretical basis of laser cooling is not so much related to the principles of thermodynamics. However, applications of this method have made, and will continue to make, great contributions to low temperature physics and statistical mechanics, therefore it is worth giving a brief description of the method here.

Laser cooling can be applied to systems of extremely dilute gaseous atoms or molecules only. Under the ordinary pressure of 1 atm, all gases will be liquefied or solidified below 3.2 K. Therefore in order that a system can remain in the gaseous state at temperatures as low as 10^{-5} K or even down to 10^{-9} K, the density of the atoms (molecules) must be extremely low, such that each atom (molecule) can travel a long distance in a straight line before colliding with another atom (molecule). Suppose the energy difference between the ground state and the first excited state of the atom is $\Delta\varepsilon$. A laser beam of frequency ν is focused on the moving atoms. The frequency ν is chosen to satisfy $h\nu < \Delta\varepsilon$, such that for the atoms moving in the same direction as, or perpendicular to, the laser beam are unable to absorb the photons of the laser beam. Therefore the motion of these atoms will not be affected by the laser beam. However, for atoms moving in the direction opposite to the beam, the frequency of the laser will be ν', which is larger than ν because of the Doppler[8] effect. If ν is properly chosen so that $h\nu' = \Delta\varepsilon$ is satisfied, then the atoms moving opposite to the beam can absorb the photons, and be excited to the first excited state. Each photon carries momentum, therefore an atom which absorbs a photon will experience a force in the direction of the photon. The excited atoms will immediately emit a photon with energy $\Delta\varepsilon$ to return to the ground state. During the emitting process the atom also suffers a recoil force in the opposite direction of the emitted photon. Therefore there are two forces acting on an atom which absorbs and re-emits a photon. However these two forces do not cancel each other completely, because the direction of the absorbed photons is fixed, but the directions of the emitted photons are random, i.e., there is an equal probability for the photon to be emitted in any direction. Therefore summing over all the atoms in the system, there is a net force on the system which is in the direction of the laser beam. This reduces the kinetic energy of the system, because the atoms which absorb the laser photons are moving in the direction opposite to the laser beam. The temperature of the system is therefore lowered.

We may also look at the absorption and emission of the photons from the view point of the laboratory frame. For an atom which absorbs a photon from the laser beam the energy increase is $h\nu$, but when it re-emits a photon it loses an energy $h\nu' = \Delta\varepsilon$. Because $h\nu < h\nu'$, therefore during the absorption and emission processes the atoms lose energy, and the total kinetic energy, and therefore the temperature of the system, is lowered. In order to increase the efficiency of cooling, six laser beams may be used simultaneously. In each of the x, y and z directions, two opposite laser beams may be used, so that no matter in what direction an atom moves there are always photons available for absorption.

Since the system considered is a gaseous system, if there is no *container* to contain the atoms, they will all escape during the cooling process. However a conventional container is not a suitable one, because

[8]Christian Doppler, Austrian mathematician (1803–1853).

the heat capacity of the container will be much larger than the gaseous atoms, and the cooling effect will be limited by the container. Therefore *a container without walls* is the best choice for the cooling process. Scientists found that, if one applies a *magneto-optic trap* in the region where all six laser beams are focused, the atoms will be trapped in that region for a sufficiently long time. A magneto-optic trap is an inhomogeneous magnetic field which acts like a potential well in mechanics.

When the temperature is lowered, the mean kinetic energy of the atoms becomes smaller and smaller. The number of atoms whose kinetic energy is equal to or greater than $\Delta\varepsilon$ becomes smaller too, this makes the efficiency of laser cooling very low, and the method is no longer useful. If one wants to lower the temperature further, then other methods such as evaporative cooling, may be used. The method of evaporative cooling, as its name implies, is to *evaporate* the atoms which have a kinetic energy larger than average away from the system, and thus lower the mean energy per atom. The procedure is to turn off the laser beams and let the atoms be trapped in the magneto-optic trap. One may then gradually reduce the height of the trapping well, letting the atoms with higher energy evaporate (i.e., to escape the trap).[9] In this way the temperature can be lowered down to around 10^{-9} K.

One of the most important applications of laser cooling and evaporative cooling is the observation of a gaseous **Bose-Einstein**[10] **condensate** in the laboratory in the mid 1990's. On the theoretical side, the theory of Bose-Einstein condensation has been known since the 1920's. It predicts that, for an ideal Bose gas, there is a non-zero critical temperature below which a Bose-Einstein condensate exists. This is a condensation in *momentum space*, which is rather different from a condensation in real space (i.e., to become liquid). Experimentally it took about 70 years to confirm the prediction of the existence of the gaseous Bose-Einstein condensate. The experimental difficulty is mainly due to the difficulty of lowering the temperature, and yet preserving the system in a gaseous state. For a system to remain in a gaseous state down to 10^{-5} K or even lower, the number density of the system must be very small. This requires roughly that the mean distance between the atoms (molecules) should be about ten or more times the size of an atom (molecule). With this distance the mutual interaction between the atoms is very small,

[9]For an elementary review of laser cooling and its applications, see Steven Chu, "Laser Trapping of Neutral Particles," *Scientific American* **266**, 71–76, 1992.

[10]Satyendranath Bose, Indian physicist (1894–1974); Albert Einstein, German physicist (1879–1955), Nobel prize laureate in physics in 1921.

thus they can remain in a gaseous state without condensation to a liquid or solid. The lower number density implies that the condensation occurs at a lower critical temperature. The critical temperature for the Bose-Einstein condensation is proportional to the 2/3 power of the number density of the gas. It was estimated that in order for a gaseous Bose-Einstein condensate to be observed in the laboratory the critical temperature should be at least as low as 10^{-6} K or even 10^{-7} K. The advancement in the technique of laser cooling made it possible to observe the gaseous Bose-Einstein condensate in the laboratory in the mid 1990's. We will introduce and discuss the theory of Bose-Einstein condensation in Chapter 12.

6.6 The Third Law of Thermodynamics

We have introduced the zeroth, first, and second laws of thermodynamics in the previous chapters. These three laws are the fundamental laws of thermodynamics. The third law concerns the thermodynamic properties of all substances near absolute zero, and thus may not be as fundamental as the three previous laws. For phenomena which are not related to absolute zero, the third law may be irrelevant. However in the limit of very low temperatures, the third law may become important. There are restrictions on some physical properties of a substance in order that the third law is not violated. The best-known example is that all the heat capacities must be equal to zero at absolute zero.

As with the second law, there are several different statements for the third law. To be exact, there are three different statements: Planck statement and two different Nernst statements. Unlike the second law, these three statements are not all equivalent. The Planck statement from which the two statements of Nernst[11] can be derived is the strongest. However from Nernst's two statements one can not derive Planck statement. The **Planck statement** is:

The entropy of every system in the state of internal equilibrium at absolute zero is zero., that is

$$\lim_{T\to 0} S = 0. \tag{6.15}$$

This is now often referred to as the third law of thermodynamics.

[11] Walter H. Nernst, German chemist (1864–1941), Nobel prize laureate in chemistry in 1920.

There are two different Nernst statements of the third law. These two statements are equivalent, which means that if one of the statements were false, then the other one would be also false. The two **Nernst statements** are:

1. **In the neighborhood of absolute zero, all reactions in a liquid or solid in internal equilibrium take place with no change in entropy,**[12] i.e.,

$$\lim_{T\to 0} \Delta S = \lim_{T\to 0} (S_2 - S_1) = 0. \tag{6.16}$$

2. **It is impossible to reduce the temperature of a system to absolute zero by any finite number of operations.**

The second statement is often referred to as the *unattainability* statement of the third law.

From the historical point of view, Nernst observed many chemical reaction experiments, and in 1906 he gave his first statement ($\Delta S \to 0$ as $T \to 0$). Later in 1912 he gave his second statement as a supplement to his first statement. In fact the two statements are equivalent. We will prove this later. In 1911 (before Nernst gave his second statement) Planck extended Nerst's first statement to give the entropy an absolute, rather than just a relative value, and presented his statement of the third law ($S \to 0$ as $T \to 0$). Obviously the Planck statement is stronger than Nernst statements. This means that if the Planck statement is true, then Nernst statements are also true. On the other hand, if Nernst statements are true, the Planck statement is not necessarily true.

Now we study the first Nernst statement of the third law. Nernst observed many chemical reaction experiments, and noted that at very low temperatures not only $\Delta G = \Delta H$ (as $T \to 0$), but also

$$\lim_{T\to 0} \left(\frac{\partial \Delta G}{\partial T}\right)_P = 0, \quad \lim_{T\to 0} \left(\frac{\partial \Delta H}{\partial T}\right)_P = 0, \tag{6.17}$$

where ΔG is the change in the Gibbs free energy before and after the chemical reaction, and ΔH is the change in enthalpy before and after the chemical reaction. Since $G = H - TS$, therefore $\Delta G = \Delta H$ as $T \to 0$ is a result of thermodynamic relations. Equation (6.17) is a postulate, called the Nernst postulate. From many experimental results, Nernst noticed

[12] **This statement is also known as the Nernst-Simon statement.**

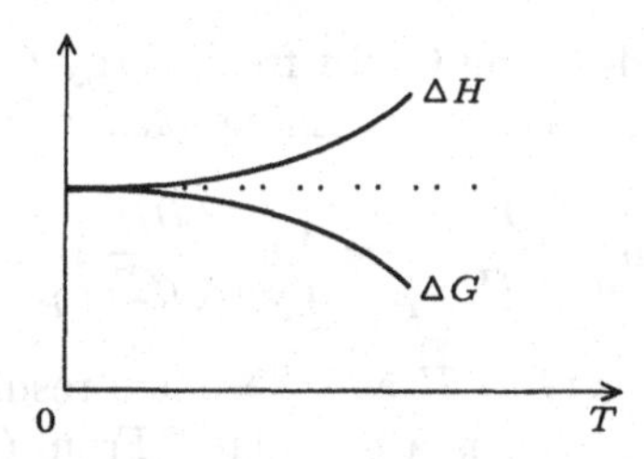

Figure 6.5: The plot of ΔG and ΔH vs. T at very low T.

that ΔG approaches ΔH, even when the temperature is not very low.[13] He used the following relation[14]

$$\begin{aligned}\Delta G &= \Delta H - T\Delta S \ (T \text{ constant}),\\ &= \Delta H + T\left(\frac{\partial \Delta G}{\partial T}\right)_P,\end{aligned} \tag{6.18}$$

to obtain the postulate of Eq. (6.17). Due to this postulate, the second term on the right hand side of the second equality of Eq. (6.18) approaches zero more rapidly as the temperature is lowered. According to Nernst's postulate, if we plot ΔG vs. T and ΔH vs. T on the same diagram, we will find that both ΔG and ΔH have a zero slope as $T \to 0$, as shown in Fig. 6.5.

Since $(\partial G/\partial T)_P = -S$, then from Nerst's postulate, Eq. (6.17), one obtains Nernst's first statement of the third law,

$$\lim_{T\to 0} \Delta S = \lim_{T\to 0} [S_2(T) - S_1(T)] = 0. \tag{6.19}$$

This means that, in the neighborhood of absolute zero, the entropy of a system depends on the temperature only, independent of all other parameters (e.g., magnetic field, chemical reactions, etc.).

Later Planck made a stronger statement of the third law. He noticed that in the neighborhood of absolute zero, when a system is in thermal

[13]The first time ^{4}He was liquefied (below 4.2 K) was in 1908 by the Dutch physicist Kamerlingh Onnes. This was two years after Nernst published his first statement of the third law. Therefore at that time the lowest temperature Nernst could reach was above 5 K. H. Kamerlingh Onnes, Dutch physicist (1853–1926), Nobel prize laureate in physics in 1913.

[14]The equation $G = H + T\left(\dfrac{\partial G}{\partial T}\right)_T$ is known as the Gibbs-Helmholtz equation.

equilibrium, not only does the Gibbs free energy $G(T)$ approach $H(T)$, but their derivatives are also approaching each other, i.e.,

$$\lim_{T\to 0}\left(\frac{\partial G}{\partial T}\right)_P = \lim_{T\to 0}\left(\frac{\partial H}{\partial T}\right)_P . \tag{6.20}$$

Since $G = H - TS$, thus $G \to H$ as $T \to 0$ is a result of thermodynamic relations, however Eq. (6.20) is a postulate. From $G = H - TS$ we also have

$$\left(\frac{\partial G}{\partial T}\right)_P = \left(\frac{\partial H}{\partial T}\right)_P - T\left(\frac{\partial S}{\partial T}\right)_P - S, \tag{6.21}$$

therefore Planck's proposed Eq. (6.20) leads to his statement of the third law, Eq. (6.15) $\lim_{T\to 0} S = 0$. By comparing Nernst's postulate Eq. (6.17) and Planck's proposed Eq. (6.20), we see that Planck replaced the differences ΔG and ΔH, proposed by Nernst, by the absolute values of G and H. Therefore Planck statement is stronger than Nernst first statement.

6.7 Equivalence of Nernst Statements

Now we want to prove that Nernst's two statements are equivalent. We use the method of cooling by a paramagnetic salt as an example of the proof. As in the proof of the equivalence of different statements of the second law, we suppose one of the statements were false, and prove that this would lead to a violation of the other statement. Suppose Nernst's first statement were false, this would imply that $\lim_{T\to 0}(S_2 - S_1) \neq 0$, and that Fig. 6.4 would become Fig. 6.6. At $T = 0$, $S_1 = S_{01}$, $S_2 = S_{02}$, and we suppose $S_{01} < S_{02}$. From the definition of the heat capacity,

$$C_X = T\left(\frac{\partial S}{\partial T}\right)_X , \tag{6.22}$$

where X is a state variable which can be controlled in the laboratory. For example it may be volume V, pressure P, etc. In the adiabatic demagnetization experiment, X is the applied magnetic field B. The temperature is reduced in the reversible adiabatic process (S=constant), i.e., the process $a \to b$ in Fig. 6.6. The temperature is reduced from T_1 to T_2. From Eq. (6.22) we have

$$S_a = S_{01} + \int_0^{T_1} \frac{C_{X_1}}{T} dT, \tag{6.23}$$

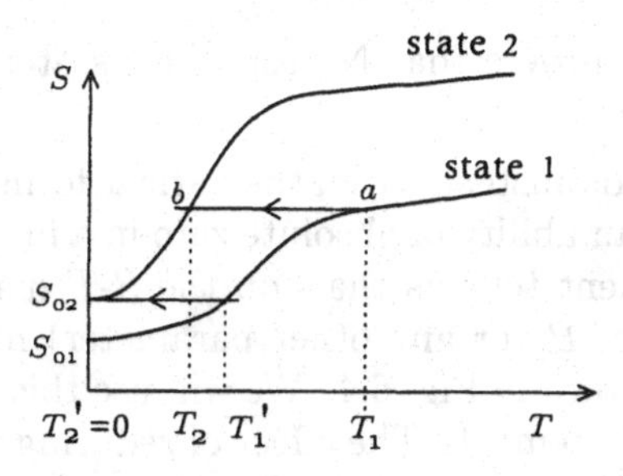

Figure 6.6: Entropy-temperature diagram for a hypothetical system in two different states 1 and 2.

$$S_b = S_{02} + \int_0^{T_2} \frac{C_{X_2}}{T} dT. \tag{6.24}$$

Since $S_a = S_b$ therefore

$$S_{02} - S_{01} = \int_0^{T_1} \frac{C_{X_1}}{T} dT - \int_0^{T_2} \frac{C_{X_2}}{T} dT. \tag{6.25}$$

Now suppose that the left hand side of the above equation is greater than zero,[15] then because $C_X > 0$ (the condition of stable equilibrium derived in Chapter 5), we can find a temperature $T_1 > 0$ such that

$$S_{02} - S_{01} = \int_0^{T_1} \frac{C_{X_1}}{T} dT. \tag{6.26}$$

Substituting this result into Eq. (6.25) we get $T_2 = 0$. This means that if the temperature of the initial state is T_1, then one needs only two steps (isothermal magnetization and adiabatic demagnetization) to get to absolute zero ($T_1 \to T_2 = 0$). This violates Nernst's second statement.

Now we want to prove the reverse situation. Suppose Nernst's second statement were false, we will prove that his first statement would also be false. Suppose one can get to absolute zero in a finite number of steps. We still use Fig. 6.6 as the reference diagram, but pay no attention to whether S_{01} and S_{02} are equal. In this situation Eqs. (6.23)–(6.25) still hold. Suppose we can get to absolute zero in a finite number of steps, we may take the step $T_1 \to T_2 = 0$, with $T_1 \neq 0$. Substituting $T_2 = 0$ into Eq. (6.25), we get Eq. (6.26). Since $T_1 > 0$ and $C_X > 0$, we have

[15] If we suppose $S_{01} > S_{02}$, we will get the same result.

$S_{02} - S_{01} > 0$. This proves that Nernst's first statement would also be false.

We may use words instead of mathematical formulas to explain the reason for the unattainability of absolute zero in a finite number of steps. Nernst's first statement tells us that, on the S-T diagram, the S curve for different values of B (or any other parameter) must have the same value at $T = 0$, as shown in Fig. 6.4. We will use this diagram to explain the unattainability statement. The effect of reducing temperature comes from the reversible adiabatic process (S=constant), which is represented by a horizontal line in Fig. 6.4. The intersection of the horizontal line with the lower S curve is the initial temperature T_i, and the intersection of the horizontal line with the higher S curve is the final temperature T_f for the adiabatic process, because $T_f < T_i$ we have a cooling effect. When T approaches zero, in each of the adiabatic processes the ratio of the final temperature T_f and the initial temperature T_i will roughly approach a constant, i.e., T_{f_n}/T_{i_n} is roughly a constant, where n means the nth adiabatic process. Suppose the ratio is 0.5, then after n adiabatic processes the final temperature is $T_f = (0.5)^n T_i$. This will not be equal to zero for a finite number n. This is the theoretical limitation of the attainability of absolute zero. In practice there are also many technological limitations in approaching absolute zero. These may include whether there is a suitable substance which can be used as the heat reservoir (necessary for the isothermal processes), the problem of insulation (for adiabatic processes), the difficulty of controlling the internal magnetic field (the substance becomes non-paramagnetic), etc.

6.8 Consequences of the Third Law

The existence of the third law will impose some restrictions on the behavior of some physical quantities. We give three examples which are quite general and are applicable to most of the thermodynamic systems. There may be other restrictions which are applicable only to some specialized systems, such as magnetic systems, surfaces and/or thin films, etc. We will not discuss these specialized systems here. All three examples we give here can be derived from Nernst's first statement, and therefore they are also consistent with the Planck statement.

1. Heat capacity:

   From the definitions of the constant volume heat capacity C_V and the constant pressure heat capacity C_P,

$$C_V = T\left(\frac{\partial S}{\partial T}\right)_V, \quad C_P = T\left(\frac{\partial S}{\partial T}\right)_P,$$

we obtain

$$S - S_0 = \int_0^T \frac{C_V}{T} dT, \tag{6.27}$$

$$S - S_0 = \int_0^T \frac{C_P}{T} dT, \tag{6.28}$$

where S_0 is the entropy at[16] $T = 0$. When $T = 0$, both of the left hand sides of the above two equations are zero, therefore the integrals of the right hand side of the above two equations must also be zero when $T = 0$. We therefore get the restrictions[17] on the heat capacities at $T = 0$:

$$C_P \geq C_V \to 0 \quad \text{as } T \to 0. \tag{6.29}$$

In general we may write C_V as

$$C_V = T^\alpha \quad (\alpha > 0) \quad \text{as } T \to 0. \tag{6.30}$$

In all known thermodynamic systems $\alpha > 0$, therefore the third law is satisfied.

2. Expansion coefficient:

   The definition of the expansion coefficient (or the expansivity) β is, Eq. (1.41),

$$\beta = \frac{1}{V}\left(\frac{\partial V}{\partial T}\right)_P = -\frac{1}{V}\left(\frac{\partial S}{\partial P}\right)_T,$$

   where the second equality comes from the Maxwell relation (5.28). When T approaches zero, S approaches a constant independent of the pressure P, therefore the condition on the expansivity is

$$\beta \to 0 \quad \text{as } T \to 0. \tag{6.31}$$

[16]According to Planck statement $S_0 = 0$, but we will see that as long as S_0 is a finite number, the following conclusions hold.

[17]The condition of $C_V = 0$ at a *single temperature* $T = 0$ does not violate the stability condition of a substance (5.61) which requires $C_V > 0$.

3. The slope of the two-phase coexistence curve:

   On the P-T phase diagram, usually there are three two-phase coexistence curves, i.e., the liquid-vapor, solid-liquid, and solid-vapor coexistence curves, as shown in Fig. 1.1. Near $T = 0$, the slope of the solid-vapor [18] coexistence curve, which will be derived in the next chapter, must satisfy

$$\frac{dP}{dT} = \frac{\Delta S}{\Delta V} \to 0 \text{ as } T \to 0, \tag{6.32}$$

   where ΔS and ΔV are, respectively, the difference of entropy and volume for the two phases. Therefore at $T = 0$ the slope of the coexistence curve must be zero.

6.9 Problems

6.1. Show that

$$\mu = -\frac{1}{c_P}\left(\frac{\partial h}{\partial P}\right)_T = -\frac{v}{c_P}(1 - \beta T),$$

where μ is the Joule-Thomson coefficient and β is the expansivity.

6.2. Show that $\left(\frac{\partial T}{\partial P}\right)_h - \left(\frac{\partial T}{\partial P}\right)_s = -\frac{v}{c_P}.$

6.3. Assuming that helium (^{4}He) obeys the van der Waals equation of state, determine the change in temperature when one kilomole of helium gas undergoes a Joule expansion at 20 K to atmospheric pressure. The initial volume of the helium is 0.12 m^3. See Table 6.1 for the values of the van der Waals constants for helium; c_v can be taken to be $1.5R$.

6.4. One kilomole of an ideal gas undergoes a throttling process from a pressure of 4 atm to 1 atm. The initial temperature is 50°C.

   (a) How much work could have been done by the ideal gas had it undergone a reversible process to the same final state at constant temperature?

[18] For helium, in addition to the solid-vapor coexistence curve, there is also a liquid-vapor coexistence curve near $T = 0$. See Fig. 7.5.

(b) How much does the entropy of the universe increase as a result of the throttling process?

6.5. Consider a van der Waals gas with a specific internal energy of $u = u_0 + c_v T - a/v$, where u_0 is a constant.

(a) Find an expression for the specific enthalpy h as a function of T and v.

(b) From $h = u + Pv$, show that the Joule-Thomson coefficient is given by

$$\mu = \frac{\kappa}{c_P} \frac{RTv}{v - b} - \frac{v}{c_P}.$$

(c) Find the isothermal compressibility κ as a function of T and v, and confirm the result of Eq. (6.11).

6.6. Compute η and μ for a gas whose equation of state is given by (a) $P(v - b) = RT$ and (b) $(P + b)v = RT$, where b is a constant. Assume that c_v and c_p are constants.

6.7. Assume that helium (^{4}He) obeys the van der Waals equation of state. Calculate the maximum inversion temperature of helium. See Table 6.1 for the values of the van der Waals constants for helium.

6.8. An apparent limit on the temperature achievable by laser cooling is reached when an atom's recoil energy from absorbing or emitting a single photon is comparable to its total kinetic energy. Make a rough estimate of this limiting temperature for rubidium atoms that are cooled using laser light with a wavelength of 780 nm. The mass number of a rubidium atom may be taken to be 85.

6.9. Show that the heat capacity at constant volume can be written as

$$C_V = \left(\frac{\partial S}{\partial \ln T}\right)_V.$$

From this show that $C_V \to 0$ as $T \to 0$. Note that the same argument can be applied to show that $C_X \to 0$ as $T \to 0$, where X is any controllable quantity in thermodynamics other than T.

6.10. Show that, in view of the third law, neither the ideal gas equation nor the van der Waals equation can hold at $T = 0$.

6.11. (a) The isothermal bulk modulus B is the reciprocal of the isothermal compressibility κ, i.e., $B = -V(\partial P/\partial V)_T$. Show that

$$\lim_{T\to 0}\left(\frac{\partial B}{\partial T}\right)_V = 0.$$

(b) By using T and V as the independent variables, show that at very low temperatures κ can be written as

$$\kappa(T,V) = a(V) + b(V)T^{\alpha},$$

where $a(V)$ and $b(V)$ are unknown functions of V (independent of T), and α is a constant. Are there any restrictions of $a(V)$, $b(V)$ and α?

6.12. Consider a solid whose equation of state is $PV + f(V) = a\,U$, where U is the internal energy, $f(V)$ is a function of the volume only, and a is a constant. It is known that the expansivity $\beta \to 0$ as $T \to 0$. Show that $C_V \to 0$ as $T \to 0$.

6.13. Use the third law to show that $\left(\frac{\partial S}{\partial X}\right)_{T=0} = 0$, where X is any controllable quantity in thermodynamics other than T. From this show that $\left(\frac{\partial T}{\partial X}\right)_{S=0} = 0$. Note that this implies that it is impossible to reach absolute zero in a finite number of steps.

6.14. The heat capacity of a nonmagnetic dielectric material in the temperature range 0.01 and 0.5 K is found experimentally to vary as $aT^{1/2} + bT^3$. Calculate the entropy of the system as a function of T. Is this finding consistent with the third law?

6.15. A polymer, held at constant tension J, shrinks as the temperature is increased. Sketch a curve of the length L of the polymer as a function of temperature near 0 K, and give reasons for all pertinent parts of your sketch. [*Hint*: The Maxwell relations are also applicable to a string, by proper substitutions of the state variables: P (pressure) replaced by $-J$, and V (volume) by L. These replacements come from the form of dW, as given in Table 1.1]

Chapter 7

Phase Equilibrium and Binary Mixtures

7.1 Introduction

If two equilibrium states of a system have rather different thermodynamic properties, then we say that these two states belong to two different *phases* of the given substance. The most common example is the three states of a substance, the gas state, the liquid state, and the solid state. These three states belong to three different *phases.* For example, we have vapor,[1] water, and ice which are the three phases of the same substance H_2O. In addition to this common example *ferromagnetism* and *paramagnetism* observed in magnetic materials is another common example of different phases of the same substance. The transition from paramagnetism to ferromagnetism, or vice versa, is one of the most studied examples in the theory of *phase transitions* (or *phase transformations*). In the process of phase transition, the two phases involved must coexist under the same external conditions, e.g., the same temperature and pressure, or the same temperature and magnetic field. There are two classes of phase transitions, the *first-order* phase transitions, and the *second-order* (or higher-order) phase transitions. The first-order phase transition is also known as the *discontinuous* phase transitions, because the process of changing from one phase to another is discontinuous. An

[1] The conversion of a substance from liquid to gas is called vaporization, and hence the name vapor for the gas phase.

example is the transformation between ice and water. The second-order phase transition is also known as the *continuous* phase transitions, because the coexistent phases have the same physical properties at the phase transition. An example is the ferromagnetic phase transition at zero magnetic field.

In this chapter we are going to discuss the conditions for phase equilibrium, in which two phases can coexist. In Chapter 5 we derived the conditions for thermodynamic equilibrium, which require that the temperature and pressure must be equal everywhere in the system. If the temperature is not equal everywhere, then heat will flow from the places with higher T to the places with lower T, until a uniform T is attained. If the pressure is not equal everywhere, then the volume will expand from the places of higher P to the places of lower P, until P is the same everywhere. Therefore we may say that both temperature T and pressure P play the role of "force" in thermodynamics. If these forces are not balanced, then the system is not in equilibrium. We have introduced a concept of the generalized force in Sec. 1.7, where P is the generalized force in a PVT system, with PdV as its associated work. From the relation $dU = TdS - PdV$, we may generalize this concept further to include T as a generalized force, and TdS as its associated work. If we generalize this equation to an open system, then from Eq. (5.35) (for a system with only type of particle) we have

$$dU = TdS - PdV + \mu\, dn,$$

where $\mu\, dn$ is the chemical work. This equation tells us that the chemical potential μ plays the same role as the temperature T and pressure P play, and thus can be considered as one of the generalized forces. When μ is not the same everywhere in the system, then the particles will move from the places with higher μ to the places with lower μ. If the system is composed of the same type of molecule, which can be in two phases, then, if the chemical potentials are not equal for these two phases, molecules will move from the phase with higher μ to the phase with lower μ. The movement of the molecules will stop only when the two phases have the same chemical potential. When this happens the two phases can coexist, and we say that the system is in phase equilibrium. In the first half of this chapter we will give a more rigorous derivation of the conditions for phase equilibrium, using a PVT system as an example to discuss thermodynamic properties of phase equilibrium.

Binary mixtures are thermodynamic systems which we encounter quite frequently in daily life and which have many potential applications.

These systems also possess many interesting phase diagrams which can be analyzed and understood by using the conditions of phase equilibrium that we derive in this chapter. This is the subject that will be discussed in the second half of this chapter.

7.2 Two-Phase Coexistence for a One-Component System

We use a PVT system as an example for the study of the condition of two-phase coexistence. In this section we discuss the simplest case, where there is only one component in the system, which may have two phases. If a system is composed of more than one component, then it is a mixture. We will consider binary mixtures in the latter part of this chapter.

For a PVT system, the most convenient independent state variables are the temperature T and pressure P. Therefore the proper thermodynamic potential to use in the study of equilibrium conditions is the Gibbs free energy G. For a system with only one component but two phases, the expression for G is given by Eq. (5.48):

$$G(T,P) = n_1\mu_1(T,P) + n_2\mu_2(T,P), \tag{7.1}$$

where n_i and μ_i are, respectively, the number of kilomoles and the chemical potential per kilomole for the molecules in the ith phase ($i = 1, 2$). We want to discuss the condition for coexistence of the two phases at constant T and P. From Sec. 5.5 we know that when T and P are kept constant, the equilibrium condition of the system is a state such that the Gibbs free energy G is a minimum. We also note that, when both T and P are constants, the chemical potentials μ_1 and μ_2 are also fixed numbers.[2] Therefore the condition that G is a minimum is

$$\delta G(T,P) = \mu_1(T,P)\delta n_1 + \mu_2(T,P)\delta n_2 = 0. \tag{7.2}$$

Suppose the system is a closed system. Then the total number of kilomoles of the molecules n_0 is a constant, thus $n_0 = n_1 + n_2$ =constant. However a molecule can be in either phase, therefore $\delta n_1 + \delta n_2 = \delta n_0 = 0$, i.e., $\delta n_2 = -\delta n_1$. The above equation becomes

$$\delta G(T,P) = (\mu_1 - \mu_2)\delta n_1 = 0. \tag{7.3}$$

[2] Note that for a one-component system, μ_1 and μ_2 are independent of n_1 and n_2. However for a mixture, the μ_i's (or g_i's) may depend on the n_i's, see Eq. (7.55).

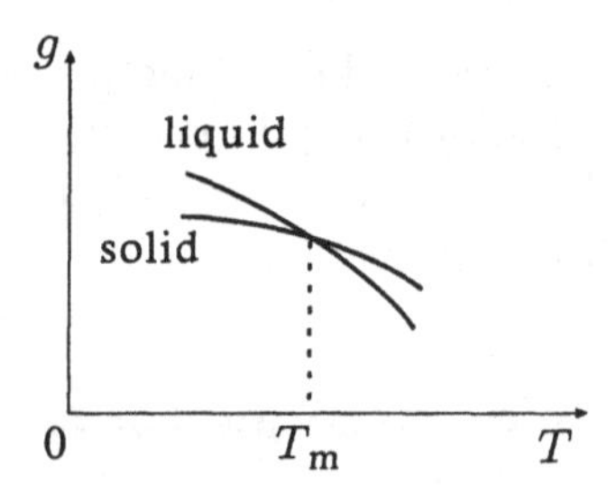

Figure 7.1: The plot of g-T curves for solid-liquid coexistence at constant pressure P_0. The two g curves intersect at $T = T_m$, which is the melting temperature at pressure P_0.

This leads to the two-phase coexistence condition as

$$\mu_1(T, P) = \mu_2(T, P). \tag{7.4}$$

When the above equation is satisfied, then the two phases can coexist.

When a system has only one component and only one phase, then its chemical potential per kilomole μ is equal to its specific Gibbs free energy g, i.e., $g(T, P) = \mu(T, P)$ (cf. Eq. (5.49)). Hence we can study two-phase coexistence from a g-T diagram (P is kept constant), where g is the ***single phase*** specific Gibbs free energy. For example if we want to study liquid-solid coexistence, then we plot the g-T curve for both the solid phase and the liquid phase on the same diagram, as shown in Fig. 7.1, where P is kept at P_0. If these two g curves intersect at temperature T_m, then T_m is the melting (or solidification) temperature when the pressure is P_0, since at the temperature T_m and pressure P_0, the g values for the liquid phase and the solid phase are equal, i.e., $g_l(T_m, P_0) = g_s(T_m, P_0)$ (l denotes liquid, and s denotes solid), or equivalently $\mu_l(T_m, P_0) = \mu_s(T_m, P_0)$. Therefore the solid and liquid can coexist at temperature T_m and pressure P_0. Since $(\partial g/\partial T)_P = -s < 0$ the two g curves in Fig. 7.1 will decrease as T increases. We know that the solid phase is stable for $T < T_m$, and the liquid phase is stable for $T > T_m$, therefore we have the following relations:

$$\begin{aligned} g_s &< g_l \quad \text{for} \quad T < T_m, \\ g_s &> g_l \quad \text{for} \quad T > T_m. \end{aligned} \tag{7.5}$$

Since $(\partial g/\partial T)_P = -s < 0$, from Fig. 7.1 we see that at the coexistence point (T_m, P_0) $s_l > s_s$, i.e., the entropy for the solid phase is smaller than that of the liquid phase. This is well-known. The molecules in the solid phase are rather ordered, while the molecules in the liquid phase are more or less randomly distributed, therefore the multiplicity of the solid phase is smaller than that of the liquid phase.

Since $g = h - Ts$, at constant T, $\Delta g = \Delta h - T\Delta s$. When the liquid and solid phases coexist, Δg is the difference between the g values of the two phases, thus $\Delta g = g_l - g_s = 0$ (because $g_l = g_s$). Therefore we have $\Delta h - T\Delta s = 0$, i.e.,

$$\Delta h = h_l - h_s = T\Delta s = T(s_l - s_s) \equiv \ell_{ls}. \tag{7.6}$$

Here $\ell_{ls} = \Delta h = T(s_l - s_s)$ $(T = T_m)$ is the heat that one kilomole of the substance absorbs when it converts from the solid phase (s is the initial state) to the liquid phase (l is the final state), $\Delta q = \Delta h = \ell_{ls} > 0$ and is called the **heat of fusion.** It is positive because the specific entropy of the liquid phase is larger than that of the solid phase, therefore the substance must absorb heat to increase its entropy ($\Delta s = \Delta q/T$) during the fusion process. The heat absorbed or released during a phase transformation is called the **latent heat**, or **heat of transformation.** There are also the **heat of vaporization** (liquid to vapor) and the **heat of sublimation** (solid to vapor). The existence (i.e., being non-zero) of a latent heat is one of the characteristics of the **first-order phase transitions**, because there is an entropy difference for the two coexistent phases. Consequently this class of phase transitions is also called the **discontinuous phase transitions**, because of the discontinuity of s, the first order derivative of g $(s = -(\partial g/\partial T)_P)$. There are also phase transitions in which the entropy is the same (continuous) for the coexistent phases, and therefore there is no latent heat for these transitions. They are known as the **second-order** (or **higher-order**) phase transitions or **continuous phase transitions.**

7.3 The Clausius-Clapeyron Equation

In this section we will go one step further to study the properties of two-phase coexistence. We use a PVT system as an example and assume that phase 1 and phase 2 coexist at temperature T and pressure P. For a PVT system there are two independent variables, and there is one condition for the two-phase coexistence. Therefore for two-phase coexistence there is only one independent variable left. This means that the *region* where two phases can coexist is a one-dimensional region, i.e., a curve, called the coexistence curve. If we plot the two-phase coexistence curve ($g_1 = g_2$) on the P-T diagram, the curve will be a ***phase boundary***, such that one side of the curve is the region where phase 1 is stable ($g_1 < g_2$); the other side is the region where phase 2 is stable ($g_1 > g_2$). This kind of diagram is called a ***phase diagram***. From this diagram we know which phase is stable for a given temperature and pressure. We denote the coexistence curve on the P-T phase diagram as $P_{\text{coex}} = P_{\text{coex}}(T)$.

Now our task is: how to obtain the function $P_{\text{coex}}(T)$? In order to simplify the notation, we drop the subscript coex in P. Suppose (T, P) is a point on the coexistence curve, and we choose another point $(T+dT, P+dP)$ which is also on the coexistence curve and very close to (T, P). The question is then: what is the relation between dT, dP and (T, P)? Since both (T, P) and $(T + dT, P + dP)$ are on the coexistence curve, we should have $g_1(T, P) = g_2(T, P)$ and $g_1(T + dT, P + dP) = g_2(T + dT, P + dP)$. Therefore

$$dg_1(T, P) = dg_2(T, P), \tag{7.7}$$

where $dg_i = g_i(T + dT, P + dP) - g_i(T, P)$ $(i = 1, 2)$. From Eq. (5.12))

$$dg = -s\,dT + v\,dP,$$

so Eq. (7.7) becomes

$$-s_1 dT + v_1 dP = -s_2 dT + v_2 dP. \tag{7.8}$$

Rearranging the above equation, we obtain

$$\left(\frac{dP}{dT}\right)_{21} = \frac{s_2 - s_1}{v_2 - v_1} = \frac{\ell_{21}}{T(v_2 - v_1)}, \tag{7.9}$$

where $\ell_{21} = T(s_2 - s_1)$ is the heat of transformation for converting from phase 1 to phase 2 (cf. Eq. (7.6)). Equation (7.9) is called the **Clausius-Clapeyron[3] equation**, which is a differential equation for the first-order phase transition coexistence curve. The term $(dP/dT)_{21}$ on the left hand side of the equality is the slope of the coexistence curve for phase 1 and phase 2. The subscripts 1 and 2 may respectively denote the vapor phase, the liquid phase, or the solid phase. If we are considering a system other than a PVT system, then 1 and 2 may denote different phases for the system under consideration. However, it should be noted

[3]Benoit P. E. Clapeyron, French chemist (1799–1864).

that Eq. (7.9) cannot be applied to second-order (or higher-order) phase transitions, because the right hand side will become ambiguous (because $s_1 = s_2$ and/or $v_1 = v_2$). In this case one needs to consider the expansion of s and v to obtain $(dP/dT)_{21}$ for the second-order phase transitions (see Problem 7.16).

7.4 Vapor Pressure of a Liquid or Solid

In this section we use 1 to denote the solid phase, 2 the liquid phase, and 3 the vapor phase in a PVT system. We want to obtain the formula for the vapor pressure of the liquid phase and the solid phase. This is equivalent to studying the problem of liquid-vapor coexistence and solid-vapor coexistence. The pressure of the vapor phase is called the *vapor pressure*. Here we use the term ***vapor*** to denote the gas phase, which emphasizes the fact that the gas coexists with a liquid or solid, but these two terms (vapor and gas) have the same meaning. Equation (7.9) which contains phase 3 (vapor) is

$$\left(\frac{dP}{dT}\right)_{3i} = \frac{\ell_{3i}}{T(v_3 - v_i)} \approx \frac{\ell_{3i}}{Tv_3}, \quad i = 1, 2, \tag{7.10}$$

where the second approximation sign is due to the fact that $v_3 \gg v_i$ $(i = 1, 2)$. If we approximate the vapor as an ideal gas, then

$$v_3 = \frac{RT}{P}.$$

Substitute the above equation into Eq. (7.10) and drop the subscript on the left hand side; we then have

$$\frac{dP}{dT} \approx \frac{\ell_{3i}}{R}\frac{P}{T^2}, \quad i = 1, 2. \tag{7.11}$$

Rearrange the terms and use the equal sign instead of the approximation sign; then

$$\frac{dP}{P} = \frac{\ell_{3i}}{R}\frac{dT}{T^2}, \quad i = 1, 2. \tag{7.12}$$

Integrate the above equation assuming that ℓ_{3i} is independent of T, then

$$P(T) = P_0 \exp\left[-\frac{\ell_{3i}}{R}\left(\frac{1}{T} - \frac{1}{T_0}\right)\right], \quad i = 1, 2. \tag{7.13}$$

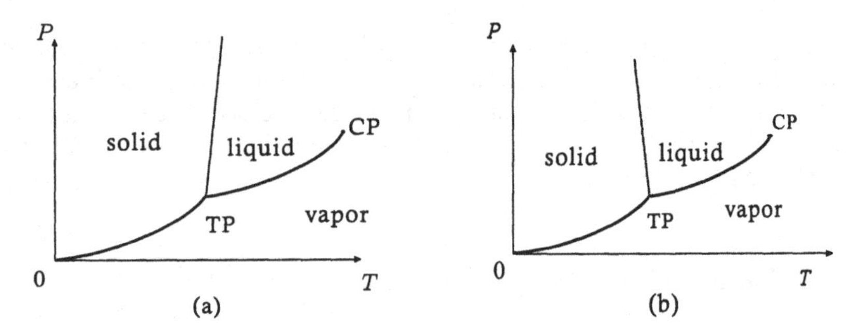

Figure 7.2: Sketched phase diagram for a PVT system, where TP denotes the triple point and CP the critical point for the liquid-vapor coexistence curve. (a) Substances for which the slope of the solid-liquid coexistence curve is positive. (b) Substances for which the slope of the solid-liquid coexistence curve is negative.

In the above equation P_0 is the vapor pressure at temperature T_0, and therefore (T_0, P_0) is a point on the coexistence curve. In Eq. (7.13), the pressure $P(T)$ is the vapor pressure of the liquid or solid at temperature T. This means that (T, P) is also a point on the coexistence curve, and $P(T)$ is known as the ***saturated vapor pressure*** at temperature T. The above equation holds only if the temperature T is close enough to T_0, because in general the latent heat ℓ_{3i} is a function of T.

7.5 Phase Diagram for a *PVT* System

Except for ^{3}He and ^{4}He, the phase diagrams in the P-T plane for all other PVT systems have about the same behavior, as shown in Fig. 7.2(a) or 7.2(b). The solid, liquid, and vapor phases are stable in separate regions in the plane, and can coexist only along the three coexistence curves. The main features of the diagram are the following.

1. The behavior near the absolute zero:

   We have proved in Sec. 6.8 that, because of the third law, the slope of the two-phase coexistence curve for any substance must approach zero as the temperature T approaches absolute zero (cf. Eq. (6.32)), i.e.,

$$\left(\frac{dP}{dT}\right)_{\text{coex}} = \frac{\Delta s}{\Delta v} \rightarrow 0 \quad \text{as } T \rightarrow 0.$$

The two-phase coexistence at $T = 0$ is a solid-vapor coexistence for all substances other than helium.

2. The triple point TP:

   At the **triple point** TP all three phases, vapor, liquid, and solid coexist. This is a fixed point on the P-T diagram for a given substance, as there is no independent variable for this point. We use $(T_{\text{tr}}, P_{\text{tr}})$ to denote the temperature and pressure for that point. The pair of values of $(T_{\text{tr}}, P_{\text{tr}})$ is one of the characteristics for a given substance. Only when the temperature is equal to T_{tr} and *simultaneously* the pressure is equal to P_{tr}, can the three phases coexist. There is no other point on the diagram where three phases can coexist. Therefore the triple point of a substance can be used as a reference temperature for temperature measurements. For example, the triple point for H_2O is at $T_{\text{tr}}= 273.16$ K and $P_{\text{tr}} = 6 \times 10^2$ Pa. Since $\ell_{ji} = T(s_j - s_i)$, near the triple point the three heats of transformation have the relation

$$\ell_{31} = \ell_{32} + \ell_{21}. \tag{7.14}$$

3. The critical point CP:

   The liquid-vapor coexistence curve will terminate at one point as T increases. This point is called the **critical point**. Beyond this point (i.e., higher T and higher P), there is no distinction between the gas phase and the liquid phase. That is the reason why there is no liquid-gas coexistence curve beyond the critical point.[4] At the critical point, the liquid-gas transition is a second-order phase transition, because the liquid and gas have the same properties (including specific volume and entropy) at this point. The temperature at the critical point is called the **critical temperature** T_c, and the pressure is called the **critical pressure** P_c. The critical point is also a fixed point for a given a substance, and it is also one of the characteristics of the substance. For H_2O the critical point occurs at $T = 374°\text{C}$ and $P = 218$ atm.

[4]This implies that it is possible to *transform* an amount of gas to liquid without going through the process of *condensation* (i.e., without going through the vapor to liquid phase transition). Consider a system which is initially in the gas state, with $T = T_0 > T_c$ and $P = P_0 < P_c$. P is then increased while T_0 is kept constant, until $P = P_1 > P_c$. T is then lowered to $T_1 < T_c$, while P_1 is kept constant. P is then lowered from P_1 to $P_2 < P_c$. The final state (T_1, P_2) will be a liquid state if T_1 and P_2 are properly chosen. In the processes described above, the system undergoes no phase transition, yet it changes from a gas state to a liquid state.

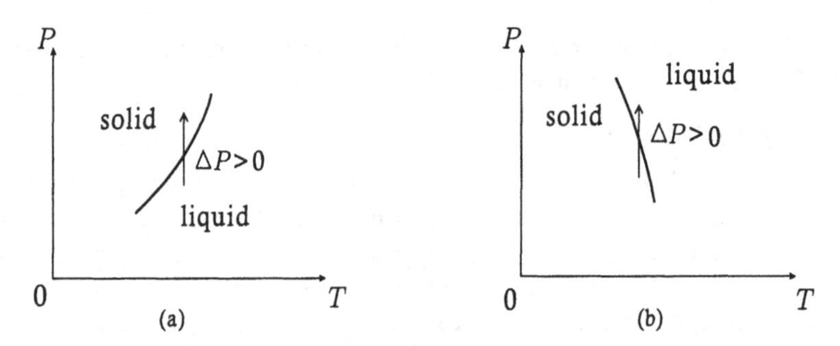

Figure 7.3: The slope of the solid-liquid coexistence curve. (a) The slope is positive, an amount of liquid may solidify to solid by applying pressure. Most of the substances belong to this case (cf. Fig. 7.2(a)). (b) The slope is negative, a solid may melt to liquid by applying pressure. An example of this kind of substance is H_2O (cf. Fig. 7.2(b)).

4. The solid-liquid coexistence curve:

   This coexistence curve will extend to infinite pressure, therefore there is no critical point for the liquid-solid coexistence curve. For most substances, when a liquid solidifies to a solid the volume decreases. Therefore for these substances, an amount of liquid, whose state is close to the coexistence curve, will solidify to a solid if the pressure is increased (at constant temperature), because the volume decreases under pressure. Therefore the slope of the coexistence curve must be positive, as shown in Fig. 7.3(a). That is

$$\left(\frac{\partial P}{\partial T}\right)_{s\text{-}l} > 0, \tag{7.15}$$

   where the subscript s denotes solid, and l denotes liquid. However there are substances which behave differently. For example when water solidifies to ice, the volume increases. Therefore for an amount of ice, whose state is close to the water-ice coexistence curve, the ice will melt to water when the pressure is increased (while the temperature is kept constant), because the volume become smaller under pressure. The slope of the solid-liquid coexistence curve should be negative for H_2O, as shown in Fig. 7.3(b). If we plot the g-P curves (T is kept constant) for both the solid and liquid phases on the same diagram, we will get two different figures in which the g's behave differently, as shown in Figs. 7.4(a) and

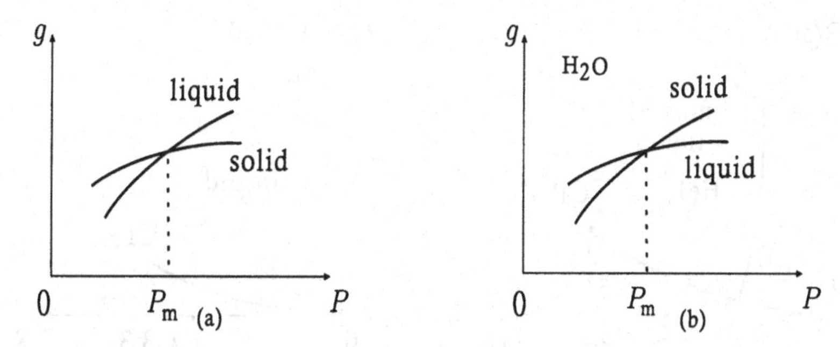

Figure 7.4: The g-P diagram (T constant) for both the solid and liquid phases, where g is the specific Gibbs free energy. The pressure P_m is the melting pressure at the temperature T which is kept constant. (a) The case where the liquid is stable for $P < P_m$. (b) The case where the solid is stable for $P < P_m$.

7.4(b). In the diagrams, P_m is the melting pressure at temperature T which is kept constant. Figure 7.4(a) shows that liquid is the stable phase for $P < P_m$, because $g_l < g_s$ in this pressure range. A substance whose g-P diagram behaves like Fig. 7.4(a), will have a phase diagram like Fig. 7.3(a), where $(\partial P/\partial T)_{s\text{-}l} > 0$. Most of the substances belong to this category. Figure 7.4(b) shows that solid is the stable phase for $P < P_m$, because $g_s < g_l$ in this pressure range. This corresponds to substances whose phase diagram has the characteristic shown in Fig. 7.3(b), where $(\partial P/\partial T)_{s\text{-}l} < 0$. The most well-known example belonging to this category is H_2O.

7.6 Phase Diagrams for ^{4}He and ^{3}He

Helium is the only PVT system whose phase diagram is different from Fig. 7.2(a) or 7.2(b). In nature there are two isotopes of helium atoms, ^{3}He and ^{4}He. For ^{3}He there are two protons and one neutron in the nucleus, and for ^{4}He there are two protons and two neutrons in the nucleus. The most important characteristic for helium is that, under the pressure of 1 atm, helium remains a liquid even when the temperature approaches absolute zero, while all other substances are solidified at this low temperature. Only when a high pressure is applied, will liquid helium be solidified. Most of the isotopes found in nature are ^{4}He atoms, and liquid ^{4}He is the most important coolant used in the laboratory. We

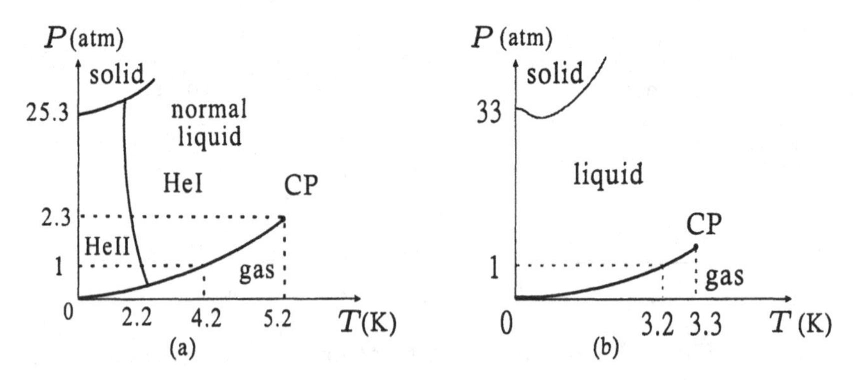

Figure 7.5: Phase diagrams (not to scale) for ^{4}He (a) and ^{3}He (b).

plot the phase diagram for ^{4}He in Fig. 7.5(a). In addition to the fact that, under normal pressure, ^{4}He remains in the liquid state as $T \to 0$, there is another unique property of liquid ^{4}He. Namely, it possesses the phenomenon of superfluidity for temperatures below the critical temperature. A superfluid behaves in a quite different fashion from a normal fluid. For example, a superfluid can flow through a capillary freely without friction, while a normal fluid can not pass through a capillary because of friction between the fluid and the wall. There are many other unique properties which can only be observed in a superfluid, and not in a normal fluid. Therefore superfluid ^{4}He and normal fluid ^{4}He are considered as two different phases, although they both belong to the liquid phase. For simplicity of terminology, normal fluid ^{4}He and superfluid ^{4}He are frequently called He I and He II, respectively. Their respective stable regions are sketched in Fig. 7.5(a). The phase transition between He I and He II is a second-order one, known as the λ transition. The slopes of the liquid-vapor and the solid-liquid coexistence curves both approach zero as T approaches zero, which is consistent with the third law.

The phase diagram of ^{3}He has about the same shape as that of ^{4}He; it is shown in Fig. 7.5(b). However, there are two important differences between the phase diagrams for ^{3}He and ^{4}He. For ^{3}He the solid-liquid coexistence curve has a minimum at about 0.3 K instead of 0 K, due to an anomalous relative entropy [5] between liquid and solid ^{3}He. Thus the

[5]Each ^{3}He has a spin $\hbar/2$, and the ^{3}He atoms are fermions. The spin entropy dominates in solid ^{3}He at low T (except for a T very close to 0 K) because the atoms are localized, but liquid ^{3}He behaves like a Fermi gas, in which the atoms obey the Pauli principle. The behavior of an ideal Fermi gas will be discussed in Chapter 12.

slope of the coexistence curve is negative for $0 < T < 0.3$ K. Yet the slope must be zero at $T = 0$, because of the third law. However, the most important difference is that for ^{3}He there is no superfluid phase. This is due to the fact that ^{4}He atoms are **Bose particles** which obey **Bose-Einstein statistics**, while the ^{3}He atoms are **Fermi**[6] **particles** which obey **Fermi-Dirac statistics**. We have mentioned in Chapter 6 that for an ideal Bose gas, the system will undergo a of Bose-Einstein condensation phase transition when the temperature is below a critical temperature. It is believed that the **superfluid** phase of liquid ^{4}He is related to the Bose-Einstein condensation, although it is much more complicated because it happens in the liquid phase. This theory seems well supported by the fact that ^{3}He (a fermion) does not possess a superfluid phase, but the composite atom ^{3}He-^{3}He (which is a boson) does exhibit superfluid phases, which were observed in the early 1970's. We will discuss the statistical properties of Bose particles and Fermi particles in Chapter 12.

7.7 Van Der Waals Gases

A gas which obeys the **van der Waals equation of state** is called a **van der Waals gas**. This equation of state takes into account that the volume of a molecule is nonzero,[7] and that there is a small yet non-negligible attractive force between molecules at large distances. These two factors are neglected in the ideal gas equation of state. Therefore this is a more realistic equation of state for *real gases*. One of the interesting properties of this equation of state is that, although it is only a mathematical equation, it can be used to analyze the vapor-liquid phase transition. The reason is that the equation contains a portion of an *unphysical* region which separates the vapor states from the liquid states. In this section we will use this equation of state to study the details of the vapor-liquid transition of a PVT system.

The van der Waals equation of state is (Eq. (1.35))

$$\left(P + \frac{a}{v^2}\right)(v - b) = RT; \quad a > 0, \quad b > 0,$$

[6] Enrico Fermi, Italian physicist (1901–1954), Nobel prize laureate in physics in 1938.

[7] This simulates the short distance repulsion between molecules.

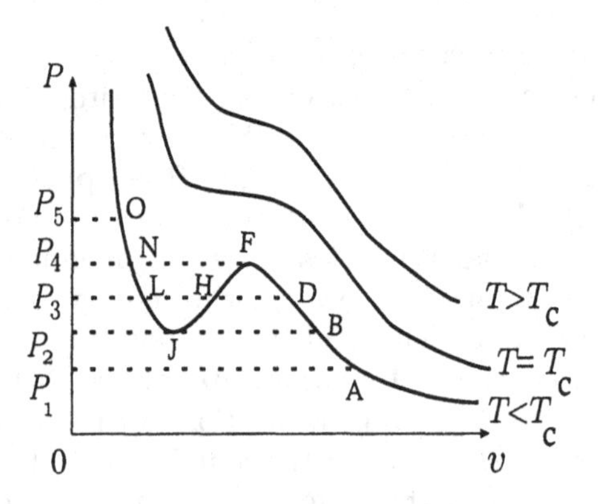

Figure 7.6: Isotherms on the P-v diagram. For the isotherms with $T > T_c$, one P corresponds to only one v. For the isotherms with $T < T_c$, one P corresponds to three v's for the pressure range $P_2 < P < P_4$.

which can be converted (by multiplying each term by v^2 and a rearrangement of the terms) to

$$Pv^3 - (Pb + RT)v^2 + av - ab = 0. \tag{7.16}$$

This is a cubic equation for v, in which a, b, and R are constants, and T and P are adjustable parameters. For a given set of values of T and P there are three roots for v. The three roots may be all real numbers, or one real number and two complex numbers (complex conjugates). Because all the quantities T, P, a, b, and R are positive numbers, therefore all the real roots of v must be positive (i.e., there is no real root of v which is smaller than 0). This means that the real roots of v are all physical solutions, but the complex roots are unphysical. Therefore for a given set of values for (T, P), Eq. (7.16) will have at least one physical root for v (v real and > 0), and it may possibly have three physical roots for v.

In order to understand the properties of Eq. (7.16), we plot the P-v diagram while keeping T a constant. We will get a curve on the diagram for a given T, which is called an *isotherm*. By taking T at different values we will get different curves, and this is shown in Fig. 7.6. We find that for temperatures T greater than a certain temperature T_c, which will be identified as the *critical temperature*, every horizontal line (i.e., P=constant line) will intersect the isotherm ($T > T_c$) at only one point. This means that there is only one real root for v, which implies that there is no phase transition. However for temperatures T less than T_c, i.e., $T < T_c$, in a certain range of pressure, the horizontal lines (P constant) may intersect the isotherm at three places, i.e., there are three real positive roots for v, as shown in Fig. 7.6. At the critical temperature T_c and critical pressure P_c, the three roots of v are all equal, i.e., it is a ***triple root***. And the root $v = v_c$ is called the critical specific volume (or simply the critical volume).

The critical temperature and critical pressure are therefore defined as the temperature and pressure such that Eq. (7.16) has a triple root for v. The set of values (T_c, v_c, P_c) denote the temperature, specific volume, and pressure at the ***critical point***.

We are interested in the isotherm such that for a given P there are three real roots of v, which is the case $T < T_c$. In the following analysis we use the letters labeled in Fig. 7.6 to represent the states. On an isotherm with $T = T_1 < T_c$, the states J and F are, respectively, the minimum and the maximum point of the curve. The slope $(\partial P/\partial v)_T$ at these two points are both zero. In the section JHF the slope is positive, i.e.,

$$\left(\frac{\partial v}{\partial P}\right)_T = \left(\frac{\partial P}{\partial v}\right)_T^{-1} > 0.$$

This means that the isothermal compressibility for the states corresponding to the segment JHF is negative, i.e. $\kappa < 0$, which implies that these states are unstable, because it violates the condition of stable equilibrium $\kappa > 0$ (cf. inequality (5.62)). Therefore although the segment JHF satisfies the van der Waals equation of state Eq. (7.16), it has no physical meaning, i.e., no physical states correspond to this segment. Only the segments ABDF and JLNO correspond to real physical states. This, however, means that for each P (and T) there are two different values of v, which implies that there are two different states for each set of the independent variables T and P. Apparently these two states belong to different phases. The segment ABDF has the property that the specific volume v is large and the pressure P is small, which is the characteristic of a gas. Therefore we may consider it as the vapor phase; the segment JLNO represents the liquid phase (smaller v and larger P). The question now is how, or under what conditions, the system changes from the vapor phase to the liquid phase, and vice versa? This is a problem of phase equilibrium. We will solve it by a graphical method, known as the *Maxwell construction*.

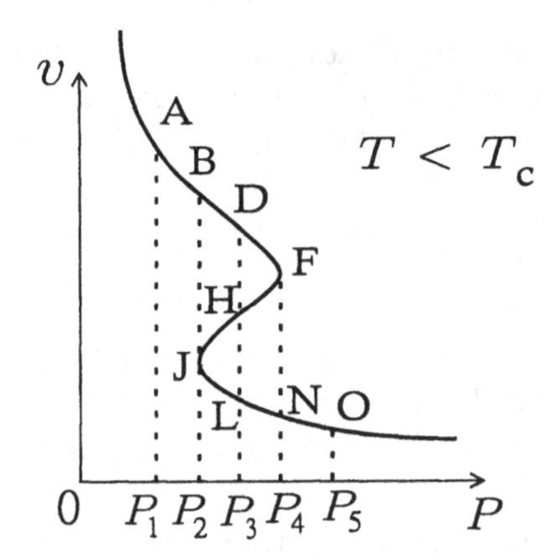

Figure 7.7: The v-P Curve for the isotherm with $T < T_c$.

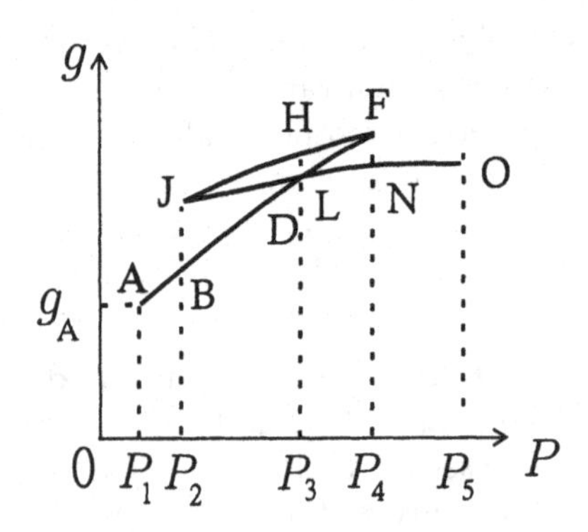

Figure 7.8: The variation of g when P is integrated along an isotherm.

7.8 The Maxwell Construction

We would like to know the answer of the following question: At a given temperature $T < T_c$, i.e., for a given isotherm, at what pressure P will the two phases coexist? In order to study this problem, we plot the isotherm ($T < T_c$) on the v-P diagram. This diagram can be obtained from Fig. 7.6 by rotating it counterclockwise by 90° and then making a mirror reflection with respect to the vertical axis, as shown in Fig. 7.7. In this diagram we emphasize that the pressures are equal to each other at B and J, which is P_2, and the pressures at D, H and L are all equal, which is denoted as P_3; also the pressures at F and N are equal to each other, which is denoted as P_4. It is to be noted that the pressure P_3, which corresponds to the points D, H, and L, is not chosen arbitrarily. These three points have special properties which will be discussed in the following. On the isotherm with $T = T_1 < T_c$,

$$dg = -s\,dT + v\,dP = v\,dP \;\; (T = T_1). \tag{7.17}$$

Integrate the above equation from point A ($P_A = P_1$) with respect to P to obtain

$$g(T_1, P) - g(T_1, P_1) = \int_{P_1}^{P} v\,dP. \tag{7.18}$$

In the following we omit T_1 in the expression because it is a constant. The above equation can be written as

$$g(P) - g_A = \int_{P_1}^{P} v\, dP = \text{area under } v(P) \text{ in Fig. 7.7,} \tag{7.19}$$

where $g_A = g(T_1, P_1)$ is the g value at A. The value of $g(P)$ is a function of P as we integrate along the isotherm, which is plotted in Fig. 7.8 with P as the horizontal axis.

By looking at the v vs. P curve in Fig. 7.7, we may easily understand how the value of g varies as we integrate along the $v(P)$ curve. The variation of g is shown in Fig. 7.8. While integrating along the segment A→B→D→F the pressure P increases, thus g increases, reaching a maximum value at point F. In the segment F→H→J the pressure P decreases, so g decreases too. Then in the segment J→L→N→O the pressure P increases again, and the value of g is increasing. Therefore the point J is a turning point. We see that the curve of g for the segment JLNO will intersect the curve of g for the segment ABDF at the point D or L. D is on the segment ABDF and L is on the segment JLNO. The point H is on the segment JHF which has the same pressure as D or L. From Fig. 7.8, and also from the integration in Fig. 7.7, we have the following results:

1. The pressure of J and B are equal, i.e., $P_J = P_B = P_2$, and $g_J > g_B$.
2. The pressure of H, L and D are equal $P_H = P_L = P_D = P_3$, and $g_H > g_L = g_D$.
3. The pressure of F and N are equal $P_F = P_N = P_4$, and $g_F > g_N$.

Now return to Fig. 7.6. We know that the segment ABDF is the vapor phase, the segment JLNO is the liquid phase, and the segment JHF is *unphysical* (because of $\kappa < 0$). At the same temperature and pressure, the state with a smaller g is the stable phase, therefore from Fig. 7.8 we know that at a fixed temperature $T = T_1$, and fixed pressure $P < P_3$, the vapor phase (corresponding to ABD) is the stable phase. For a larger pressure $P > P_3$, the liquid phase (corresponding to LNO) is the stable phase. Therefore the phase transition occurs when the pressure P is equal to P_3 (while $T = T_1$). At this pressure $g_L = g_D$, i.e., the chemical potentials of the vapor phase and the liquid phase are equal, $\mu_{\text{liquid}} = \mu_{\text{vapor}}$. This is the condition of two-phase coexistence.

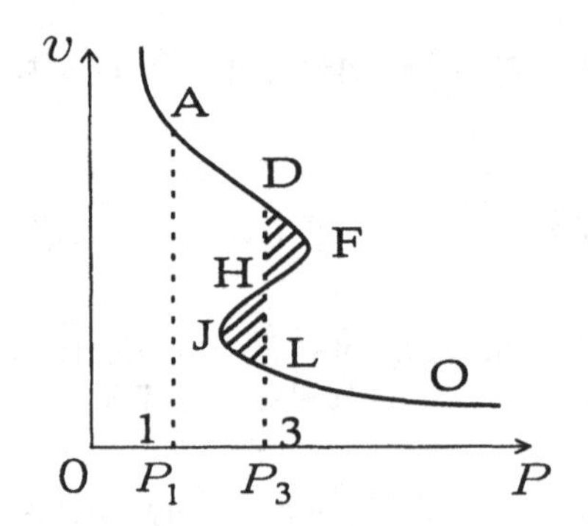

Figure 7.9: The Maxwell construction for two-phase coexistence. The pressure P_3 makes area(DFH)=area(HJL).

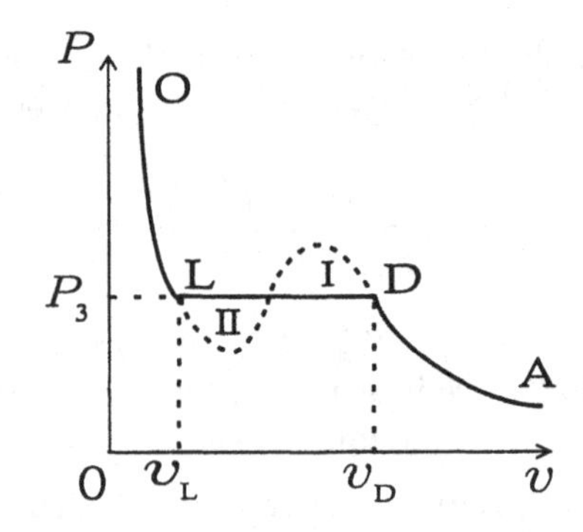

Figure 7.10: An isotherm on the P-v diagram for liquid-vapor coexistence. The coexistent line DL ($P = P_3$) makes area I=area II.

It is desirable if we can determine the coexistence pressure P_3 from the isotherm shown in Fig. 7.7. We simplify the labels and lines of this diagram and reproduce it, as shown in Fig. 7.9. Since the specific Gibbs free energy $g = \int_{P_1} v dP$ =area under the $v(P)$ curve in Fig. 7.9, we get the respective g values at D and L as (take all the areas as positive),

$$g_{\mathrm{D}} = g_{\mathrm{A}} + \text{area}\,(\text{AD31}),$$
$$g_{\mathrm{L}} = g_{\mathrm{A}} + \text{area}\,(\text{AD31}) + \text{area}\,(\text{DFH}) - \text{area}\,(\text{HJL}).$$

Therefore the condition for phase equilibrium at D and L, $g_{\mathrm{D}} = g_{\mathrm{L}}$, becomes

$$\text{area}\,(\text{DFH}) = \text{area}\,(\text{HJL}). \tag{7.20}$$

Now we return back to Fig. 7.6. We see that in the pressure range for which we have three roots of v for a given P, we move the horizontal isobaric line up and down until this line intersects the van der Waals curve such that two equal areas are formed (one above and the other one below the isobaric line), then this is the pressure for two-phase coexistence. Using the condition of Eq. (7.20) to obtain the coexistence pressure is called the **Maxwell construction**.

We redraw Fig. 7.6, but keep only one isotherm $T = T_1 < T_c$, and use the solid lines to denote the stable states and the dashed lines to denote the unstable (unphysical) states. The new diagram is shown in Fig. 7.10. From the Maxwell construction, if the constant pressure line

LD (with $P = P_3$) denotes the two-phase coexistence pressure, then area I=area II, as shown in the figure. At D, the system is 100% vapor with the specific volume $v = v_{\rm D}$. At L, the system is 100% liquid, with the specific volume $v = v_{\rm L}$. Now we would like to study the process of how the system evolves from a vapor state to a liquid state when the temperature of the system is kept constant at $T_1 < T_c$. The system starts from a low P and large v state, which is a vapor state (with $P < P_3$ and $v > v_{\rm D}$). We refer to Fig. 7.10 and suppose the system starts from the state A in the figure. The pressure on the vapor is then increased slowly (while T_1 is kept constant); the state of the system will change along the isotherm moving in the direction of D. In this process, pressure is increased and the specific volume v is decreased. The system remains in the vapor phase until the state D is reached. At state D ($P = P_3$, $v = v_{\rm D}$) the system is 100% vapor. The pressure P_3 is the two-phase coexistence pressure, therefore when the external pressure is increased slightly beyond P_3, the volume v decreases rapidly along the line DL to keep the pressure constant at P_3. The rapid decrease of v is because some of the vapor condenses to liquid to keep the pressure constant. In this process the state of the system changes along the horizontal line DL (P=constant line); this is the two-phase coexistence region. With the external pressure kept applied, the volume v continues to decrease to keep $P = P_3$, until the state L is reached, where $v = v_{\rm L}$. At L the system is 100% liquid, and there is no vapor in this state. Therefore if more external pressure is applied, the pressure of the system will increase rapidly along LO where v changes only slightly. This is a characteristic of a liquid state.

Now we would like to know when the specific volume v is in the range $v_{\rm L} < v < v_{\rm D}$, what is the mole fraction of liquid (or vapor)? If there are totally n_0 kilomoles of molecules in the system, in which $n_{\rm v}$ kilomoles are in the vapor phase, and n_l kilomoles are in the liquid phase, we have $n_v + n_l = n_0$. We define two mole fractions x_l and $x_{\rm v}$,

$$x_l = \frac{n_l}{n_l + n_{\rm v}} \ \text{(liquid)}, \quad x_{\rm v} = \frac{n_{\rm v}}{n_l + n_{\rm v}} \ \text{(vapor)}. \tag{7.21}$$

Obviously $x_l + x_{\rm v} = 1$, we obtain

$$v = x_{\rm v} v_{\rm D} + x_l v_{\rm L} = (1 - x_l) v_{\rm D} + x_l v_{\rm L}, \tag{7.22}$$

$$v_{\rm D} - v = x_l (v_{\rm D} - v_{\rm L}).$$

The solutions are

$$x_l = \frac{v_{\rm D} - v}{v_{\rm D} - v_{\rm L}} \quad \text{and} \quad x_{\rm v} = \frac{v - v_{\rm L}}{v_{\rm D} - v_{\rm L}}. \tag{7.23}$$

Therefore the mole fraction for the liquid phase x_l and that for the vapor phase x_V, can be obtained from the geometry of Fig 7.10. Specifically x_l and x_V can be obtained from the lengths $v_D - v_L$, $v_D - v$, and $v - v_L$. Equation (7.23) is known as the **lever rule**, because the equation for the ratio x_V/x_l is the same as that of a lever formula if we interpret it geometrically.

7.9 The Critical Point and the Reduced Equation of State

In the phase diagram for a *PVT* system, Fig. 7.2, we note that there is a **critical point** CP in the vapor-liquid coexistence curve. The temperature, pressure, and specific volume at this point are called the **critical temperature** T_c, the **critical pressure** P_c, and the **critical volume** v_c. For $T > T_c$, there is no distinction between the liquid phase and the gaseous phase; therefore on the P-v plot in Fig. 7.6, for any value of the pressure P there corresponds only one v for an isotherm with $T > T_c$, but for an isotherm with $T < T_c$, there may be three v's corresponding to a single P, for a certain range of P. When T increases from below T_c, the three v's which correspond to the same P will get closer and closer. At $T = T_c$ the three v's will coincide and become a ***triple root*** for Eq. (7.16). Therefore at the critical point we may rewrite Eq. (7.16) as

$$(v - v_c)^3 = 0 \quad \text{for} \quad T = T_c,\ P = P_c. \tag{7.24}$$

The critical point has another important property. It is an ***inflection point*** of the isotherm, which means that on the left side of the critical point the curve is concave upward, and on the right side the curve is concave downward. Therefore, at the critical point, the first order and the second order derivatives of P with respect to v are both zero:

$$\left(\frac{\partial P}{\partial v}\right)_T = 0 \ \text{and} \ \left(\frac{\partial^2 P}{\partial v^2}\right)_T = 0 \ \text{at CP}. \tag{7.25}$$

From the van der Waals equation of state (with critical point values) and Eq. (7.25), we have the following three coupled equations:

$$P = \frac{RT_c}{v_c - b} - \frac{a}{v_c^2}, \tag{7.26}$$

$$\left(\frac{\partial P}{\partial v}\right)_T = -\frac{RT_c}{(v_c - b)^2} + \frac{2a}{v_c^3} = 0, \tag{7.27}$$

$$\left(\frac{\partial^2 P}{\partial v^2}\right)_T = \frac{2RT_c}{(v_c - b)^3} - \frac{6a}{v_c^4} = 0. \tag{7.28}$$

Solving these three coupled equations, we obtain

$$P_c = \frac{a}{27b^2}, \quad v_c = 3b, \quad T_c = \frac{8a}{27Rb}. \tag{7.29}$$

These three equations are frequently used to determine the van der Waals constants a and b for different gases. By substituting the experimental values at the critical point (P_c, v_c, T_c) into Eq. (7.29), a and b can be determined. However we have three equations but only two undetermined constants, therefore an inconsistency may occur. In order to avoid this inconsistency one could use only P_c and T_c to determine a and b, since there is a larger uncertainty in the experimental measurement of v_c.

From Eq. (7.29) we obtain

$$\frac{P_c v_c}{RT_c} = \frac{3}{8}, \tag{7.30}$$

which means that $P_c v_c / T_c$ is independent of the values of a and b. All the van der Waals gases obey Eq. (7.30). Therefore this condition may be used to test whether a real gas obeys the van der Waals equation of state. A comparison of Eq. (7.30) with all the known experimental critical values of the real gases shows that the discrepancies in most cases are greater than 20%. Hence the van der Waals equation of state is not an ideal equation of state for most real gases. However this equation of state is relatively simple, and it takes into account the characteristics of real gas molecules (molecules possess a finite volume, and there is a weak attractive interaction at large distance). The equation also possesses a liquid-vapor phase transition behavior; therefore it is useful in studying real gases, at least as a good approximate model. There is one more interesting property about the critical point of the van der Waals gas which is worth mentioning. We may use the critical pressure P_c, the critical volume v_c, and the critical temperature T_c to define the **reduced variables**:

$$P_r = \frac{P}{P_c}, \quad v_r = \frac{v}{v_c}, \quad T_r = \frac{T}{T_c}, \tag{7.31}$$

where P_r, v_r and T_r are called the *reduced pressure*, the *reduced volume*, and the *reduced temperature*, respectively. By using P_r, v_r, and T_r as variables, the van der Waals equation of state (1.35) becomes

$$\left(P_r + \frac{3}{v_r^2}\right)(3v_r - 1) = 8T_r. \tag{7.32}$$

This is called the **reduced equation of state** for a van der Waals gas. The characteristic of this equation of state is that it does not contain the constants a and b, and thus it contains no system-dependent constants. Therefore it applies to all van der Waals gases, i.e., all van der Waals gases satisfy the same equation of state (7.32).

7.10 Order of Phase Transitions

In the introduction section, we mentioned that there are two classes of phase transitions: the first-order and second-order (or higher-order) phase transitions. Now we want to give the theoretical basis for this classification. We use 1 and 2 to denote the different phases. At temperature T and pressure P, the condition for phase 1 and 2 to coexist is, by Eq. (7.4),

$$g_1(T,P) = g_2(T,P). \tag{7.33}$$

The classification of the phase transitions is according to whether the first order derivatives of g_1 and g_2 with respect to T and to P are equal or unequal. Now the first order derivatives of g with respect to T and to P are, respectively, equal to $-s$ and v, i.e., $(\partial g/\partial T)_P = -s$ and $(\partial g/\partial P)_T = v$, therefore if in the two-phase coexistence s and v are not equal for the two phases, i.e.,

$$s_1(T,P) \neq s_2(T,P), \quad v_1(T,P) \neq v_2(T,P), \tag{7.34}$$

then this belongs to a first-order phase transition. The term **first-order** comes from the fact that the first order derivatives of g are not equal to each other for the two phases. It is also called a **discontinuou** phase transition, because s and v are discontinuous (unequal) in the phase transition. The most common examples are the phase transitions between the three phases of a PVT system, such as the vapor⇌liquid, liquid⇌solid, and vapor⇌solid transitions.

On the other hand if, in the two-phase coexistence, the first order derivatives of g are equal for the two phases i.e., $s_1 = s_2$, and $v_1 = v_2$, it is called a **second-order** (or **higher-order**) phase transition.[8] Since at the two-phase coexistence, $s_1 = s_2$, and $v_1 = v_2$, so s and v are

[8]If the second order derivatives of g are also equal, then at least one of the higher order derivatives must be unequal, otherwise there will be no phase transition. This is why they are also called higher-order phase transitions.

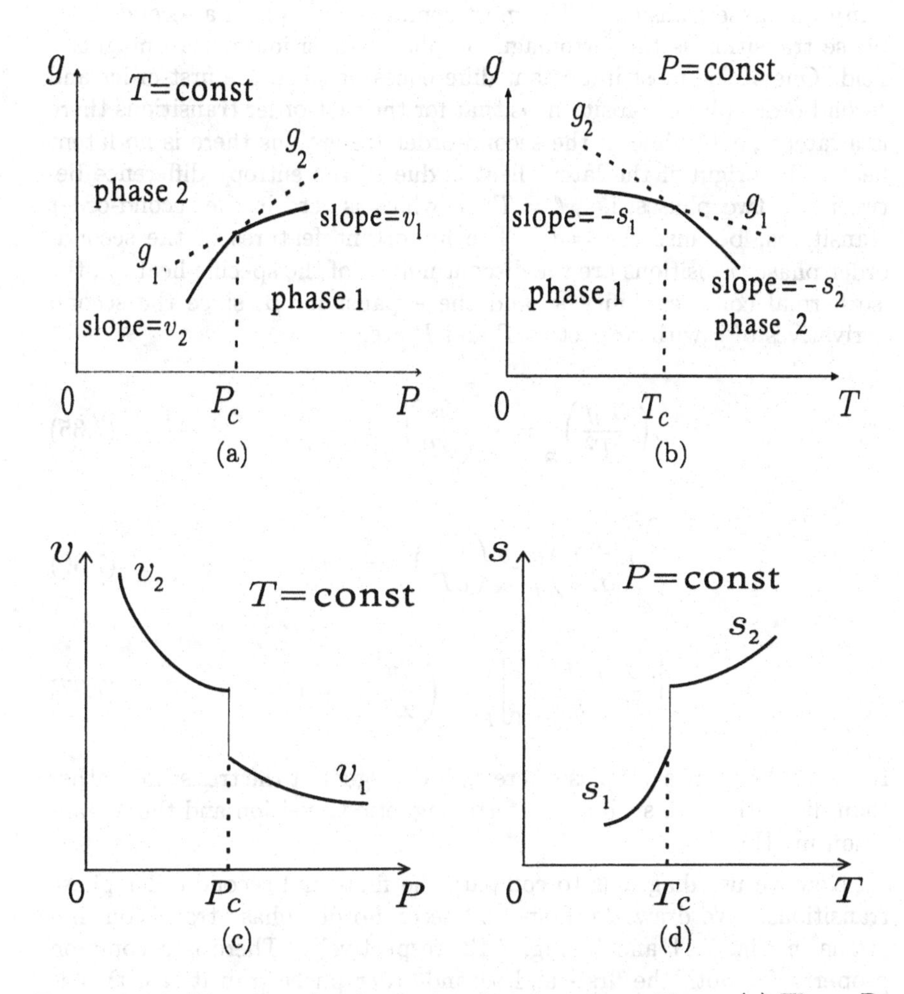

Figure 7.11: Diagrams for the first-order phase transitions. (a) The g-P plot (T constant). The solid line is the g line for the stable phase, and the dashed line is the g line for the unstable phase. (b) The g-T plot (P constant). The solid line is the g line for the stable phase, and the dashed line is the g line for the unstable phase. (c) The v-P plot (T constant). (d) The s-T plot (P constant).

continuous in the phase transition, therefore this is also called a **continuous** phase transition. The most common example of a second-order phase transition is the ferromagnetic phase transition at zero magnetic field. One of the most important differences between the first-order and second-order phase transitions is that for the first-order transitions there is a latent heat, while in the second-order transitions there is no latent heat. The origin of the latent heat is due to the entropy difference between the two phases, i.e., $\ell = T\Delta s$, which is zero in the second-order transitions, because $\Delta s = 0$. The important features of the second-order phase transitions are the discontinuities of the specific heat c_P, the isothermal compressibility κ, and the expansivity β, since the second derivatives of g with respect to T and P are,

$$\left(\frac{\partial^2 g}{\partial T^2}\right)_P = -\left(\frac{\partial s}{\partial T}\right)_P = -\frac{c_P}{T}, \tag{7.35}$$

$$\left(\frac{\partial^2 g}{\partial P^2}\right)_T = \left(\frac{\partial v}{\partial P}\right)_T = -v\kappa, \tag{7.36}$$

$$\left[\frac{\partial}{\partial T}\left(\frac{\partial g}{\partial P}\right)_T\right]_P = \left(\frac{\partial v}{\partial T}\right)_P = v\beta. \tag{7.37}$$

However there are examples where c_p is divergent at the transition rather than discontinuous, such as the ferromagnetic transition and the λ transition in ^{4}He.

Now we use diagrams to compare the first- and second-order phase transitions. We draw the first- and second-order phase transition diagrams in Fig. 7.11 and in Fig. 7.12, respectively. There is a common property for both the first- and second-order phase transitions, that is that the specific Gibbs free energy g are equal for the coexistence phases, which is shown in all the diagrams in Figs. 7.11(a)–(b) and in Fig. 7.12(a). The first order derivatives of g may be unequal, as shown in Figs. 7.11(c)–(d) which belong to the first-order phase transitions. An example where the first order derivative of g is equal for the two coexistent phases is shown in Fig. 7.12(a), which belongs to a second-order phase transition. In this case we have to go to the second order derivatives, as shown in Figs. 7.12(c)–(d), which are examples of a discontinuity in the specific heat c_p.

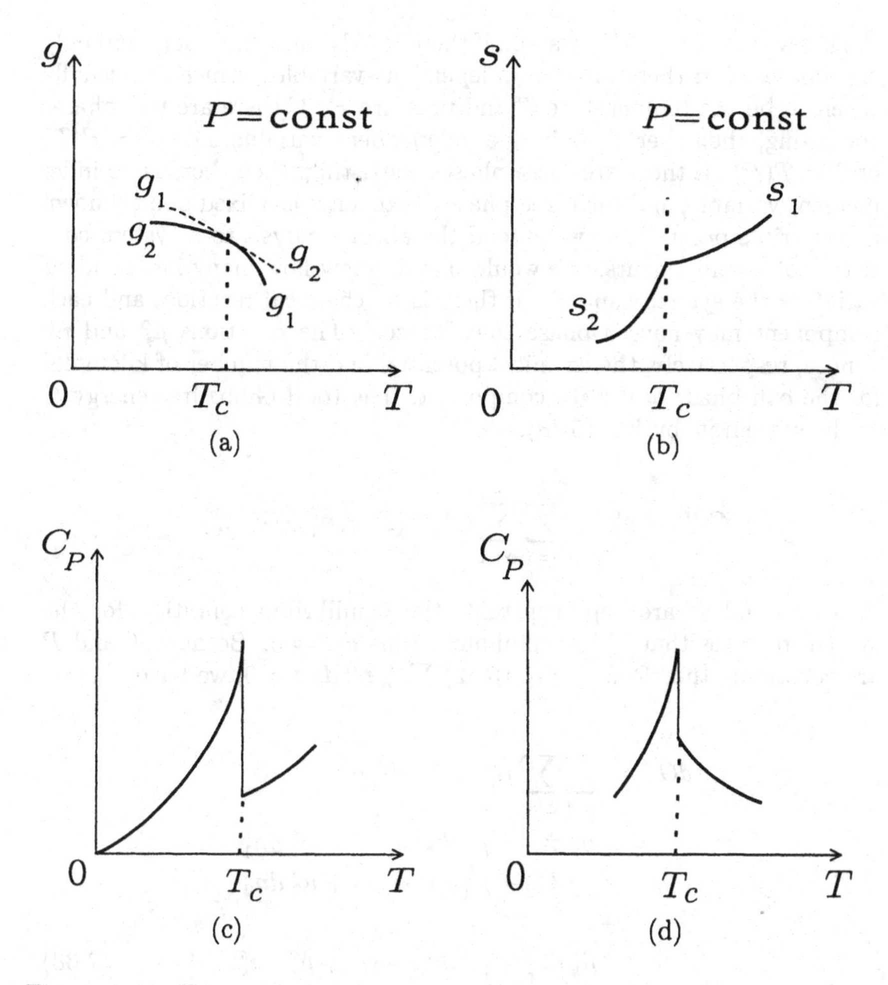

Figure 7.12: Diagrams for the second-order phase transitions. (a) The g-T plot (P constant). T_c is the temperature of the phase transition. (b) The s-T plot (P constant). At T_c, s is continuous. (c) An example where c_p is discontinuous at the phase transition, such as a superconductivity phase transition. (d) An example where c_p is divergent at the phase transition, such as a ferromagnetic phase transition.

7.11 The Gibbs Phase Rule

We know that for a PVT system, if there is only one component and only one phase, then there are two independent variables, which are usually taken to be the temperature T and pressure P. If there are two phases coexisting, then there is only one independent variable, i.e., $P = P(T)$ or $T = T(P)$. If there are three phases coexisting, then there is no independent variable, and the three-phase coexistence is a fixed point, known as the triple point. Now we extend the above analysis to a system consisting of k components. We would like to know how many independent variables the system can have if there is no chemical reaction, and each component may have π-phase coexistence.[9] The notations μ_i^α and n_i^α denote, respectively, the chemical potential and the number of kilomoles for the αth phase of the ith component. The total Gibbs free energy G of the system is, by Eq. (5.48),

$$G(T, P, n_i^\alpha) = \sum_{i=1}^{k} \sum_{\alpha=1}^{\pi} n_i^\alpha \, \mu_i^\alpha(T, P, n_1^\alpha, \cdots, n_k^\alpha).$$

When T and P are kept constant, the equilibrium condition for the system requires that G is a minimum, thus $dG = 0$. Because T and P are constants, therefore by Eq. (5.52) $\sum_{i,\alpha} n_i^\alpha \, d\mu_i^\alpha = 0$, we have

$$\begin{aligned} dG &= \sum_{i=1}^{k} \sum_{\alpha=1}^{\pi} \mu_i^\alpha \, dn_i^\alpha = 0 \\ &= \mu_1^1 \, dn_1^1 + \mu_1^2 \, dn_1^2 + \cdots + \mu_1^\pi \, dn_1^\pi \\ &\quad + \mu_2^1 \, dn_2^1 + \mu_2^2 \, dn_2^2 + \cdots + \mu_2^\pi \, dn_2^\pi \\ &\quad + \cdots \\ &\quad + \mu_k^1 \, dn_k^1 + \mu_k^2 \, dn_k^2 + \cdots + \mu_k^\pi \, dn_k^\pi . \end{aligned} \tag{7.38}$$

We are considering a closed system, therefore the total number of kilomoles for each component is a constant, i.e., $\sum_{\alpha=1}^{\pi} n_i^\alpha$ =constant. Thus

$$\sum_{\alpha=1}^{\pi} dn_i^\alpha = 0, \quad \text{or} \quad dn_i^1 = -\sum_{\alpha=2}^{\pi} dn_i^\alpha . \tag{7.39}$$

[9] In this section we use a Roman letter i ($i = 1, 2, \cdots, k$) to denote the components, and a Greek letter α ($\alpha = 1, 2, \cdots, \pi$) to denote the phases.

Substituting Eq. (7.39) into Eq. (7.38), we have

$$\begin{aligned}
&- \mu_1^1(dn_1^2 + dn_1^3 + \cdots + dn_1^\pi) + \mu_1^2\, dn_1^2 + \cdots + \mu_1^\pi\, dn_1^\pi \\
&- \mu_2^1(dn_2^2 + dn_2^3 + \cdots + dn_2^\pi) + \mu_2^2\, dn_2^2 + \cdots + \mu_2^\pi\, dn_2^\pi \\
&- \cdots \\
&- \mu_k^1(dn_k^2 + dn_k^3 + \cdots + dn_k^\pi) + \mu_k^2\, dn_k^2 + \cdots + \mu_k^\pi\, dn_k^\pi \\
&= 0. \qquad (7.40)
\end{aligned}$$

Rearranging the terms, we obtain

$$\begin{aligned}
&(\mu_1^2 - \mu_1^1)\, dn_1^2 + (\mu_1^3 - \mu_1^1)\, dn_1^3 + \cdots + (\mu_1^\pi - \mu_1^1)\, dn_1^\pi \\
&+ (\mu_2^2 - \mu_2^1)\, dn_2^2 + (\mu_2^3 - \mu_2^1)\, dn_2^3 + \cdots + (\mu_2^\pi - \mu_2^1)\, dn_2^\pi \\
&+ \cdots \\
&+ (\mu_k^2 - \mu_k^1)\, dn_k^2 + (\mu_k^3 - \mu_k^1)\, dn_k^3 + \cdots + (\mu_k^\pi - \mu_k^1)\, dn_k^\pi \\
&= 0. \qquad (7.41)
\end{aligned}$$

In the above equation, all the differentials dn_i^α ($i = 1, 2, \cdots, k$, and $2 \le \alpha \le \pi$) are arbitrary small quantities, which can vary independently. Therefore their respective coefficients must be equal to zero, in order that Eq. (7.41) holds. This implies that

$$\mu_i^1 = \mu_i^2 = \mu_i^3 = \cdots = \mu_i^\pi, \quad i = 1, 2, 3 \cdots, k. \qquad (7.42)$$

This means that for each component, the conditions for the different phases to coexist are that the chemical potentials for the different phases must all be equal to one another. Note that there are no relations between the chemical potentials for different components.

We have derived the conditions for multi-phase coexistence for each component, and from this we can determine how many independent variables there are for a system consisting of k components, each with π phases. At first, let us forget about the conditions of coexistence, and just count the total number of independent variables of the system. The answer is $2 + \pi(k-1)$. The number 2 represents temperature T and pressure P, and the number $\pi(k-1)$ comes from the fact that each phase has $k-1$ independent variables. The reason is that for each phase there are k components, and there are k mole fractions which can be defined, $x_i^\alpha = n_i^\alpha / \sum_{j=1}^k n_j^\alpha$ ($i = 1, 2, \cdots, k$), but not all the x_i^α's are independent variables, they must satisfy the relation

$$x_1^\alpha + x_2^\alpha + \cdots + x_k^\alpha = 1. \qquad (7.43)$$

Therefore there are $k - 1$ independent variables for each phase. There are π phases, therefore the total number of independent variables for the π phases are $\pi(k-1)$, to which we add T and P to get the total number $2 + \pi(k-1)$.

Now we consider the conditions for the π-phase coexistence. From Eq. (7.42) we see that there are $\pi - 1$ equalities in the equation, therefore there are $\pi - 1$ conditions for the π-phase coexistence for each component. There are k components in the system, therefore the total number of constraints is $k(\pi - 1)$. We define the **variance** f as the the number of *degrees of freedom* of a system, then f is equal to the total number of independent variables minus the total number of constraints, i.e.,

$$f = 2 + \pi(k-1) - k(\pi - 1) = 2 + k - \pi. \tag{7.44}$$

This is known as the **Gibbs phase rule**. It gives the degrees of freedom for a system consisting of k components, where each component can have π-phase coexistence. We list several examples:

1. A system consists of one component and only one phase, thus $k = 1$, $\pi = 1$, and $f = 2 + 1 - 1 = 2$. This is the most common system we have discussed in previous chapters. It is a one-component homogeneous system with two independent variables, which are usually taken to be the temperature T and pressure P.

2. A homogeneous system consists of two different gaseous molecules. In this case $k = 2$, $\pi = 1$, and $f = 2 + 2 - 1 = 3$. In addition to T and P, there is one more independent variable which may be either the mole fraction of the first kind of molecule x_1, or the mole fraction of the second kind of molecule x_2 ($x_i = n_i/(n_1 + n_2)$, $i = 1, 2$).

3. A system consists of one component, and there is a two-phase coexistence. An example is water-vapor coexistence. In this case $k = 1$, $\pi = 2$, and $f = 2 + 1 - 2 = 1$. Therefore there is only one independent variable. If we choose temperature T as the independent variable, then the coexistence pressure is a function of T i.e., $P = P(T)$, which may be interpreted as the (saturated) vapor pressure.

4. A system consists of one component, but there is a three-phase coexistence. This is the triple point for a PVT system. In this case $k = 1$, $\pi = 3$, and $f = 2 + 1 - 3 = 0$. Therefore there is no

independent variable. The triple point is a fixed point, no other point can have three-phase coexistence.

7.12 Mixing of Ideal Gases

In this section we discuss the properties of a homogeneous mixture of several different ideal gases. Suppose there are k different ideal gases, and n_j denotes the number of kilomoles of the jth gas ($j = 1, 2, \cdots, k$). All the gas molecules are contained in the same container of volume V. The temperature of the whole system is T, and all the gas molecules obey the ideal gas equation of state, as if there were only one kind of molecule occupying the entire volume V. Therefore

$$P_j = \frac{n_j RT}{V}, \quad j = 1, 2, \cdots, k, \tag{7.45}$$

where P_j is the pressure of the jth gas if there is no other kind of gas in the volume V. This equation still holds even when there are other kinds of molecules in the volume V. In the latter case P_j is the pressure on the wall of the container due to the jth type of gas; it is called the *partial pressure* of the jth gas. We add all the P_j's from $j = 1$ to $j = k$,

$$P = \frac{nRT}{V}, \quad n = n_1 + n_2 + \cdots + n_k, \tag{7.46}$$

where $P = P_1 + P_2 + \cdots + P_k$ is the sum of all the partial pressures, and is the total pressure on the wall of the container. From Eqs. (7.45)–(7.46) we have

$$P_j = x_j P, \quad x_j = \frac{n_j}{n}, \tag{7.47}$$

where x_j is the mole fraction of the jth type of gas. This equation is known as **Dalton's[10] law of partial pressure**. For the jth gas, the partial pressure P_j is proportional to the mole fraction x_j.

We are now going to study what is the difference between the specific Gibbs free energy for a one-component ideal gas $g(T, P)$ and that of an ideal gas mixture $g_j(T, P)$. From Eq. (5.11), $g = u - Ts + Pv = h - Ts$ therefore if we know h and s then we know g. For an ideal gas, h is a function of T only ($h = u + Pv = u(T) + RT$), thus

$$dh = \left(\frac{\partial h}{\partial T}\right)_P dT + \left(\frac{\partial h}{\partial P}\right)_T dP = c_p dT. \tag{7.48}$$

[10]John Dalton, British chemist (1766–1844).

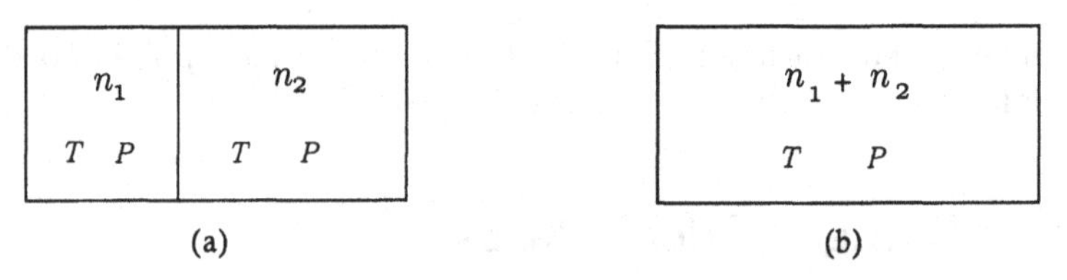

Figure 7.13: Mixing of two ideal gases, which have the same temperature T and pressure P. (a) Before mixing. (b) After mixing.

For most cases, c_p for an ideal gas can be considered as a constant independent of T, then we have

$$h = c_p T + h_0. \tag{7.49}$$

The specific entropy s for a one-component ideal gas as a function of T and P is given in Eq. (4.21):

$$s(T, P) = c_p \ln T - R \ln P + s_0.$$

Thus the specific Gibbs free energy g for a one-component ideal gas is

$$g(T, P) = c_p T - c_p T \ln T + RT \ln P - T s_0 + h_0. \tag{7.50}$$

The above equation can be written as

$$g = RT[\ln P + \phi(T)], \tag{7.51}$$

where $\phi(T)$ is a function which depends on T only.

Next we consider the mixture of two ideal gases. We use 1 and 2 to denote the two different gases. The gases before mixing are shown in Fig. 7.13(a), where gas 1 and gas 2 have the respective number of kilomoles n_1 and n_2. Both gases have the same temperature T and pressure P. Remove the partition between 1 and 2 so that the two gases are mixed together, as shown in Fig. 7.13(b). The total number of kilomoles after mixing is $n = n_1 + n_2$, the total pressure P is the same as the pressure of either gas before mixing. From Dalton's law of partial pressure, the partial pressure for the jth gas after mixing is

$$P_j = x_j P, \quad x_j = \frac{n_j}{n_1 + n_2} \quad (j = 1, 2).$$

From Eq. (5.51), the total Gibbs free energy before mixing G_i is

$$G_i(T, P, n_1, n_2) = n_1 g_{1i} + n_2 g_{2i},$$

where g_{1i} and g_{2i} are the specific Gibbs free energies of gas 1 and gas 2 before mixing. From Eq. (7.51) we have

$$g_{1i} = RT[\ln P + \phi_1(T)], \quad g_{2i} = RT[\ln P + \phi_2(T)], \tag{7.52}$$

$$G_i = n_1 RT[\ln P + \phi_1(T)] + n_2 RT[\ln P + \phi_2(T)]. \tag{7.53}$$

After mixing, both gas 1 and gas 2 occupy the whole container, as shown in Fig. 7.13(b). The gases have the same temperature T as before mixing, but the pressures of the gases are reduced from P to the respective partial pressures P_1 and P_2. Both gas 1 and gas 2 are still ideal gases, therefore the total Gibbs free energy after mixing G_f is

$$G_f = n_1 g_{1f} + n_2 g_{2f}, \tag{7.54}$$

where g_{1f} and g_{2f} are the specific Gibbs free energies of gas 1 and gas 2 after mixing. From Eq. (7.51) and $P_1 = x_1 P$, $P_2 = x_2 P$, we have

$$\begin{aligned} g_{1f} &= RT[\ln P_1 + \phi_1(T)] = g_{1i} + RT \ln x_1, \\ g_{2f} &= RT[\ln P_2 + \phi_2(T)] = g_{2i} + RT \ln x_2. \end{aligned} \tag{7.55}$$

The difference of the total Gibbs free energy G before and after mixing can now be evaluated,

$$\begin{aligned} \Delta G = G_f - G_i &= n_1(g_{1f} - g_{1i}) + n_2(g_{2f} - g_{2i}) \\ &= RT(n_1 \ln x_1 + n_2 \ln x_2) \\ &= nRT(x_1 \ln x_1 + x_2 \ln x_2) < 0. \end{aligned} \tag{7.56}$$

The final inequality follows because both x_1 and x_2 are positive numbers but smaller than 1, and $\ln x < 0$ for $0 < x < 1$. This means that G has decreased after mixing, and thus this is an irreversible process. The mixed state is more stable than the separated state. We can also understand this phenomenon from the consideration of entropy. The entropy change between before and after mixing is

$$\begin{aligned} \Delta S &= -\left[\frac{\partial(\Delta G)}{\partial T}\right]_P \\ &= -nR(x_1 \ln x_1 + x_2 \ln x_2) > 0. \end{aligned} \tag{7.57}$$

Since $\Delta S > 0$ this is an irreversible process, as we have proved from the consideration of the change of the Gibbs free energy. The above result for ΔS is the entropy increase due to mixing of two ideal gases, and is called the **entropy of mixing.** If we take $x_1 = x_2 = 1/2$, then we get a result which is identical to that obtained in Eq. (3.46).

7.13 Binary Mixtures

In this section we will study the problem of the phase equilibrium of a binary mixture, that is a system composed of two different kinds of molecules. If the mixture is in the gas phase (excluding vapor-liquid coexistence), then it can only be mixed homogeneously, no matter whether the gases are ideal or non-ideal. Then the system can have only one phase, the gas phase, which is relatively simple, so we will not discuss it any further. In this section we are mainly interested in the mixture of two liquids. There are two possibilities for these kinds of mixture: (1) The two liquids can be mixed homogeneously for any mole fraction of the components. This is called a **miscible mixture**. An example is the mixture of water and alcohol. In this case, the situation is rather like the mixture of two gases. (2) Only when the mole fraction of one of the components is less than a certain number, can the mixture be mixed homogeneously. When the mole fraction exceeds that number, then the mixture cannot be mixed homogeneously. The system will separate into two spatial regions, such that the two regions have two different mole fractions. These two regions are considered as two different phases; this phenomenon is called a **phase separation**,[11] and the system is said to have has a **solubility gap**. Such a system is called an **immiscible mixture**. An example is the mixture of water and gasoline. These two liquids cannot be mixed freely; a phase separation occurs. The mass density of gasoline is smaller than that of water, therefore gasoline floats on the surface of water. In fact the gasoline that floats on the surface of water is not pure, it is a *homogeneous* mixture of water and gasoline with a very small mole fraction of water. The water under the gasoline is not pure water either, it is also a *homogeneous* mixture of water and gasoline with a very small mole fraction of gasoline. This is to say that in general any two liquids will be miscible when the mole fraction of one of the components is small enough. Only when the mole fraction exceeds a certain number, will phase separation occur. This is the main topic we will discuss in this section.

1. Homogeneous mixtures

First we discuss the case where the two liquids can be mixed to form a homogeneous mixture. Suppose the temperature is T and the pressure is P, and there are n_1 kilomoles of component 1 and n_2 kilomoles of

[11] Phase separation may also be found in solid mixtures.

component 2. The total number of kilomoles is $n = n_1 + n_2$. The total Gibbs free energy G and its derivative dG of the system are (Eqs. (5.48) and (5.30)):

$$G(T, P, n_1, n_2) = n_1\mu_1 + n_2\mu_2, \tag{7.58}$$

$$dG = -SdT + VdP + \mu_1 dn_1 + \mu_2 dn_2. \tag{7.59}$$

We define the specific Gibbs free energy g as

$$g \equiv \frac{G}{n} = x_1\mu_1 + x_2\mu_2, \tag{7.60}$$

where $x_i = n_i/n$ is the mole fraction of the ith component ($i = 1, 2$), $x_1 + x_2 = 1$. When the system consists of only one component, g is a function of T and P, but when the system consists of two components, there is one additional independent variable. We choose x_2 as the third independent variable. In order to simplify the notation we rename x_2 as x, i.e., x is the mole fraction of the second component, and $x_1 = 1 - x$.

The specific Gibbs free energy g is a function of T, P and x, thus

$$g = g(T, P, x), \quad x = \frac{n_2}{n_1 + n_2}. \tag{7.61}$$

From Eq. (7.59) we have

$$\begin{aligned}
\mu_1 &= \left(\frac{\partial G}{\partial n_1}\right)_{T,P,n_2} = \frac{\partial}{\partial n_1}(ng) \\
&= g + n\left(\frac{\partial g}{\partial x}\right)_{T,P}\left(\frac{\partial x}{\partial n_1}\right)_{n_2} \\
&= g - x\left(\frac{\partial g}{\partial x}\right)_{T,P},
\end{aligned} \tag{7.62}$$

$$\mu_2 = \left(\frac{\partial G}{\partial n_2}\right)_{T,P,n_1} = g + (1 - x)\left(\frac{\partial g}{\partial x}\right)_{T,P}, \tag{7.63}$$

where we have used

$$\left(\frac{\partial x}{\partial n_1}\right)_{n_2} = -\frac{n_2}{n^2} = -\frac{1}{n}x,$$

and

$$\left(\frac{\partial x}{\partial n_2}\right)_{n_1} = \frac{1}{n} - \frac{n_2}{n^2} = \frac{1}{n}(1 - x).$$

Equations (7.62)–(7.63) are useful formulas for a homogeneous binary mixture; they will be used in the latter discussions. From these two equations we obtain

$$\left(\frac{\partial \mu_1}{\partial x}\right)_{T,P} = -x\left(\frac{\partial^2 g}{\partial x^2}\right)_{T,P}, \tag{7.64}$$

$$\left(\frac{\partial \mu_2}{\partial x}\right)_{T,P} = (1-x)\left(\frac{\partial^2 g}{\partial x^2}\right)_{T,P}. \tag{7.65}$$

2. Phase separation

When phase separation occurs, it means that two homogeneous mixtures with different mole fractions coexist. We consider *the two homogeneous mixtures with different mole fractions as two different phases.* These two different phases occupy different spatial regions and they are *spatially separated.* We use phase I to denote the homogeneous mixture with $x = x_{\text{I}}$, and phase II to denote the homogeneous mixture with $x = x_{\text{II}}$; in general $x_{\text{I}} \neq x_{\text{II}}$. If $x_{\text{I}} < x_{\text{II}}$, then we say that phase I is the component 1-rich phase, and II the component 2-rich phase. Suppose the mass density for a pure component 1 liquid is ρ_1, and that for a component 2 liquid is ρ_2. If $\rho_2 > \rho_1$ then the mass density of phase II will be larger than that of phase I. In phase separation, because the mixture of phase II is heavier than the mixture of phase I, therefore phase II will sink to the bottom of the container, and phase I will float forming the upper layer of the container, as shown in Fig. 7.14. This is the phenomenon of phase separation.

The problem now is how to determine x_{I} and x_{II}? Suppose the temperature is T and the pressure is P, the conditions that phase I and phase II coexist are

$$\mu_1(T,P,x_{\text{I}}) = \mu_1(T,P,x_{\text{II}}), \tag{7.66}$$

and

$$\mu_2(T,P,x_{\text{I}}) = \mu_2(T,P,x_{\text{II}}), \tag{7.67}$$

where $\mu_i(T,P,x_{\text{I}})$ and $\mu_i(T,P,x_{\text{II}})$ are, respectively, the chemical potential for the ith component ($i = 1,2$) in phase I and phase II. In the following we drop the variables T and P in the functional arguments, because they are kept constant in the analysis. Note that in both phases I

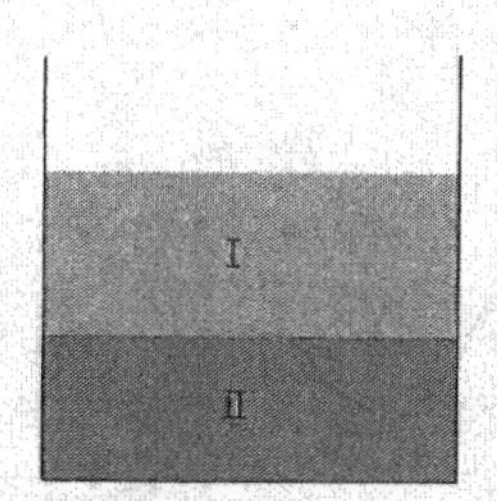

Figure 7.14: The phase separation of a liquid binary mixture. The mass density of phase II is larger than that of phase I, therefore the phase I mixture floats on top of the phase II mixture.

and II, the system is a homogeneous mixture, therefore Eqs. (7.62)–(7.63) are applicable. The coexistence conditions Eqs. (7.66)–(7.67) become

$$g(x_{\rm I}) - x_{\rm I}\left(\frac{\partial g}{\partial x}\right)^{\rm I}_{T,P} = g(x_{\rm II}) - x_{\rm II}\left(\frac{\partial g}{\partial x}\right)^{\rm II}_{T,P}, \tag{7.68}$$

$$g(x_{\rm I}) + (1 - x_{\rm I})\left(\frac{\partial g}{\partial x}\right)^{\rm I}_{T,P} = g(x_{\rm II}) + (1 - x_{\rm II})\left(\frac{\partial g}{\partial x}\right)^{\rm II}_{T,P}. \tag{7.69}$$

In these equations, the superscripts I and II of the partial differentiations indicate that the quantities are evaluated at $x = x_{\rm I}$ and at $x = x_{\rm II}$, respectively. Subtracting Eq. (7.69) from Eq. (7.68) we have

$$\left.\left(\frac{\partial g}{\partial x}\right)_{T,P}\right|_{x=x_{\rm I}} = \left.\left(\frac{\partial g}{\partial x}\right)_{T,P}\right|_{x=x_{\rm II}}. \tag{7.70}$$

Substituting this equation into Eq. (7.68) or (7.69), we obtain

$$\left.\left(\frac{\partial g}{\partial x}\right)_{T,P}\right|_{x=x_{\rm I}} = \frac{g(x_{\rm II}) - g(x_{\rm I})}{x_{\rm II} - x_{\rm I}}. \tag{7.71}$$

We draw the g-x (T, P constant) plot for a *homogeneous binary mixture* in Fig. 7.15. At $x = 0$, the mixture is 100% component 1 liquid (there is no component 2), therefore $g = \mu_1^0$, where the superscript 0 denotes that it is a quantity for a one-component system. At $x = 1$, the mixture is 100% component 2 liquid (no component 1), and $g = \mu_2^0$. On

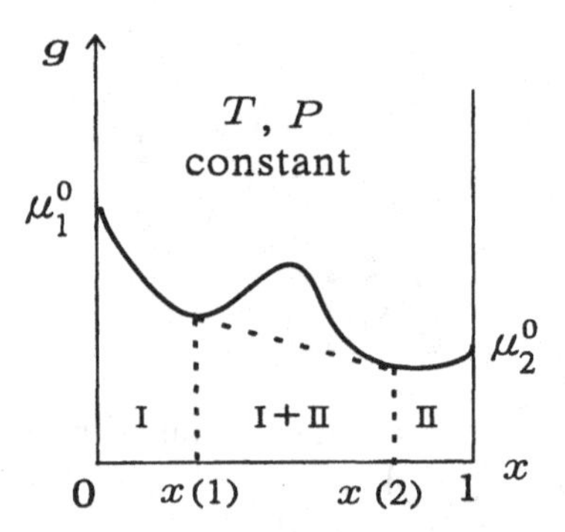

Figure 7.15: The g-x plot at constant T and P for a homogeneous binary mixture. If there are two concave-up sections, then there exists a straight line which will be tangent to the g curve at two points, $x = x(1)$ and $x = x(2)$. We may identify $x(1) = x_{\rm I}$ and $x(2) = x_{\rm II}$. Phase separation occurs for $x_{\rm I} < x < x_{\rm II}$. The mixture is stable and homogeneous for both $x < x_{\rm I}$ and $x > x_{\rm II}$.

the g-x plot, if the g curve has only one *concave-up* section, then phase separation will not occur.[12] This means that the system will be mixed homogeneously for any mole fraction of x $(0 \le x \le 1)$. If the curve of g vs. x has *two or more* concave upward sections, then phase separation will occur. In Fig. 7.15 there are two regions where the g curve is concave upward. This can be used as an example to explain why phase separation will occur, and how one can determine the mole fraction of phase I and II, i.e., $x_{\rm I}$ and $x_{\rm II}$, by a graphical method. The explanation is given in the following.

We see that Eqs. (7.70) and (7.71) can be interpreted geometrically. Equation (7.70) indicates that the slope of g at $x = x_{\rm I}$ and at $x = x_{\rm II}$ must be equal to each other. The meaning of Eq. (7.71) is that the slope of the straight line connecting the two points $(x_{\rm I}, g(x_{\rm I}))$ and $(x_{\rm II}, g(x_{\rm II}))$ is equal to the slope of g at $x = x_{\rm I}$, which is also equal to the slope of g at $x = x_{\rm II}$ by Eq. (7.70). Therefore the graphical method for determining $x_{\rm I}$ and $x_{\rm II}$ is to draw a straight line, which is tangent to the g vs. x curve at two different points, say at $x = x(1)$ and at $x = x(2)$ with $x(1) < x(2)$, as shown in Fig. 7.15. It is easy to see that we may identify

[12]This case will be discussed in Sec. 7.15. The case where there is no concave-up section can occur only at $T = 0$, which will also be discussed in p. 211.

$x_{\rm I} = x(1)$ and $x_{\rm II} = x(2)$, because $x(1)$ and $x(2)$ satisfy the coexistence conditions Eqs. (7.70)–(7.71).

Now we will study the behavior of the mixture when the mole fraction x of component 2 is increased from 0 to 1. Referring to Fig. 7.15, there are three regions of x in which the mixture behaves in a quite different fashion.

(a) For $0 \leq x \leq x_{\rm I}$: single phase (homogeneous mixture).
In this region every mole fraction x corresponds to one g value only. Therefore there is only one phase, we may called it the component 1-rich phase, since in this region component 1 dominates the composition.

(b) For $x_{\rm II} \leq x \leq 1$: single phase (homogeneous mixture).
In this region, the situation is the same as in (a), each x corresponds to one g value. It is a single phase region, but now it is the component 2-rich phase.

(c) For $x_{\rm I} < x < x_{\rm II}$: two-phase coexistence (phase separation).
The most important feature in this region is that, in addition to the original g curve, there is one more possible g curve for the system. The new g curve is a straight line connecting $(x_{\rm I}, g(x_{\rm I}))$ and $(x_{\rm II}, g(x_{\rm II}))$, which is the common tangent of the original g curve at $x_{\rm I}$ and $x_{\rm II}$. We will show in a moment that the g value of the common tangent is the specific Gibbs free energy for the case of phase I-II coexistence. Therefore in this region of x, each x will correspond to two possible g values. One is the original g value, the concave-down section in Fig. 7.15, which is the g value for the homogeneous mixture. The other one is the common tangent, shown in Fig. 7.15, which is the g value for the two-phase coexistence. It is easy to see from Fig. 7.15 that in the region $x_{\rm I} < x < x_{\rm II}$ the g value of the common tangent is always less than the g value of the homogeneous mixture. Therefore the state corresponding to the common tangent, i.e., phase separation, is the stable phase. The reason is that at constant T and P, the state with a smaller g is the stable state. Therefore the *homogeneous mixture* is unstable against phase separation. In the phase separate state, two homogeneous mixtures with different mole fractions, one with $x = x_{\rm I}$ and the other with $x = x_{\rm II}$ coexist; these are separated spatially as shown in Fig. 7.14.

Now we have to prove that the common tangent line is the g value of the phase separation state. For a mole fraction x in the range $x_{\rm I} < x < x_{\rm II}$, we assume that the mole fractions for phase I and II are $y_{\rm I}$ and $y_{\rm II}$, respectively, with $y_{\rm I} + y_{\rm II} = 1$. Therefore the total g value for phase I-II coexistence is

$$g = y_{\rm I} g(x_{\rm I}) + y_{\rm II} g(x_{\rm II}).$$

We may choose

$$y_{\rm I} = \frac{x_{\rm II} - x}{x_{\rm II} - x_{\rm I}}, \quad y_{\rm II} = \frac{x - x_{\rm I}}{x_{\rm II} - x_{\rm I}},$$

then the total g value for phase I-II coexistence is

$$\begin{aligned} g &= \left(\frac{x_{\rm II} - x}{x_{\rm II} - x_{\rm I}}\right) g(x_{\rm I}) + \left(\frac{x - x_{\rm I}}{x_{\rm II} - x_{\rm I}}\right) g(x_{\rm II}) \\ &= \frac{g(x_{\rm II}) - g(x_{\rm I})}{x_{\rm II} - x_{\rm I}} (x - x_{\rm I}) + g(x_{\rm I}). \end{aligned} \tag{7.72}$$

From the second equality of the above equation we know that g vs. x is a straight line, which passes the point $(x_{\rm I}, g(x_{\rm I}))$ with a slope $[g(x_{\rm II}) - g(x_{\rm I})]/(x_{\rm II} - x_{\rm I})$. The slope of the g line is the same as the common tangent as expressed in Eq. (7.71). This proves that the common tangent line is the g line for the state of phase separation.

The main feature of the graphical method, which we introduced above, is to draw a common tangent to the g vs. x curve (with T and P constant). There are two points which are worth mentioning in regards to this method. (1) If there is no straight line which can be a tangent for g at two different points, then it means that phase separation will not occur.[13] (2) The common tangent must lie below the g curve for the homogeneous mixture for a phase separation to occur. If the common tangent lies above the g curve of the homogeneous mixture, then we do not draw this common tangent, it is not the one we want. The reason is simple, because the g value of the common tangent is greater than the g value of the homogeneous mixture. In this case we can usually draw at least two common tangent lines which both lie below the g curve of the homogeneous mixture, because there will be three or more concave-up sections for the g curve. Therefore the mixture may have three or more phases.

[13]However there is an exception at $T = 0$. See p. 211 for the explanation.

Obviously the shape of the g-x curve depends sensitively on the temperature T (if P is kept constant). When T increases slowly from low temperatures, the values of $x_{\rm I}$ and $x_{\rm II}$ will get closer and closer, until $T = T_c$, then $x_{\rm I} = x_{\rm II}$. For $T > T_c$, $x_{\rm I}$ and $x_{\rm II}$ no longer exist. This means that for $T > T_c$, the g-x curve has only one concave-up section, and no phase separation occurs. We call the temperature T_c the **critical temperature**, because for $T < T_c$ there is a phase separation, and for $T > T_c$ there is no phase separation.

7.14 Ideal Solutions and Regular Solutions

A solution has the same meaning as a mixture. In this section we consider only two-component solutions, therefore it has the same meaning as a binary mixture. However, in a solution, we consider one of the components as being the primary (the **solvent**), and the other component as being the secondary (the **solute**). If the mole fraction of the solute is very small, then it is called a **dilute solution**. In this section we use a solution as an example of a binary mixture to understand the behavior of the mixture, such as when there will be a phase separation, and when there is no phase separation. Before we go on, we would like to introduce two terminologies, the *ideal solution* and the *regular solution*.

We consider a two-component system, where the number of kilomoles of component 1 and component 2 are, respectively, n_1 and n_2. Initially the two components are separated, and there is no interaction between them. The two components are then mixed together. The external temperature T and pressure P are kept unchanged before and after mixing. Before mixing the total Gibbs free energy G_0 of the system is (Eq. (5.48))

$$G_0 = n_1\mu_1^0(T,P) + n_2\mu_2^0(T,P),$$

where μ_1^0 and μ_2^0 are the chemical potentials of component 1 and component 2, respectively, before mixing. After mixing the total Gibbs free energy of the system becomes G; therefore the change in the total Gibbs free energy is ($G = H - TS$)

$$\Delta G = G - G_0 = \Delta H - T\Delta S \quad (T, P \text{ constant}), \tag{7.73}$$

where $\Delta H = H - H_0$ is the change in total enthalpy H from before to after mixing. It is called the **heat of mixing**, because at constant pressure ΔH is the heat absorbed or released. $\Delta S = S - S_0$ is the entropy change from before to after mixing; it is called the *entropy of mixing*.

1. A solution is called an **ideal solution** (or an **ideal mixture**) if the heat of mixing is zero, i.e., $\Delta H = 0$, and the entropy of mixing ΔS is taken to be the entropy of mixing for two ideal gases (Eq. (7.57)). Therefore for an ideal solution, the following two conditions must be satisfied,

$$\begin{aligned} \Delta H &= 0, \quad \text{and} \\ \Delta S &= -n_1 R \ln x_1 - n_2 R \ln x_2, \end{aligned} \tag{7.74}$$

   where x_1 and x_2 are the mole fractions of component 1 and component 2, respectively.

2. A solution is called a **regular solution** if the heat of mixing is not zero, i.e., $\Delta H \neq 0$, and the entropy of mixing ΔS is taken to be the entropy of mixing for two ideal gases (Eq. (7.57)). Therefore for a regular solution the following two conditions must be satisfied,

$$\begin{aligned} \Delta H &\neq 0, \quad \text{and} \\ \Delta S &= -n_1 R \ln x_1 - n_2 R \ln x_2. \end{aligned} \tag{7.75}$$

The thermodynamic properties of an ideal solution are the same as those of an ideal gas mixture, which we have studied in Sec. 7.12. In this section we are interested in the properties of a regular solution. We use a simple model to study under what conditions phase separation may occur. It is obvious that the form of the heat of mixing ΔH plays an important role in the properties of a regular solution. The physical meaning of the heat of mixing is the energy change of the system when some of the solvent molecules are replaced by solute molecules, while the total number of kilomoles is kept constant. This is a measure of the difference between the solvent-solvent interaction energy and the solvent-solute interaction energy. The model we take is

$$\Delta H = \lambda \frac{n_1 n_2}{n} = \lambda n x_1 x_2, \quad n = n_1 + n_2. \tag{7.76}$$

This means that the heat of mixing is proportional to the number of kilomoles of each component, with the proportionality constant λ. We assume the simplest case that λ is a constant, independent of either n_1

or n_2.[14] From Eqs. (7.73), (7.57) and (7.76) the total Gibbs free energy after mixing is

$$G = n_1\mu_1^0 + n_2\mu_2^0 + \lambda n x_1 x_2 + RT(n_1 \ln x_1 + n_2 \ln x_2). \tag{7.77}$$

As in the previous section, we rename x_2 as x ($x_1 = 1 - x$), therefore the specific Gibbs free energy g of the mixture is given by

$$\begin{aligned} g = \frac{G}{n} &= x_1\mu_1^0 + x_2\mu_2^0 + \lambda x_1 x_2 + RT(x_1 \ln x_1 + x_2 \ln x_2) \\ &= (1-x)\mu_1^0 + x\mu_2^0 + \lambda x(1-x) \\ &\quad + RT[(1-x)\ln(1-x) + x \ln x]. \end{aligned} \tag{7.78}$$

We could use the above equation for g to plot the g-x diagram, and use the graphical method introduced in the previous section to study whether there is a phase separation or not. However this is not such a good approach to do it, because the shape of the g vs. x curve depends sensitively on the value of λ (μ_1^0 and μ_2^0 may be considered as constants). Hence, in order to get a general conclusion, we need to plot many diagrams, each for a given value of λ. Therefore it is desirable to find a better technique. In the following we use mathematical analysis with the help of the graphical method.

We noted that for a binary mixture to possess a phase separation, the g vs. x curve must have at least two concave-up sections in order that we can draw a straight line which is a tangent to two points of the curve. We denote the two tangent points as $x_{\rm I}$ and $x_{\rm II}$. Therefore from Eq. (7.78)

$$\begin{aligned} \left(\frac{\partial g}{\partial x}\right)_{T,P}\bigg|_{x=x_{\rm I}} &= \mu_2^0 - \mu_1^0 + RT\left(\ln \frac{x_{\rm I}}{1-x_{\rm I}}\right) + \lambda(1-2x_{\rm I}) \\ &= \mu_2^0 - \mu_1^0 + RT\left(\ln \frac{x_{\rm II}}{1-x_{\rm II}}\right) + \lambda(1-2x_{\rm II}) \\ &= \left(\frac{\partial g}{\partial x}\right)_{T,P}\bigg|_{x=x_{\rm II}} \end{aligned} \tag{7.79}$$

Both the form of ΔH in Eq. (7.76) and that of ΔS in Eq. (7.57) are symmetrical with respect to the exchange of x_1 and x_2, therefore $x_{\rm II} =$

[14]The constant λ may be a function of temperature, but this does not affect our analysis.

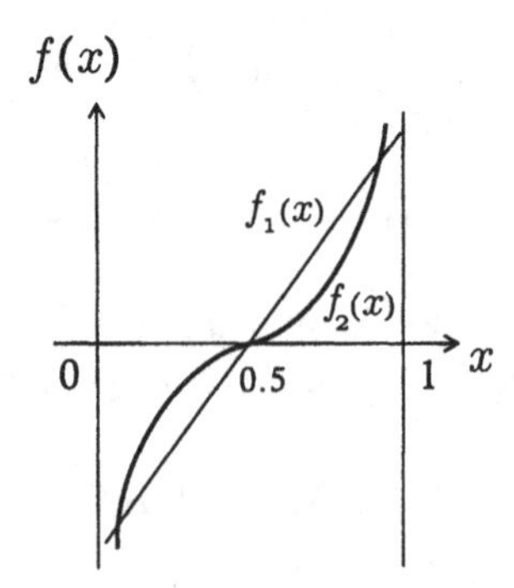

Figure 7.16: Solving $f_1(x) = f_2(x)$ by the graphical method, where f_1 and f_2 are defined in Eqs. (7.81)–(7.82).

$1 - x_{\rm I}$, because $x_1 = 1 - x_2$. Physically we can say that the model we consider is a system in which the roles played by component 1 and 2 are equivalent. Substituting $x_{\rm II} = 1 - x_{\rm I}$ into Eq. (7.79), we have[15]

$$RT\left(\ln\frac{x_{\rm I}}{1-x_{\rm I}}\right) + \lambda(1 - 2x_{\rm I}) = 0. \tag{7.80}$$

Obviously $x_{\rm I} = 1/2$ is a solution, but this is a special point, i.e., $x_{\rm I} = x_{\rm II} = 1/2$. We would like to get more general solutions, which can be obtained by using the graphical method to solve Eq. (7.80). Define

$$f_1(x) = \frac{2\lambda}{RT}\left(x - \frac{1}{2}\right), \tag{7.81}$$

$$f_2(x) = \ln\frac{x}{1-x}. \tag{7.82}$$

Plot f_1 vs. x and f_2 vs. x on the same graph; the intersections of these two curves are the solutions of Eq. (7.80), $x_{\rm I}$. The function $f_1(x)$ is a very simple curve, which is a straight line passing through zero at the point $x = 1/2$ with a slope $2\lambda/RT$. The function $f_2(x)$ is also not

[15]The condition in Eq. (7.80) is equivalent to saying that the slope of the common tangent is $(\partial g/\partial x)_{T,P} = \mu_2^0 - \mu_1^0$. This slope is equal to that of the straight line which connects the two points $(x = 0, g = \mu_1^0)$ and $(x = 1, g = \mu_2^0)$ on the g vs. x curve. From Eq. (7.78) if we take $x_{\rm II} = 1 - x_{\rm I}$, then $[g(x_{\rm II}) - g(x_{\rm I})]/(x_{\rm II} - x_{\rm I}) = \mu_2^0 - \mu_1^0$, therefore the second condition of phase separation, Eq. (7.71) is also satisfied.

complicated. At $x = 1/2$, $f_2 = 0$, when $x \to 0^+$, $f_2 \to -\infty$, and when $x \to 1^-$, $f_2 \to \infty$. It is easy to see that $f_2(x)$ has an inversion symmetry with respect to $x = 1/2$, which is shown in Fig. 7.16. We can see that if the slope of $f_1(x)$ at $x = 1/2$ is greater than that of $f_2(x)$ at $x = 1/2$, then, in addition to $x = 1/2$, Eq. (7.80) will have two other solutions. Therefore we need to calculate the slope of $f_2(x)$ at $x = 1/2$ to determine how many solutions Eq. (7.80) has. The slope is

$$\frac{df_2}{dx} = \frac{1}{x} - \frac{-1}{1-x} = \frac{1}{x(1-x)} = 4, \quad \text{for } x = \frac{1}{2}.$$

Therefore when

$$\frac{2\lambda}{RT} > 4 \text{ or } T < \frac{\lambda}{2R}, \tag{7.83}$$

Eq. (7.80) has three solutions. It is not necessary that all these three solutions satisfy the condition of phase separation. We have to check whether the obtained common tangent lies below or above the g curve. We found that the tangent at $x = 1/2$ lies above the g curve, therefore it is not the solution we want. This is because when the inequality (7.83) is satisfied, we have

$$\begin{aligned} \left(\frac{\partial^2 g}{\partial x^2}\right)_{T,P} &= \frac{RT}{x(1-x)} - 2\lambda \\ &= 4RT - 2\lambda < 0, \quad \text{for } x = \frac{1}{2}. \end{aligned} \tag{7.84}$$

Therefore the point $x = 1/2$ is in the concave-down section of the g curve, whose tangent (at $x = 1/2$) will lie above the curve. This is not the solution we are looking for. Therefore there are only two unequal solutions which will satisfy the condition for phase separation,[16] which are x_{I} and x_{II}. Inequality (7.83) is therefore the criterion for the existence of phase separation.

Before using inequality (7.83) to discuss the existence of phase separation, we would like to give a more general look at this problem. Whether the solute molecules can mix well with the solvent molecules depends on whether the change in the specific Gibbs free energy Δg is negative or positive. If $\Delta g < 0$, then the different molecules will mix well, and if

[16]The solutions x_{I} ($< 1/2$) and x_{II} ($> 1/2$) must be in the concave-up sections of the g curve, whose tangents will lie below the g curve. The reason is simple, because at $x = 1/2$ the g curve is concave downward, so it must be concave upward at x_{I} and x_{II}.

$\Delta g > 0$, they will not mix well, but will tend to separate. For the model discussed, we have (from Eqs. (7.73), (7.75) and (7.76)),

$$\Delta g = g - g_0 = \Delta h - T\Delta s, \tag{7.85}$$

$$\Delta h = \lambda x(1-x),$$

$$\Delta s = -R[x \ln x + (1-x)\ln(1-x)].$$

We see that if $T\Delta s > \Delta h$, then $\Delta g < 0$, and the solute tends to mix well with the solvent. Therefore for high enough temperatures the solution will mix well, becoming a homogeneous mixture, no matter what Δh is. However the condition $T\Delta s > \Delta h$ also depends on the mole fraction x of the solute. For the model we used, $\Delta h = \Delta s = 0$ at $x = 0$,[17] and $d\Delta h/dx = \lambda$ (at $x = 0$) is finite, but $d\Delta s/dx = R\ln[(1-x)/x] \to \infty$ as $x \to 0$; therefore the solution tends to mix well for small x. The same property will be found for x close to 1, because our model is symmetric under the interchange of x and $1-x$. In fact, the property that the two different kinds of molecules will mix well when x is close to 0 or 1 is quite general; it will hold for any realistic model.

Now we use inequality (7.83) to discuss the behavior of the solution.

1. For $\lambda \leq 0$, there is no phase separation, because $\Delta g < 0$ holds for any T and any x. The solution tends to mix well for any T and any x. Physically, the condition $\lambda < 0$ implies that the interaction energy between the different kinds of molecules is smaller than that of the same kind of molecules. Therefore there is a tendency for different kinds of molecules to get closer to each other, and no phase separation will occur. For the case $\lambda = 0$, the solution becomes an ideal solution, which behaves like an ideal gas mixture, thus there is no phase separation. The stable state for the solution is a homogeneous mixture for any x and any T.

2. For $\lambda > 0$, we can define a critical temperature T_c,

$$T_c = \frac{\lambda}{2R}. \tag{7.86}$$

The temperature T_c is proportional to the magnitude of λ. The larger λ, the higher the critical temperature T_c. The physical significance of the critical temperature T_c is the following.

[17] Any model should satisfy this condition, because $x = 0$ implies a one-component system.

(a) For temperature $T > T_c$, inequality (7.83) is not satisfied. Therefore there is no phase separation. At this high temperature range, the energy due to the entropy of mixing term, $T\Delta s$, is larger than the heat of mixing term Δh, therefore $\Delta g < 0$. Hence the solution tends to mix well, and the stable state is a homogeneous solution for all x.

(b) For temperature $T < T_c$, inequality (7.83) is satisfied, and phase separation occurs. At these low temperatures, the energy due to the entropy of mixing term, $T\Delta s$, is not large enough to overcome the increased mixing energy Δh. Therefore the molecules tend not to mix with the other kind of molecules. However, as we have pointed out above, homogeneous mixing always occurs for x close to 0 and 1, except for $T = 0$. For $T \neq 0$, phase separation occurs only for x close to 1/2, i.e., in the range $x_{\mathrm{I}} < x < x_{\mathrm{II}}$.

(c) For temperature $T = T_c$, $x_{\mathrm{I}} = x_{\mathrm{II}}$. Phase I and phase II become identical at this temperature. For $T > T_c$ and $T < T_c$ the solution behaves in a quite different fashion, but the change of behavior is continuous at $T = T_c$. The temperature T_c is therefore called the **critical temperature.**

7.15 Phase Diagrams for Binary Mixtures

In this section we use the graphical method to explain the possible phase diagrams that a binary mixture may have. The diagrams we plot here are only sketchy ones; they may not be scaled correctly, but the main features of the curves are correct. For a binary mixture, in addition to T and P, there is a third independent variable x, the mole fraction of one of the components. In addition to vapor, liquid, and solid phases, a binary mixture may exhibit phase separation as we have studied in the previous section; therefore the phase diagram for a binary mixture may be much more complicated than a one-component PVT system. In most cases the phase diagram is a T-x plot, with the pressure P as an adjustable constant, which will be taken to be 1 atm in our following discussions. We will discuss the following four cases, using the term *state* to distinguish the vapor, liquid and solid phases. (1) one *state*, no phase separation; (2) one *state*, with phase separation; (3) two-*state* coexistence, no phase separation for both *states*; and (4) two-*state* coexistence, and there is a phase separation in one of the *states*.

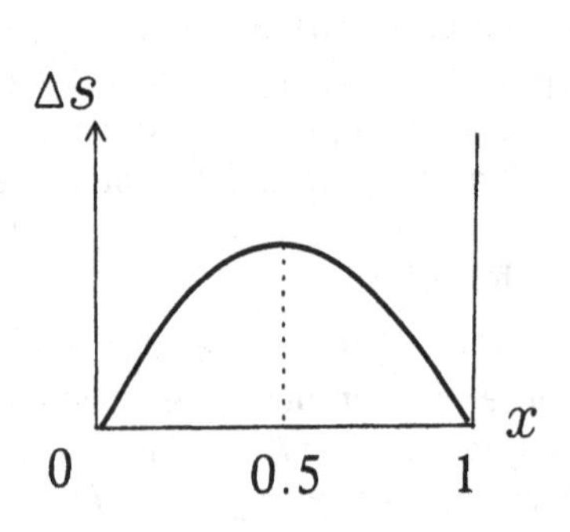

Figure 7.17: Entropy of mixing Δs vs. x for an ideal solution.

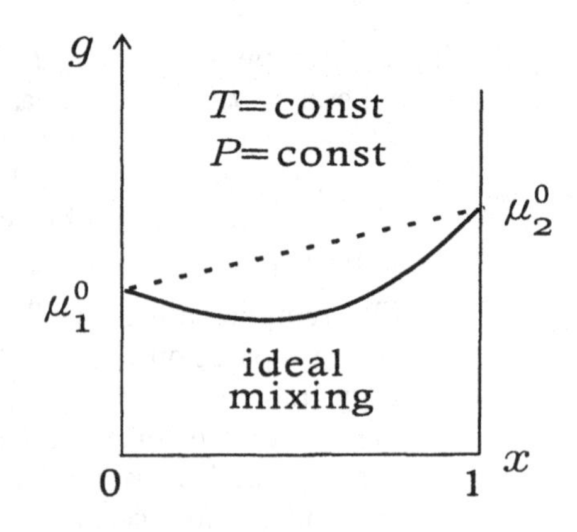

Figure 7.18: The g vs. x plot for an ideal solution.

1. One *state* and no phase separation

This is a miscible mixture, such that the system will mix homogeneously for any mole fraction x. Examples are the mixing of two gases, ideal solutions, and regular solutions with a negative heat of mixing. We use an ideal solution as an example in the following discussion. An ideal solution satisfies the conditions of Eq. (7.74); therefore the specific Gibbs free energy g of the homogeneous mixture is (cf. Eq. (7.78) with $\lambda = 0$)

$$g = (1 - x)\mu_1^0 + x\mu_2^0 - T\Delta s,$$

where μ_1^0 and μ_2^0 are constants for constant T and P, and Δs is the specific entropy of mixing for two ideal gases,

$$\Delta s = \frac{\Delta S}{n} = -R[(1 - x)\ln(1 - x) + x\ln x]. \tag{7.87}$$

The diagrams of Δs vs. x and g vs. x are plotted in Figs. 7.17 and 7.18. We find that the g curve is everywhere concave upward for all x between 0 and 1. Therefore the homogeneous mixture state is the stable state (g is minimum) for all mole fractions x ($0 \le x \le 1$). There is no phase separation. Therefore in the T-x diagram, there is only one phase and we have no curve to draw.

2. One *state* and a phase separation

This is an immiscible mixture, and it may happen in a liquid or in a solid, but not in a gas mixture. We use a regular solution with a positive specific heat of mixing ($\Delta h > 0$) as an example. The change in specific Gibbs free energy from before to after the mixing is, by Eq. (7.85),

$$\Delta g = g - g_0 = \Delta h - T\Delta s,$$

where Δs is the specific entropy of mixing for two ideal gases (Eq. (7.87)). We assume that $\Delta h > 0$ and that it is independent of the temperature. The Δh-x plot is shown in Fig. 7.19(a), where $\Delta h = 0$ at $x = 0$ and $x = 1$. The shape of the g vs. x curve (T, P constant) depends sensitively on the value of the temperature T, while the pressure P is still kept fixed. We plot several isotherms of the g-x curves at different values of T in Figs. 7.19(b)–(c). We plot the $T = 0$ isotherm in Fig. 7.19(b), which shows that the $g(x)$ curve is always concave downward for all x ($0 \leq x \leq 1$), and there is no concave upward section. At first sight one might think that there is no phase separation, because one cannot draw a straight line which can be tangent to g at two points. However this conclusion is wrong, because the straight line which connects the points $(0, g(0))$ and $(1, g(1))$, is the g curve for the coexistence of phases I and II with $x_{\mathrm{I}} = 0$ and $x_{\mathrm{II}} = 1$. The g value of this line is smaller than that of the homogeneous mixture for all x ($0 < x < 1$). Therefore phase separation is the stable state while the homogeneous mixture is unstable. Thus if $\Delta h > 0$, the mixture will be separated completely at $T = 0$, i.e., the two components do not mix at all and remain as pure systems, with $x = 0$ and $x = 1$, respectively. However when T increases the situation will change greatly. We plot four isotherms of the g-x curves (P constant) at temperatures T_1, T_2, T_3, and T_4 in Fig. 7.19(c), where $0 < T_1 < T_2 < T_3 < T_4$. The important features of the isotherms are the following.

(a) The temperature T_1 is only slightly greater than 0, but the shape of the g curve is significantly different from that of the $T = 0$ curve, especially near $x = 0$ and $x = 1$. This is because $d\Delta h/dx$ =constant at $x = 0$ and $x = 1$, but $(\partial \Delta s/\partial x)_{T,P} = R\ln[(1-x)/x]$ which approaches ∞ at $x = 0^+$ and $dg/dx \to -\infty$. On the other hand at $x = 1^-$, $dg/dx \to \infty$, therefore the g curve at $x = 0^+$ and $x = 1^-$ are both concave upward, and the conditions of phase separation, Eqs. (7.70)–(7.71), can be satisfied with $x_{\mathrm{I}} > 0$ and $x_{\mathrm{II}} < 1$. This means that both phases I and II are no longer purely component 1 or purely component 2, there is a slight homogeneous mixing of the two components.

(b) The temperature T_2 is a temperature greater than T_1, but less than the critical temperature. As T increases, the contribution of the

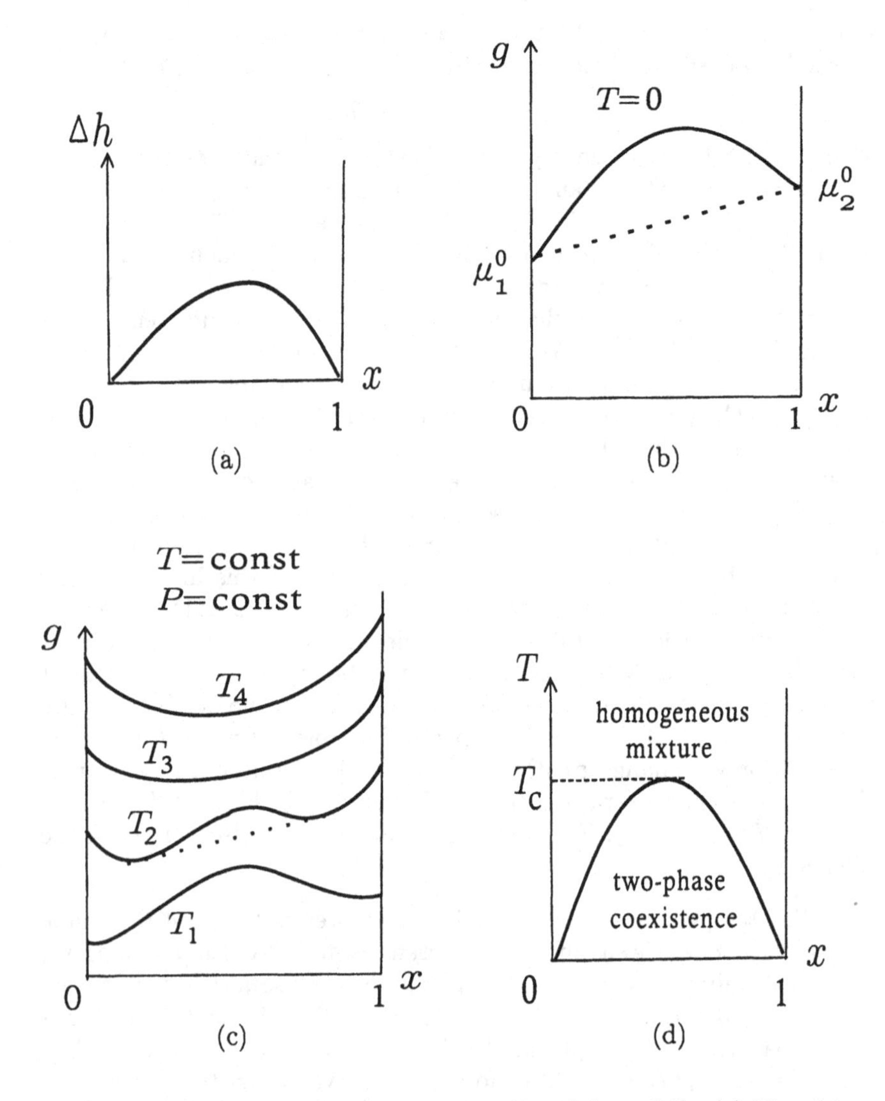

Figure 7.19: Diagrams for an immiscible liquid (or solid). (a) The Δh-x plot, where $\Delta h > 0$ is the heat of mixing and x is the mole fraction. (b) The g-x plot for $T = 0$ (P constant). (c) The g-x plot (P constant). Each curve is an isothermal curve, with $0 < T_1 < T_2 < T_3 < T_4$, and $T_3 = T_c$ is the critical temperature. (d) The phase diagram T-x (P constant).

entropy of mixing, $-T\Delta s < 0$, to the specific Gibbs free energy g (Eq. (7.85)) becomes more significant, and the concave-up sections become wider and deeper. Phase separation occurs at a larger mole fraction of x_{I} and a smaller mole fraction of x_{II}. This means that phase I is a homogeneous mixture with a larger mole fraction of component 2, and in phase II there is a larger mole fraction of component 1.

(c) The temperature T_3 is the critical temperature, i.e., $T_3 = T_c$ ($> T_2$). At this temperature the two concave-up sections merge to become one concave-up section, thus $x_{\text{I}} = x_{\text{II}}$, therefore phase I and phase II become identical. The important characteristic of T_c is that for temperatures $T < T_c$, there is a phase separation; for temperatures $T > T_c$ there is no phase separation; the mixture can be mixed homogeneously for any mole fraction x.

(d) For temperature $T_4 > T_c$, there is only one concave-up section and no phase separation. The mixture can be mixed homogeneously for any mole fraction x.

We combine the information for $T = 0$, T_1, T_2, T_3, and T_4, and plot the phase diagram T-x in Fig. 7.19(d). For every temperature $T < T_c$, there are three regions corresponding to different equilibrium stable states of the system. The two regions, one with x close to 0, the other with x close to 1, are regions of homogeneous mixtures. The region with smaller x is the mixture rich in component 1, and the region with larger x is the mixture rich in component 2. The region in the middle is the region of two-phase coexistence, i.e., the phase separation region. For the temperatures $T > T_c$ there is only one region, which is the region of a homogeneous mixture.

3. Two-*state* coexistence and no phase separation

The two coexistent *states* may be any two of the vapor, liquid, and solid *states*. We take a mixture of two liquids which coexist with its vapor as an example. Suppose the boiling temperatures for liquid 1 and liquid 2 are, respectively, T_1 and T_2 (P=1 atm), and assuming $T_2 > T_1$. In Fig. 7.20(a) we draw five different isotherms on the g-x diagram (P=1 atm), where the solid and the dashed curves denote, respectively, the specific Gibbs free energy for the ***homogeneous liquid*** g_l and that of the vapor g_{v}. Because there is no phase separation, all the g_l and the g_{v} curves are concave upward curves. Five different temperatures T are chosen in the following values or range (from high T to low T): (1) $T > T_2$, (2) $T = T_2$, (3) $T_1 < T < T_2$, (4) $T = T_1$, and (5) $T < T_1$. We explain briefly the relation between the g_l and the g_v curves at each of the five temperatures.

(a) At the temperatures $T > T_2$, both components are in a stable vapor phase, therefore for any x, g_v is smaller than g_l. The system is a homogeneous vapor.

(b) As the temperature T decreases, g_l and g_v will slowly get closer. When $T = T_2$, g_l and g_v will have the same value at $x = 1$, because the vapor and liquid coexist for pure component 2 ($x = 1$) at this temperature. However for the region with the mole fraction $x < 1$, we have $g_l > g_v$, which implies that the vapor phase is stable in this region.

(c) When the temperature T is decreased further to the range $T_1 < T < T_2$, we have $g_l < g_v$ near $x = 1$, and $g_l > g_v$ near $x = 0$. In this case we can draw a common tangent line to g_l and g_v, which implies that there is liquid-vapor coexistence. The same analysis used to explain the existence of phase separation for a binary mixture can also be applied here. In the coexistence region, the vapor and the liquid are homogeneous mixtures, but they have different mole fractions.

(d) When the temperature T is lowered to $T = T_1$, g_l and g_v will have the same value at $x = 0$. At this temperature pure liquid 1 coexists with its vapor, but in the region with $x \neq 0$, $g_l < g_v$, the system is a homogeneous liquid mixture for $0 < x \leq 1$.

(e) When the temperature T is lowered to $T < T_1$, both components condense to the liquid phase. Therefore for any x ($0 \leq x \leq 1$), g_v is greater than g_l. The system is a homogeneous liquid.

The results of the above analysis are plotted on the T-x phase diagram in Fig. 7.20(b), in which the region denoted by v+l is the region of vapor-liquid coexistence. In this region, vapor and liquid coexist but with different mole fractions. For example at $T = T_A = T_B$, the mole fraction of the vapor phase is x_B, and that of the liquid phase is x_A. How x_A and x_B are determined is explained in the figure.

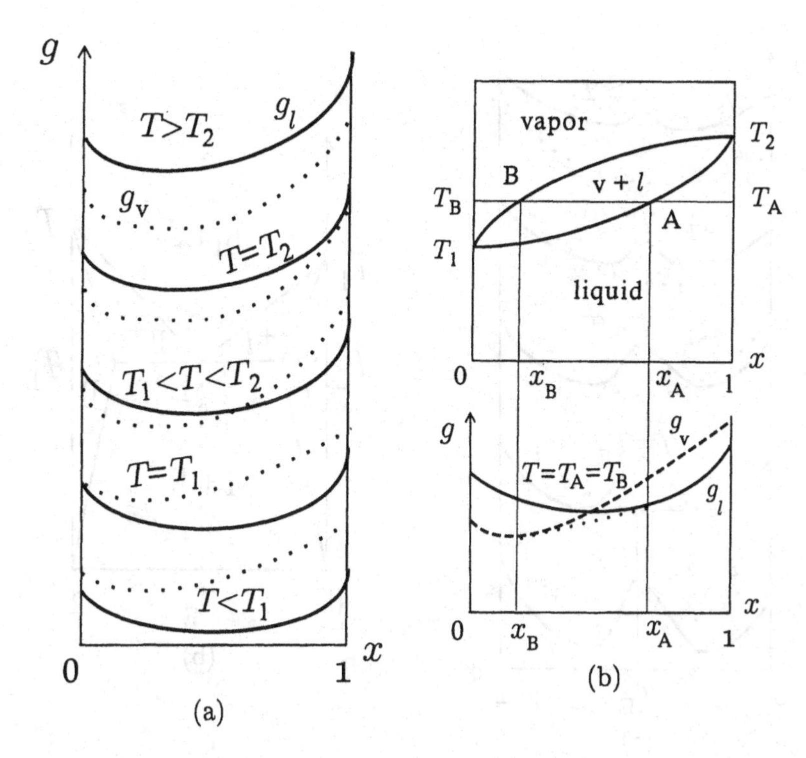

Figure 7.20: The phase diagram for a miscible binary liquid mixture which coexists with its vapor (the pressure is 1 atm). (a) The isotherms of the g-x plot for liquid and vapor at five different temperatures. The solid and dashed curves are, respectively, the g curves for a homogeneous liquid and vapor. The five temperatures are $T > T_2$, $T = T_2$, $T_1 < T < T_2$, $T = T_1$, and $T < T_1$. (b) The upper diagram is the T-x plot obtained from (a), where the region denoted by v+l is the vapor-liquid coexistence region. The lower diagram is used to explain how to get the coexistence curve on the T-x plot of the upper diagram. For example at the temperature $T = T_{\rm A}(= T_{\rm B})$, the tangent points for the common tangent to g_l and $g_{\rm v}$ are at $x = x_{\rm A}$ and $x = x_{\rm B}$; then the mole fractions for A and B on the coexistence curve in the upper diagram are, respectively, $x_{\rm A}$ and $x_{\rm B}$.

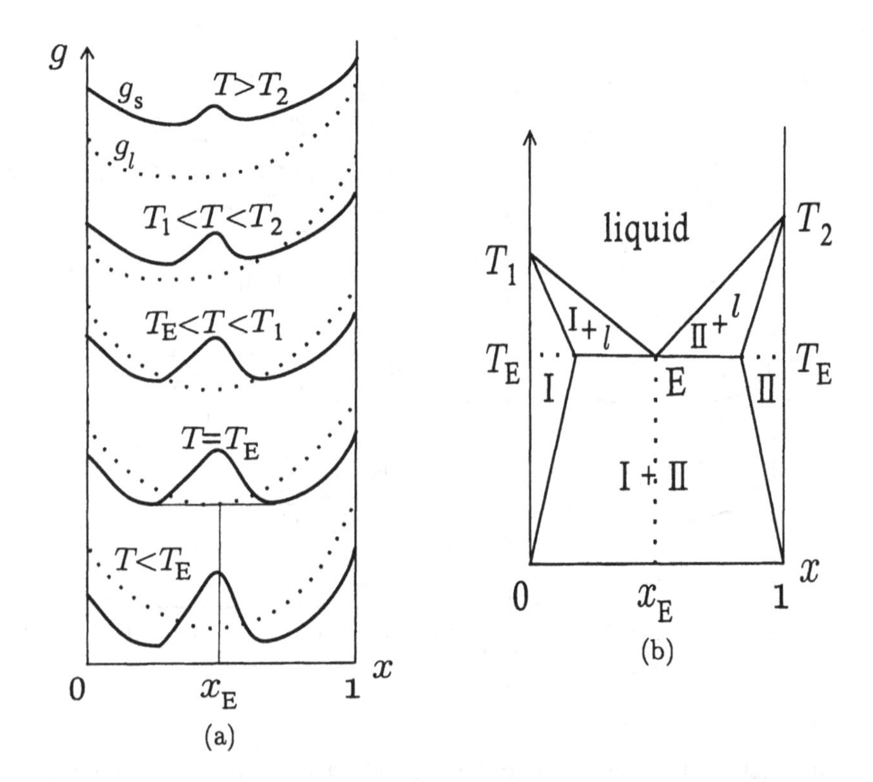

Figure 7.21: Phase diagram for a binary mixture near solid-liquid coexistence: the liquid *state* is miscible and the solid *state* is immiscible. The melting points for the component 1 and 2 are T_1 and T_2, respectively, assuming $T_1 < T_2$. (a) The isotherms for the g-x plot for five different temperatures (at constant pressure). The solid curves and the dashed curves represent, respectively, the specific Gibbs free energy g for the homogeneous solid mixture g_s and the homogeneous liquid mixture g_l. The five temperatures chosen are $T > T_2$, $T_1 < T < T_2$, $T_E < T < T_1$, $T = T_E$, and $T < T_E$. At $T = T_E$ there is a straight line which is a tangent to g_s at two points, and also a tangent to g_l at one point. (b) The phase diagram T vs. x at a constant P, where E is the eutectic point. It is a triple point and is the lowest possible freezing point, $T_E < T_1 < T_2$.

4. Two-*state* coexistence and a phase separation: an eutectic system

The example we are going to discuss is a two-component system in the liquid-solid coexistence region, such that the liquid *state* is miscible and the solid *state* is immiscible. Therefore the liquid phase will mix homogeneously for any mole fraction, but the solid *state* will have phase separation. Suppose the melting temperatures for component 1 and 2 are T_1 and T_2, respectively, and assuming $T_1 < T_2$. We plot five isotherms of the g-x curves on the same diagram in Fig. 7.21(a), where the solid curves denote the specific Gibbs free energy for a *homogeneous solid mixture* g_s, and the dashed curves denote that of a *homogeneous liquid mixture* g_l. The pressure for these curves is fixed, and the five temperatures we have chosen for the plot are: (1) $T > T_2$, (2) $T_1 < T < T_2$, (3) $T_E < T < T_1$, (4) $T = T_E$, and (5) $T < T_E$.

As we have analyzed above, if there is a straight line which can be a tangent to the g curves (including g_s and g_l) at two points, and the tangent line lies below the g curves, then phase separation will occur. We use this statement as a guideline to plot the T-x phase diagram in Fig. 7.21(b). In the diagram, there are several solid curves which divide the two-dimensional space into several regions. The letters I, II, and l in each region denote the equilibrium phase in that region. If there is a + sign between two letters, then it means two-phase coexistence. The letters I, II, and l represent the following phases:

(a) I: a homogeneous solid mixture rich in component 1;
(b) II: a homogeneous solid mixture rich in component 2;
(c) l or liquid: a homogeneous liquid mixture.

The most important characteristic to note in Figs. 7.21(a) and 7.21(b) is that there exists a special temperature T_E. At this temperature there is a tangent line which is a tangent to the g_s curve at two points, but it is also a tangent to the g_l curve at one point. If the tangent point for the g_l curve is $x = x_E$, then the point (x_E, T_E) is a triple point. Solid phases I, II and a liquid mixture with a mole fraction x_E coexist at the triple point. In addition to the triple point, the most important characteristic for T_E is that it is lower than both T_1 and T_2, i.e., $T_E < T_1 < T_2$. This implies that the liquid mixture with a mole fraction close to x_E will have a freezing temperature lower than both the pure liquid 1 and pure liquid 2. If the freezing point of a mixture of two substances is lower than the freezing point for both substances, then the system is called a **eutectic system**. The triple point (x_E, T_E) is call the **eutectic point**. An eutectic system, which can lower the freezing points of pure substances, has many applications. For example people may spread salt crystals on a snowy road, to melt the snow, because the ice-salt mixture has a much lower melting point than pure ice and salt.

7.16 Problems

7.1. Both graphite and diamond are composed of carbon atoms, but with different crystal structures. We say that they belong to different phases. Although at ordinary pressures and temperatures both graphite and diamond seem to be stable, however they are not in phase equilibrium, because the specific Gibbs free energies $g_{\mathrm{g},0}$ and $g_{\mathrm{d},0}$ are not equal (the subscripts g and d denote graphite and diamond, respectively, and 0 indicates the value at $P = 1$ atm). The fact is that one of the phases is stable and the other metastable. We have assumed that T is a constant at room temperature in all the considerations. It is known that at $P = 1$ atm the specific volume of graphite and diamond are, respectively, $v_{\mathrm{g}} = 5.31 \times 10^{-6}$ m^3 mole^{-1} and $v_{\mathrm{d}} = 3.42 \times 10^{-6}$ m^3 mole^{-1}. The isothermal compressibility of graphite and diamond are, respectively, $\kappa_{\mathrm{g}} = 3 \times 10^{-6}$ atm^{-1}, and $\kappa_{\mathrm{d}} \approx 0$. It is also known that at $P \approx 1.61 \times 10^4$ atm, these two phases are in phase equilibrium, i.e., $g_{\mathrm{g}} = g_{\mathrm{d}}$ at $P = 1.61 \times 10^4$ atm. By using the above data, calculate $g_{\mathrm{g},0} - g_{\mathrm{d},0}$. Which one, graphite or diamond, is the stable phase for $P < 1.61 \times 10^4$ atm? And for $P > 1.61 \times 10^4$ atm?

7.2. The density of water and ice at 1 atm and 0°C are 1000 kg/m^3 and 917 kg/m^3, respectively. The heat of fusion of ice is 3.34×10^5 J/kg at 0°C and 1 atm.

(a) Find the slope of the ice-water coexistence curve at 1 atm.

(b) How much pressure would you have to put on an ice cube to make it melt at −1°C?

7.3. Use physical arguments to show that when a system is in a two-phase coexistence state, the specific heat capacity at constant pressure and the expansivity are both infinite, i.e., $c_p \to \infty$ and $\beta \to \infty$ at two-phase coexistence.

7.4. Show that, during a first-order phase transition, the change of the specific internal energy u is given by

$$\Delta u = \ell\left(1 - \frac{d\ln T}{d\ln P}\right),$$

where ℓ is the molar latent heat of the transition.

7.5. Let ℓ be the molar latent heat of transformation from phase 1 to phase 2. Show that

$$\frac{d\ell}{dT} = c_{P2} - c_{P1} + \ell\left[\frac{1}{T} - \frac{v_2\beta_2 - v_1\beta_1}{v_2 - v_1}\right],$$

where β is the expansivity. Simplify the above result when phase 2 is vapor and phase 1 is either liquid or solid. The vapor can be approximated as an ideal gas, note that $v_2 \gg v_1$, and $\beta_2 \gg \beta_1$.

7.6. Prove that the molar heat capacity c of a saturated vapor, i.e., the molar heat capacity corresponding to a heating process in which the vapor is kept saturated, is given at a temperature T by

$$c = c_P^{(g)} - \frac{\ell}{T} = c_P^{(l)} + T\frac{d}{dT}\left(\frac{\ell}{T}\right),$$

where ℓ is the molar heat of vaporization; $c_P^{(g)}$ and $c_P^{(l)}$ are, respectively, the molar heat capacity at constant pressure for the vapor and the liquid. Note that in obtaining the second equality, the result of the previous problem may be helpful, where the vapor is considered as an ideal gas.

7.7. Below the triple point ($-56.2°$C) the vapor pressure of solid CO_2 is given as

$$\ln P = -\frac{3116}{T} + 16,$$

where P is given in atmospheres, and T is in K. The molar latent heat of melting of CO_2 is 8330 J, and is considered to be independent of T. Calculate the vapor pressure exerted by liquid CO_2 at 20°C, and explain why solid CO_2 is often referred to as *dry ice*.

7.8. The vapor pressure of a particular solid and of a liquid of the same material are given, respectively, by

$$\ln P = 0.04 - \frac{6}{T} \quad \text{(solid)}$$

and

$$\ln P = 0.03 - \frac{4}{T} \text{ (liquid)},$$

where P is given in atmospheres, and T in K.

(a) Find the temperature and pressure of the triple point of this material.

(b) Find the approximate values of the three latent heats at the triple point. Treat the vapor as an ideal gas.

7.9. The heat of fusion of ice is 3.34×10^5 J/kg at 0°C and 1 atm. The mass density of water and ice at the same temperature and pressure are 1000 kg/m^3 and 917 kg/m^3, respectively. The saturated vapor pressure and heat of vaporization of water at 0°C are 610 Pa and 2.51×10^6 J/kg. Estimate approximately the temperature of the triple point of H_2O by using the above data. Note that the temperature of the triple point is very close to 0°C, because dT/dP for the ice-water coexistence curve is very small (cf. Problem 7.2). For simplicity you may take 0°C to be 273 K.

7.10. Near the triple point on the P-T phase diagram, show that the slope of the solid-vapor coexistence curve is, in general, greater than that of the liquid-vapor coexistence curve.

7.11. Sketch qualitatively the curves in a g-P and a g-T plane of the phases of a substance which sublimates rather than melts.

7.12. Sketch qualitatively the curves for the three phases of H_2O (ice, water, and steam) in the g-T plane at the pressure $P = 1$ atm. Put all three curves on the same set of axes, and label the temperature $T = 0$°C and 100°C. How would the graphs differ at a pressure of 0.001 atm? Note that the triple point pressure is about 0.006 atm.

7.13. Sketch qualitatively the curves for the three phases of H_2O (ice, water, and steam) in the g-P plane at $T = -10$°C. Put all three curves on the same set of axes and label the pressure $P = 1$ atm.

7.14. For temperature $T < 0.3$ K, the slope of the ^{3}He solid-liquid coexistence curve on the P-T phase diagram is negative, i.e., 0.3 K is a minimum for the coexistence curve (see Fig. 7.5(b)).

(a) Which phase, solid or liquid, is more dense? Which phase has more specific entropy? Use $(\partial g/\partial P)_T$ and the Clausius-Clapeyron equation to explain.

(b) Show that the slope of the solid-liquid coexistence curve must be zero at $T = 0$.

(c) One may compress liquid ^{3}He adiabatically and reversibly until it becomes a solid. If just before the phase change the temperature is 0.1 K, will the temperature after the phase change be higher or lower? Explain your result by using the condition of reversible adiabatic processes and the result of part (a) of this problem.

7.15. Find the rate of change of the boiling temperature with altitude, dT_b/dz, near sea level in °C per km. Assume that the temperature of the air is 300 K, independent of the altitude, but the density of the air decreases with increasing altitude z:

$$\rho(z) = \rho(0)\, e^{-mgz/kT},$$

where m is the average mass of an air molecule. You may use the following approximate values for H_2O at 100°C: the latent heat of vaporization 2.5×10^6 J/kg, the water density 10^3 kg/m^3, and the vapor density 0.6 kg/m^3. The air density at the sea level is 1.3 kg/m^3 and the air can be treated as an ideal gas. Note that the mass density $\rho = M/V = Nm/V$.

7.16. In a second-order phase transition $s_i = s_f$ and/or $v_i = v_f$ at the two-phase coexistence, where i and f denote, respectively, the initial and the final phase.

(a) If $s_i = s_f$, show that along the coexistence curve

$$\frac{dP}{dT} = \frac{1}{T}\frac{c_{P_f} - c_{P_i}}{\beta_f v_f - \beta_i v_i}.$$

(b) If $v_i = v_f$, show that along the coexistence curve

$$\frac{dP}{dT} = \frac{\beta_f - \beta_i}{\kappa_f - \kappa_i}.$$

Note that if both $s_i = s_f$ and $v_i = v_f$ hold at the two-phase coexistence, then both equations in (a) and (b) must hold simultaneously. There is only one point that can satisfy both of the equations, which is known as the critical point.

7.17. Show that for a two-component liquid whose vapor can be treated as an ideal gas mixture,

$$\left(\frac{d \ln P_1}{d \ln x_1}\right)_{T,P} = \left(\frac{d \ln P_2}{d \ln x_2}\right)_{T,P},$$

where P_1 and P_2 are the partial pressures of the components of the vapor, and x_1 and x_2 are the mole fraction of the liquid. [*Hint*: You may use the result of Problem 5.20(b).]

7.18. Prove the following statements for an ideal solution. (a) There is no volume change. (b) The internal energy does not change.

7.19. Consider a mixture of alcohol and water in equilibrium with their vapors.

(a) Determine the number of degrees of freedom for the system and state what they are.

(b) Show that for each constituent (i=1 or 2)

$$-s_i'' dT + v_i'' dP + \left(\frac{\partial \mu_i''}{\partial x''}\right)_{T,P} dx''$$
$$= -s_i''' dT + v_i''' dP + \left(\frac{\partial \mu_i'''}{\partial x'''}\right)_{T,P} dx''',$$

where $s_i'' = -\left(\frac{\partial g_i''}{\partial T}\right)_{P,n_j}$, $v_i'' = \left(\frac{\partial g''_i}{\partial P}\right)_{T,n_j}$; x'' is the mole fraction of one of the constituents in the liquid phase; and x''' is the mole fraction of the same constituent in the vapor phase. Note that the notations with double-prime refer to quantities in the liquid state, and those of triple-prime refer to the vapor phase.

(c) Show that

$$\left(\frac{\partial P}{\partial T}\right)_{x''} = \frac{x'''(s_1''' - s_1'') + (1 - x''')(s_2''' - s_2'')}{x'''(v_1''' - v_1'') + (1 - x''')(v_2''' - v_2'')},$$

where x'' is held constant artificially. [*Hint*: You may use the results of Problem 5.20(b), and part (b) of this problem.]

7.20. Consider a dilute binary liquid solution, with $n_{\rm A}$ kilomoles of solvent and $n_{\rm B}$ kilomoles of solute ($n_{\rm B} \ll n_{\rm A}$).

(a) Starting from Eq. (7.77), show that the total Gibbs free energy of the solution can be *approximated* as

$$G(T,P) = n_{\rm A}\mu_{\rm A}^0 + n_{\rm B}f(T,P) - n_{\rm B}RT(\ln n_{\rm A} - \ln n_{\rm B} + 1),$$

where $f(T,P)$ is a function of T and P, independent of $n_{\rm A}$ and $n_{\rm B}$. Find $f(T,P)$ in terms of $\mu_{\rm A}^0$, $\mu_{\rm B}^0$ and λ.

(b) If the saturated vapor pressure of the pure solvent (i.e., $n_{\rm B} = 0$) at temperature T is P_0, show that the saturated vapor pressure P of the solution at T is given by (assuming that the solute does not evaporate at all)

$$\frac{P}{P_0} = 1 - \frac{n_{\rm B}}{n_{\rm A}}.$$

This is known as the **Raoult's**[18] **law**. This result tells us that the vapor pressure of the solvent tends to decrease when a small amount of solute molecules are added.

7.21. Consider the same dilute liquid solution as in the previous problem. If the boiling temperature of the pure solvent (i.e., $n_{\rm B} = 0$) at pressure P is T_0, show that the boiling temperature of the solution T is given by (assuming that the solute does not evaporate at all)

$$T - T_0 = \left(\frac{n_{\rm B}}{n_{\rm A}}\right)\frac{RT_0^2}{\ell},$$

where ℓ is the heat of vaporization per kilomole of the pure solvent. This result tells us that the boiling temperature of the solvent tends to increase when a small amount of solute molecules are added.

7.22. Consider an ideal liquid solution composed of substances 1 and 2 at a constant temperature T. The liquid solution coexists with its saturated vapor, which can be considered as an ideal gas mixture. If x_1 and x_1' are, respectively, the mole fractions of substance 1 in the liquid phase and in the vapor phase, show that the ratio x_1'/x_1 is not equal to 1. Under what condition will x_1'/x_1 be close to 1? [*Hint*: You may consider the condition of two-phase coexistence for pure substance 1, i.e., $x_1 = x_1' = 1$, first.]

[18] Francois Marie Raoult, French chemist (1830-1901).

7.23. Consider the vapor-liquid coexistence of a regular solution of a binary mixture with the heat of mixing given by

$$\Delta H = \lambda\,(n_1 + n_2)x_1x_2,$$

where λ is a constant, and n_i and x_i (i=1,2) are the number of moles and the mole fraction for the ith component in the liquid phase. The vapor phase of the solution can be considered as an ideal gas mixture (i.e., $\lambda = 0$ for the vapor phase). The chemical potentials of the pure liquids $\mu_i^{(l)0}$ (i=1,2) can be considered as pressure independent, i.e., $\mu_i^{(l)0}(T,P) \approx \mu_i^{(l)0}(T)$. The superscript 0 refers to the pure system 1 or 2.

(a) Find the expression $\mu_i^{(l)} = \mu_i^{(l)}(T,P,x_1)$ for the chemical potential for the ith component in the homogeneous liquid phase mixture.

(b) Find the partial vapor pressure (coexistent with the liquid phase) P_i (i=1,2) for the ith component as a function of the composition x_1 and λ/RT. The saturation vapor pressures for the pure system $P_i^0(T)$ (i=1,2) are assumed to be known functions of T.

(c) Calculate dP_i/dx_1 and find the critical temperature T_c for a given λ. (Note that for $T > T_c$ the liquid phase is a homogeneous phase for all x_1, and for $T < T_c$ phase separation occurs.) You can obtain T_c from dP_i/dx_1 (which has a special form at T_c); there is no need to solve the phase separation problem in the liquid phase.

(d) Sketch P_1, P_2, and P vs. x_1 (for T constant) on the same graph, where $P = P_1 + P_2$ is the total vapor pressure. The shape of each curve will depend on the value of λ/RT. Sketch the graphs for λ/RT=−2, 0, 1, 2, and 3 (one graph for each λ/RT). Discuss the characteristics for each graph. You may discuss the case for the ideal solution ($\lambda = 0$) first and compare it with the results of the cases $\lambda \neq 0$. The results of (c) are also important in plotting the graphs.

7.24. Consider a miscible binary mixture of A and B, with boiling points $T_{\rm A}$ and $T_{\rm B}$, respectively. The specific Gibbs free energies for the mixture (at constant T and P=1 atm) as a function of x are shown in Fig. 7.22 for seven different temperatures, where x =

$n_B/(n_A + n_B)$ is the mole fraction of B. In the figure the solid curves (labeled by g_l) are the specific Gibbs free energies for the liquid phase, and the dashed curves (labeled by g_v) are those of the vapor phase. From the free energy graphs, draw qualitatively the phase diagram T vs. x for the binary system, which should show all the vapor and liquid phase boundaries. Explain how you get the phase diagram. Note that there is a particular composition at which this mixture will condense with no change of composition. This special composition is called an **azeotrope**.

7.25. Repeat the previous problem but with a slightly different shape in the g-x curves. In this problem we consider the case that the concave-up section of the liquid's g curve is deeper than that of the vapor's g curve, which is opposite to that of the previous problem. One of the main features of the g curves in this problem is that at some temperature range, $g_l(x) > g_v(x)$ for $x = 0$ and $x = 1$, but $g_l(x) < g_v(x)$ for x in the range $0 < x_1 < x < x_2 < 1$ for some x_1 and x_2. Construct the T vs. x phase diagram qualitatively, and show that the system also has an azeotrope.

7.26. Consider a binary mixture which has a T-x phase diagram, as shown in Fig. 7.21(b). Suppose you start from some composition

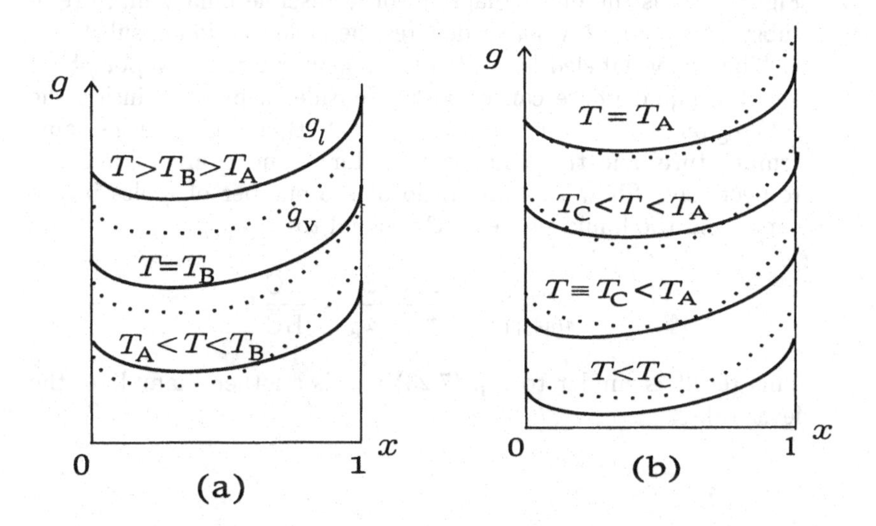

Figure 7.22: Problem 7.24.

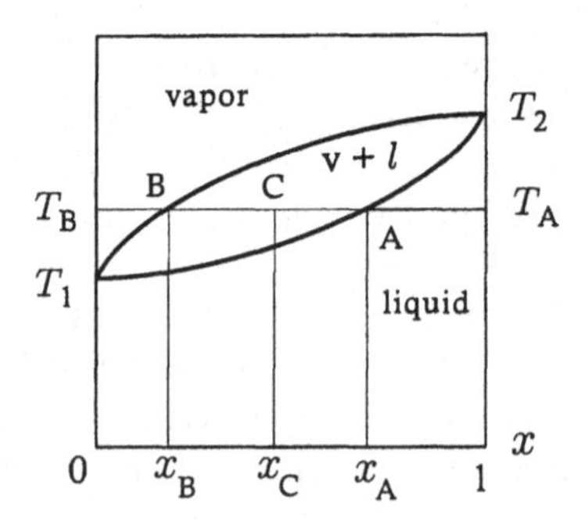

Figure 7.23: Problem 7.27.

x_0 ($0 < x_0 < 1$) at temperature $T > T_2$, and then lower the temperature T while keeping the composition x_0 fixed. Describe the processes of solidification of the solution as T is lowered to absolute zero. Apparently the solidification processes may be different for a different starting composition x_0. How many different processes (qualitatively, not quantitatively) are there? Describe all possible different processes by varying x_0.

7.27. Figure 7.23 is the phase diagram for a miscible binary mixture of substances 1 and 2, where x denotes the mole fraction of substance 2. The region labeled by v+l is the region where the vapor phase and the liquid phase can coexist. Consider a point C inside the v+l region, with $x = x_C$. In the figure, A, B, and C have the same temperature, and the mole fractions for A and B are x_A and x_B, respectively. Show that the ratio of the number of moles in the vapor and the liquid phase at C is equal to

$$\frac{n_{\text{vapor,C}}}{n_{\text{liquid,C}}} = \frac{x_A - x_C}{x_C - x_B} = \frac{\overline{\text{AC}}}{\overline{\text{BC}}},$$

This result is similar to Eq. (7.23) and is another example of the lever rule.

Chapter 8

Applications of Thermodynamics

8.1 Introduction

In this chapter we will discuss several topics involving the applications of thermodynamics. First we introduce two non-PVT thermodynamic systems, namely magnetic systems and surface (thin film) systems, which are important for both scientific research and technological applications. The discussion of these two systems will also help the reader to learn how to apply the thermodynamic relations established in PVT systems to non-PVT systems, there is no need to derive these relations all over again. Other topics we will discuss include how the temperature of the atmosphere depends on the elevation, how the presence of other molecules (such as nitrogen and oxygen) affect the saturated vapor pressure, and the formation of liquid drops. These topics are related to the thermodynamics of the atmosphere, and may be helpful in the understanding of weather conditions. We will also discuss the phenomenon of osmotic pressure in dilute solutions, which is an important problem in biology and medical science. Finally we will show that thermodynamics can be successfully applied to blackbody radiation, which is a non-material system as the mass of radiation vanishes. The theory of blackbody radiation played an important role in the development of quantum theory. In quantum theory the quanta of electromagnetic radiation are called photons. We will see that the thermodynamics of a photon gas can be used to derive many properties which are consistent with those obtained in the statistical theory of a photon gas. Application of the statistical theory to a photon gas will be given in Chapter 11.

8.2 Thermodynamics of Magnetism

Aside from PVT systems magnetic systems are the most studied systems in thermodynamics. Magnetic systems are ideal systems for pure academic research in the theory of phase transitions and other thermodynamic studies, because they can be described by simple theoretical models which are useful in statistical mechanics. Moreover, magnetic materials are also important in many technological applications in the laboratory, industry, medicine, etc. Therefore it is worth for us to study the general thermodynamic properties of magnetic systems in more detail. In fact the thermodynamic laws and relations we derived in the previous chapters for PVT systems are still valid for other thermodynamic systems, except that we have to change the PVT state variables to other state variables which are appropriate for other systems. For example, the equation of state for a PVT system involves the variables (P, V, T), and for a magnetic system the appropriate variables are (B, M, T), where B is the external magnetic field and M is the magnetic moment. The task for us is then to find a transformation rule which can be used to transform the set of variables (P, V) to the set (B, M).

We consider a magnetic material of length l and a cross-sectional area A, which is surrounded by a coil wound uniformly with a winding of N turns with negligible resistance that carries a current I, as shown in Fig. 8.1. In the time interval dt, the current is increased from I to $I + dI$, the induced emf $\mathcal{E}$ in the winding is

$$\mathcal{E} = -NA\frac{dB}{dt}, \tag{8.1}$$

where B is the magnetic induction in the winding.[1] The work done by the induced emf $\mathcal{E}$ in a time interval dt is

$$dW = \mathcal{E}Idt = -NAIdB = -V\mathcal{H}dB, \tag{8.2}$$

where $V = Al$ is the volume of the system and $\mathcal{H} = NI/l$ is the external

[1]We assume that B, $\mathcal{H}$, which will be introduced shortly, and m are all spatially homogeneous quantities, and all three vectors are parallel to one another.

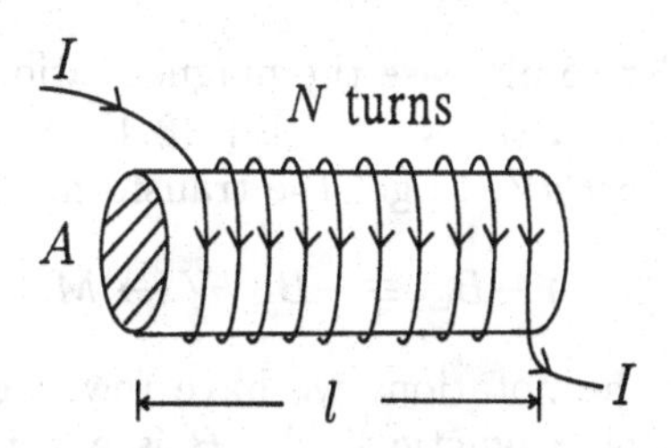

Figure 8.1: A current-carrying coil can be used to magnetize a magnetic material. The coil, having a cross-sectional area A and length l, is wound uniformly with a winding of N turns with negligible resistance that carries a current I.

magnetic field intensity. From the relation [2]

$$B = \mu_0(\mathcal{H} + m), \tag{8.3}$$

where μ_0 is the permeability of the vacuum [3] and m is the magnetization (i.e., the magnetic moment per unit volume). By substituting Eq. (8.3) into Eq. (8.2) we have

$$đW = -\mu_0 V\mathcal{H}d\mathcal{H} - \mu_0 V\mathcal{H}dm.$$

Note that the first term on the right hand side of the above equation is the work required to increase the field in a vacuum, if the material were not present (in which case m and dm would be zero). The second term is therefore the work associated with the magnetization of the material. Therefore the work of magnetization is

$$đW = -B_{\text{ex}}dM, \quad M \equiv Vm, \tag{8.4}$$

where M is the total magnetic moment of the system and $B_{\text{ex}} = \mu_0\mathcal{H}$ is the *external magnetic induction*, which does not include the contribution from M. It should be noted that in Eq. (8.2) the term $đW$ is the work done by the induced $\mathcal{E}$ in the winding, not by the external voltage. Therefore Eq. (8.4) is *the work done by the magnetic system.* This work is negative for $dM > 0$, which implies that external work must be done

[2] If Gaussian units are used, then $B = \mathcal{H} + 4\pi m$. The units of the related physical quantities are: B: gauss; $\mathcal{H}$: oersted; M: erg oersted^{-1}.

[3] In SI (MKS) units, $\mu_0 = 4 \times 10^{-7}$ tesla m ampere^{-1}, and the units for the related physical quantities are: B: tesla; $\mathcal{H}$: ampere-turn m^{-1}; M: joule tesla^{-1}.

on the system in order to increase the magnetization. We may compare the work done by a magnetic system, Eq. (8.4), with the work done by a PVT system, $đW = PdV$, to get the transformation formulas

$$P \to -B_{\text{ex}} \equiv -B, \quad V \to M. \tag{8.5}$$

In order to simplify the notations we have rewritten B_{ex} as B, which represents the external magnetic field. B is a generalized force; it is an intensive variable, independent of the size of the system. The total magnetic moment M is a generalized displacement, which is an extensive variable with a magnitude proportional to the size of the system. We should note that in the replacement of P by B, we have to introduce a minus sign in front of B. From the equation $dU = TdS - PdV$ in a PVT system, the corresponding equation for a magnetic system is

$$dU = TdS + B\,dM. \tag{8.6}$$

In a PVT system we defined a state variable the enthalpy, $H = U + PV$, therefore if we use the transformation formulas (8.5) we may define the state variable **magnetic enthalpy**

$$H_{\text{M}} = U - BM \equiv E. \tag{8.7}$$

In a magnetic system, $-BM$ is the magnetic potential energy for the magnetic moment M in a magnetic field B, therefore the magnetic enthalpy H_{M} is identical to the **total internal energy** E, which is the sum of the non-magnetic internal energy U and the magnetic internal energy $-BM$. Therefore in the following we will use the total internal energy E instead of the magnetic enthalpy H_{M}. From Eq. (8.7) we have

$$dE = TdS - MdB. \tag{8.8}$$

We may define two heat capacities C_M and C_B from Eqs. (8.6) and (8.8) ($đQ = TdS$),

$$C_M = \left(\frac{đQ}{dT}\right)_M = \left(\frac{\partial U}{\partial T}\right)_M, \tag{8.9}$$

$$C_B = \left(\frac{đQ}{dT}\right)_B = \left(\frac{\partial E}{\partial T}\right)_B, \tag{8.10}$$

where C_M and C_B are, respectively, the constant M and constant B heat capacities. The heat capacities C_M and C_B correspond to the heat capacities C_V and C_P in a PVT system. Therefore from the first and second

Tds equations (Eqs. (4.28)–(4.29)) we derived in Sec. 4.5, and the transformation formulas (8.5), we obtain the following two TdS equations for a magnetic system:

$$TdS = C_M dT - T\left(\frac{\partial B}{\partial T}\right)_M dM, \tag{8.11}$$

$$TdS = C_B dT + T\left(\frac{\partial M}{\partial T}\right)_B dB. \tag{8.12}$$

We define the **magnetic free energy** F_{M} as

$$F_{\mathrm{M}} = E - TS, \quad \text{or} \quad E = F_{\mathrm{M}} + TS; \tag{8.13}$$

its derivative to first order is

$$dF_{\mathrm{M}} = dE - TdS - SdT = -SdT - MdB. \tag{8.14}$$

Therefore we have

$$\left(\frac{\partial F_{\mathrm{M}}}{\partial T}\right)_B = -S, \tag{8.15}$$

$$\left(\frac{\partial F_{\mathrm{M}}}{\partial B}\right)_T = -M. \tag{8.16}$$

If we can obtain $F_{\mathrm{M}} = F_{\mathrm{M}}(M, B, T)$, then Eq. (8.16) may be used to obtain the equation of state for a magnetic system $M = M(B, T)$. It is worth mentioning that in Eq. (8.13), the magnetic free energy F_{M} is equal to $E - TS$ rather than $U - TS$, therefore F_{M} does not correspond to the Helmholtz free energy F in a PVT system. In a magnetic system, the variable E corresponds to the enthalpy H in a PVT system, therefore in fact the magnetic free energy F_{M} corresponds to the Gibbs free energy $G = H - TS$ in a PVT system. We call F_{M} the *magnetic free energy.* For the convenience of the reader, we tabulate the corresponding state variables for a magnetic and a PVT systems in Table 8.1.

Table 8.1 Corresponding Variables for a Magnetic and a *PVT* Systems

PVT System	Magnetic System
P (pressure)	$-B$ (magnetic field)
V (volume)	M (total magnetic moment)
T (temperature)	T (temperature)
S (entropy)	S (entropy)
H (enthalpy)	E (total internal energy)
G (Gibbs free energy)	F_{M} (magnetic free energy)

From Eq. (8.14) we obtain one of the Maxwell relations for a magnetic system,

$$\left(\frac{\partial S}{\partial B}\right)_T = \left(\frac{\partial M}{\partial T}\right)_B . \tag{8.17}$$

This equation can also be obtained directly from Eq. (5.28) derived in Chapter 5, by proper substitutions of the state variables which are tabulated in Table 8.1. Equation (8.17) is the theoretical basis showing that we can use a paramagnetic salt for the purpose of cooling, as we discussed in Chapter 6. For a paramagnetic system at constant magnetic field, the magnetization decreases as the temperature increases, i.e., $(\partial M/\partial T)_B < 0$. From Eq. (8.17), we have

$$\left(\frac{\partial S}{\partial B}\right)_T < 0. \tag{8.18}$$

Therefore for an isothermal magnetization process (increasing B) the entropy S decreases, and the subsequent adiabatic demagnetization will decrease T, as shown in Fig. 6.4. The equation of state for a paramagnetic system, to a good approximation, can be described by **Curie's[4] law**

$$M = C_{\mathrm{c}}\frac{B}{T}, \tag{8.19}$$

where C_{c} is a constant, called the Curie constant. Curie's law satisfies the condition of inequality (8.18). This equation of state violates the

[4]Pierre Curie, French physicist (1859–1906), Nobel prize laureate in physics in 1903.

third law as $T \to 0$ (see Problem 6.17), therefore paramagnetic systems no longer exist at low enough T. These systems transform to other phases, such as ferromagnetic or anti-ferromagnetic phases. We will give a more general study of the paramagnetic systems in Sec. 11.9, and a brief discussion of a ferromagnetic system in Sec. 11.11.

8.3 Surface Tension

In our previous studies of a thermodynamic system, we always assumed that the system under study is spatially homogeneous, i.e., the physical properties of the system are everywhere the same throughout the system, the ***boundary (surface) effects are neglected.*** For a gaseous system, this assumption is close to the real situation. For a liquid or a solid, however, this assumption is a good one only when the system is large enough (i.e., macroscopically large), and the phenomena under study are not related to the surface (boundary). This is because the surface molecules are in an environment which is different from that of the interior molecules, and therefore the surface molecules may have properties that are different from the interior molecules. Surface effects usually prevail only within three or four layers of molecules under the surface, called the surface layers. For liquids or solids whose molecular arrangements are in a *normal* shape, the number of molecules in the surface layer is of the order of $N^{2/3}$, where N is the total number of molecules of the system. Therefore the fraction of the surface molecules to the total number of molecules is on the order of $N^{-1/3}$, which can be considered as zero if N is macroscopically large.

Surface science is a branch of science which focuses on the properties of the surface layers, which became an active and important research area after the 1970's. Surface effects are important when: (1) the total number of molecules of a system N is not very large and $N^{-1/3}$ is not negligibly small. Thus for mesoscopic systems, surface effects are important; (2) the shape of a system is not *normal*, so that the number of the surface molecules is not of the order of $N^{2/3}$, e.g., thin films, and porous materials etc.; (3) phenomena in which the surface molecules play a major role. There are a lot of phenomena in nature belonging to this category, such as surface reconstruction, adsorption, catalysis, etc.

Although the surface molecules and the bulk molecules are of the same type, their properties may differ from one another. In some sense it is similar to their belonging to different phases, but this is just an analogy, there is no surface phase.[5] The surface molecules possess an important physical quantity called **surface tension**, which is unique to the surface molecules; the bulk molecules do not have this quantity. The number of nearest neighbor molecules for a molecule on the surface is less than that of a bulk molecule. Therefore, because molecules attract each other, the energy of a surface molecule is higher than that of a bulk molecule. At constant T and V (or constant T and P), a system is in an equilibrium state when its free energy is minimum. Therefore a liquid system tends to minimize its surface area in order to minimize its free energy, due to the existence of surface tension.[6] When we want to increase the surface area of a liquid, we must exert an external force on the system, because a liquid has a tendency to keep its surface area as small as possible. We call the force which tries to keep the surface area as small as possible the **surface tension**. However we will see that the surface tension has the units of *force per unit length*, not just force. We will now discuss the thermodynamics of a surface. For simplicity we consider a system of a liquid thin film.

First we derive the expression for the configuration work of a thin film. We consider a rectangular shaped wire which sustains a liquid thin film, as shown in Fig. 8.2. The right end of the rectangular wire has a length l, which can be moved without friction. An external force $\mathcal{F}$ pulls the right-end wire to the right to increase the area of the film. The surface tension σ will resist the action of $\mathcal{F}$, trying to keep the area of the film unchanged. When the external force $\mathcal{F}$ just overcomes the resistance of the surface tension, the right-end wire will move very slowly to the right, and the area will increase. When the process is very slow it is a quasi-static process; if the wire has no friction during the motion the process can be considered as a reversible process. When the right-end wire has a displacement dx, the area of the film is increased from $A = 2lx$ to $A + dA = 2l(x + dx)$; the factor of 2 comes from the fact that the film has an upper and a lower surface. Therefore the change of area is $dA = 2ldx$. The work done by the film (against the external force $\mathcal{F}$) in this process is given by

[5]The surface affects only three to four layers of molecules close to the surface, called the surface layers. The transition of the properties from the surface to the bulk is a gradual one, and each layer within the surface layers behaves differently. Therefore the surface layers are not a uniform system; this is quite different from the phases we discussed in Chapter 7.

[6]Surface tension also exists in solid systems, but the shape of a solid is so rigid that can not be changed by surface tension, which is too small.

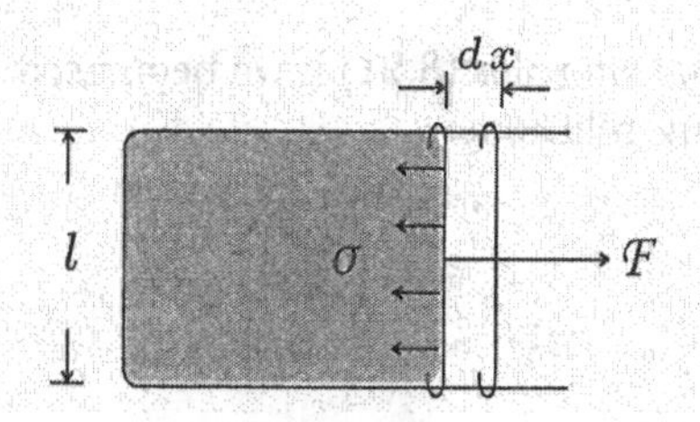

Figure 8.2: Derivation of the configuration work for a thin film. Under the action of the external force $\mathcal{F}$, the wire at the right end moves a small distance dx, which changes the area of the thin film from the original $A = 2lx$ to the final $A + dA = 2l(x + dx)$.

$$đW = -\mathcal{F}\,dx = -\sigma 2l\,dx = -\sigma\,dA, \tag{8.20}$$

where we have used the quasi-static process condition $\mathcal{F} = 2l\sigma + 0^+$. The surface tension σ therefore has the units of newtons per meter, and it is an intensive variable. Comparing the form of $đW$ in Eq. (8.20) and that of a PVT system $đW = PdV$, we find the following transformation rule for the surface system,

$$P \to -\sigma, \quad V \to A. \tag{8.21}$$

There is, however, an important difference between a surface system and a PVT system. In a PVT system we can choose any two of the three state variables P, V and T as the independent variables. For a surface system, however, the surface tension σ depends on the temperature T only, since it is found experimentally that for a given film, the external force per unit length $\mathcal{F}/l$ in Fig. 8.2 depends on T only, independent of the area A of the film. Therefore it is not proper to choose σ as the independent variable, because it is the same as choosing T as the independent variable. In a surface system we therefore choose T and A as the independent variables.

Next we consider the thermodynamics of a surface system with an area A at temperature T. From Eq. (5.10) for the Helmholtz free energy for a PVT system,

$$dF = -SdT - PdV,$$

we get the corresponding equation for the surface Helmholtz free energy:

$$dF = -SdT + \sigma\,dA, \tag{8.22}$$

where the transformation rules (8.21) have been used. From this equation we have the following relations:

$$S = -\left(\frac{\partial F}{\partial T}\right)_A, \tag{8.23}$$

and

$$\sigma = \left(\frac{\partial F}{\partial A}\right)_T. \tag{8.24}$$

Since σ is a function of T only, independent of the area A of the surface, the above equation can be integrated at a constant temperature to obtain[7]

$$F = \sigma A, \quad \text{or} \quad \sigma = \frac{F}{A}. \tag{8.25}$$

This means that the **surface tension is the surface Helmholtz free energy per unit area**. This is a simple and yet important result for a surface system. Note that, since σ cannot act as an independent variable (because it is a function of T only), there is no distinction between the surface Helmholtz and the surface Gibbs free energies. Therefore we may also say that **the surface tension is the surface Gibbs free energy per unit area.**

From Eqs. (8.23) and (8.25), we obtain the surface entropy,

$$S = -\left(\frac{\partial F}{\partial T}\right)_A = -\frac{d\sigma}{dT}A.$$

The specific entropy s, i.e., the entropy per unit area, is therefore

$$s = \frac{S}{A} = -\frac{d\sigma}{dT}. \tag{8.26}$$

Since s must be a positive quantity, we should have $d\sigma/dT < 0$. The surface tension σ decreases as T increases. The internal energy U of a surface can be obtained from the relation $U = F + TS$. From Eqs. (8.25) and (8.26) we obtain the specific internal energy of a surface $u = U/A$:

$$u = \sigma - T\frac{d\sigma}{dT}. \tag{8.27}$$

---

[7]There is a constant of integration in Eq. (8.25) which can be chosen to be 0, because $F = 0$ when $A = 0$.

Therefore u is a function of T only, independent of the surface area A. We define the heat capacity per unit area c_A as

$$c_A = \frac{du}{dT} = \frac{d\sigma}{dT} - \frac{d\sigma}{dT} - T\frac{d^2\sigma}{dT^2} = -T\frac{d^2\sigma}{dT^2}, \tag{8.28}$$

which can also be obtained from the relation

$$c_A = T\left(\frac{\partial s}{\partial T}\right)_A = -T\frac{d^2\sigma}{dT^2}. \tag{8.29}$$

For many pure liquids in equilibrium with their vapor phase, the surface tension has an equation of state of the form:

$$\sigma = \sigma_0 \left(1 - \frac{T}{T_c}\right)^n, \tag{8.30}$$

where σ_0 is some reference surface tension,[8] n is a constant and T_c is known as the critical temperature. We note that $\sigma = 0$ at $T = T_c$, and Eq. (8.30) has no meaning for $T > T_c$. This implies that for $T \geq T_c$ there is no surface tension. The reason for this is because for $T \geq T_c$ there is no distinction between the liquid and gas phases. Therefore T_c is the critical temperature in the vapor-liquid coexistence.

Experimentally it is found that the typical value for n in Eq. (8.30) lies between 1 and 2. It is interesting to note that for these values of n, $d\sigma/dT < 0$ and the specific entropy $s > 0$ as expected. However it also implies that s decreases as T increases, and finally $s = 0$ at $T = T_c$. Although $s = 0$ at $T = T_c$ may seem reasonable, because at $T = T_c$ $\sigma = 0$, the surface no longer exists and thus there is no surface entropy. However there is a rather unexpected consequences of $s = 0$ at $T = T_c$ that $ds/dT < 0$ and $c_A < 0$ for $T \leq T_c$. The fact that $c_A < 0$ is rather unusual. Nevertheless, it does not violate the requirement that C_V should be greater than zero (see Eq. (5.61)) for a thermodynamic system to be stable. The reason is that a surface system is not an isolated system, it is just part of a bulk system.

[8]From Eq. (8.30) σ_0 is the surface tension at $T = 0$. However, except for helium, all liquids solidify at temperatures well above $T = 0$ at ordinary pressure. Therefore we may consider σ_0 just a constant.

8.4 Dependence of Vapor Pressure on Total Pressure

In Eq. (7.13) we derived a formula for a liquid to coexist with its saturated vapor. In this derivation we assumed that there was only one kind of molecule both in the liquid and vapor phases. In this section we will study the case where there are other kinds of molecules in the vapor phase, how the saturated vapor pressure will be affected, i.e., how Eq. (7.13) should be modified. We consider a liquid-vapor coexistence system, in which the liquid is composed of only component 1 and the vapor phase is composed of several components. For simplicity we consider the case where the vapor phase is composed of component 1 and component 2, which may contain several constituents. Suppose at temperature T and pressure P the liquid phase coexists with the vapor phase. The condition for the coexistence is, by Eq. (7.4),

$$\mu_{1l}(T,P) = \mu_{1\mathrm{v}}(T,P), \tag{8.31}$$

where the subscript 1 denotes component 1, and l and v denote the liquid and vapor phase, respectively. In the above equation P is the total pressure, therefore $P = P_1 + P_2$, where P_1 is the partial pressure of component 1. The physical meaning of P_1 is that when there are molecules of component 2 ($P_2 \neq 0$), P_1 is the saturated vapor pressure $P_\mathrm{v}(T)$ of the liquid at temperature T. Now we keep the temperature T constant and add more component 2 molecules to the vapor phase. In this process the total pressure changes from P to $P + dP$, and the partial pressure will also change from P_1 to $P_1 + dP_1$. The condition of liquid-vapor coexistence now becomes $\mu_{1l}(T, P+dP) = \mu_{1\mathrm{v}}(T, P+dP)$, which is, from Eq. (8.31),

$$d\mu_{1l}(T,P) = d\mu_{1\mathrm{v}}(T,P) \quad (T = \mathrm{constant}). \tag{8.32}$$

We approximate the vapor as an ideal gas mixture, therefore from Eq. (7.55), we have

$$\begin{aligned} \mu_{1\mathrm{v}}(T,P) &= g_{1\mathrm{v}}(T,P) = RT\ln P_1 + RT\phi_1(T), \\ d\mu_{1\mathrm{v}} &= RT\frac{dP_1}{P_1} \quad (T = \mathrm{constant}). \end{aligned} \tag{8.33}$$

For the liquid we have

$$d\mu_{1l} = dg_{1l} = v_l dP \quad (T = \mathrm{constant}). \tag{8.34}$$

From Eqs. (8.32)–(8.34) one obtains

$$\frac{dP_1}{P_1} = \frac{v_l}{RT} dP. \tag{8.35}$$

Now we assume that the initial state i is the state that component 2 is zero, and $P_1 = P_v^0 = P$; and the final state f is the state when component 2 is not zero, with the partial pressure of component 1 as $P_1 = P_v$, $P = P_1 + P_2 > P_1$. We integrate Eq. (8.35) from state i to state f, and obtain (assuming v_l is independent of P)

$$\int_{P_v^0}^{P_v} \frac{dP_1}{P_1} = \frac{v_l}{RT} \int_{P_v^0}^{P} dP \quad (T = \text{constant}), \tag{8.36}$$

$$\ln \frac{P_v}{P_v^0} = \frac{v_l}{RT}(P - P_v^0) \quad (T = \text{constant}). \tag{8.37}$$

Equation (8.37) is the equation we are looking for. We see that when component 2 is zero, $P_v = P_v^0$ and $P = P_v^0$; thus the total pressure P is the saturated vapor pressure P_v^0. When we add some molecules of component 2 to the saturated vapor, the total pressure will be larger than the initial saturated pressure, i.e., $P > P_v^0$. From Eq. (8.37) we see that the saturated vapor pressure P_v will also be increased to a value larger than the initial saturated vapor pressure P_v^0, when there are no component 2 molecules.

We use water vapor in the atmosphere as an example to estimate the effect of other molecules on the saturated vapor pressure. The saturated vapor pressure for pure H_2O at 300 K is known to be $P_v^0 = 3.6 \times 10^3$ N/m^2, the specific volume for water is $v_l = 18 \times 10^{-3}$ m^3/kilomole, and $RT = 8.31 \times 10^3 \times 300$ J/kilomole·K. Now we add other molecules to the vapor to let the total pressure P become 10 atm, i.e., $P = 1.01 \times 10^6$ N/m^2. Substituting these values into Eq. (8.37) we obtain

$$\begin{aligned} \ln \frac{P_v}{P_v^0} &= \frac{18 \times 10^{-3}}{8.31 \times 10^3 \times 300}(1.01 \times 10^6 - 3.6 \times 10^3) \\ &= 7.29 \times 10^{-3} \ll 1. \end{aligned} \tag{8.38}$$

Therefore the value of P_v/P_v^0 is very close to 1, and we can write $P_v/P_v^0 = 1 + \Delta P_v/P_v^0$, where $\Delta P_v/P_v^0 \ll 1$. Equation (8.38) becomes

$$\ln \frac{P_v}{P_v^0} = \ln\left(1 + \frac{\Delta P_v}{P_v^0}\right) \approx \frac{\Delta P_v}{P_v^0} = 7.29 \times 10^{-3} < 1\%, \tag{8.39}$$

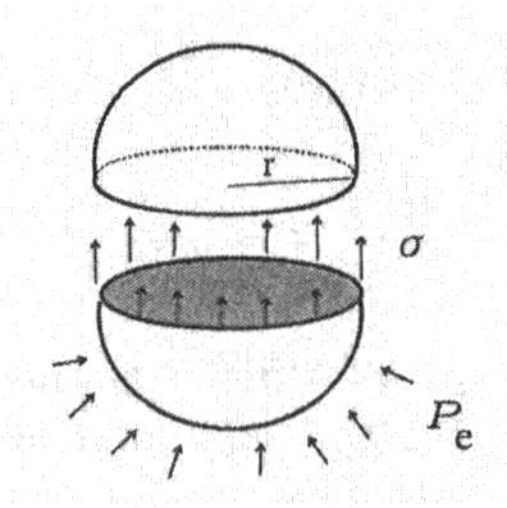

Figure 8.3: The forces acting on the lower half of a spherical drop of liquid, which include the internal pressure $P_{\rm i}$ (not shown), the external pressure $P_{\rm e}$, and the surface tension σ due to the upper half of the drop.

where $\Delta P_{\rm v}$ is the change of the saturated water vapor pressure due to the presence of other molecules. Equation (8.39) tells us that the change of the saturated vapor pressure is negligible, even when the total pressure is increased to 10 atm.

8.5 Vapor Pressure of a Liquid Drop

We have derived Eq. (7.13) for the saturated vapor pressure for a liquid (or solid) with a plane surface. In this section we will discuss the vapor pressure for a liquid drop and to see how it is different from Eq. (7.13). We consider a *spherical* liquid drop with a finite radius r, which is in equilibrium with an external pressure $P_{\rm e}$. Due to the presence of the surface tension σ, the internal pressure of the drop $P_{\rm i}$ is greater than the external pressure $P_{\rm e}$. We consider the force balance as shown in Fig. 8.3. We imagine that we cut the sphere by a horizontal plane into two halves, and consider the forces on the lower half. In addition to the internal and external pressures, there is also a force due to the surface tension. The surface tension force on the lower half sphere (due to the upper half sphere) is an upward force, whose magnitude is $2\pi r\sigma$. The sum of all the horizontal forces are zero, hence we have the equation for the force balance (vertical forces)[9]

$$2\pi r\sigma + \pi r^2 P_{\rm e} = \pi r^2 P_{\rm i}, \tag{8.40}$$

[9]We are considering a small liquid drop, therefore the gravitational force can be neglected.

$$P_\text{i} - P_\text{e} = \frac{2\sigma}{r}. \tag{8.41}$$

Therefore the internal pressure is greater than the external pressure, i.e., $P_\text{i} > P_\text{e}$. This is because there are two forces, the external pressure and the surface tension, acting on the internal molecules.

We will use Eq. (8.37) to compute the effect of the surface tension on the saturated vapor pressure. This equation gives the saturated vapor pressure when there are a different kind of molecules present in the vapor phase. We found that the effect of the foreign molecules is negligible. Now we consider that there are not different kinds of molecules in the vapor, but rather the molecules of the surface layer plays the role of the foreign molecules in the vapor phase. By Eq.(8.41), we see that the pressure due to the surface molecules is $2\sigma/r$. We use $P_\text{v}^{(r)}$ to denote the saturated vapor pressure of a spherical drop of radius r. Therefore the internal pressure of the drop P_i is

$$P_\text{i} = P_\text{e} + \frac{2\sigma}{r} = P_\text{v}^0 + \frac{2\sigma}{r}, \tag{8.42}$$

where we let P_e be equal to the saturated vapor pressure for the liquid with a plane surface P_v^0 ($P_\text{v}^0 = P_\text{v}^{(\infty)}$), this is because there are no other molecules in the vapor phase. Also the quantity P_v^0 in the above equation has the same meaning as that in Eq. (8.37). The quantity P_i can be considered as the total pressure P in Eq. (8.37). Therefore Eq. (8.37) can be rewritten as

$$\ln \frac{P_\text{v}^{(r)}}{P_\text{v}^0} = \frac{v_l}{RT}(P_\text{i} - P_\text{v}^0) = \frac{v_l}{RT}\frac{2\sigma}{r}, \tag{8.43}$$

$$r = \frac{2v_l\,\sigma}{RT\,\ln(P_\text{v}^{(r)}/P_\text{v}^0)}. \tag{8.44}$$

The above equation is the relation of the saturated vapor pressure $P_\text{v}^{(r)}$ for a liquid drop with radius r and the saturated vapor pressure P_v^0 for a liquid with a plane surface when the temperature of the system is T.

However the above relation between r and $P_\text{v}^{(r)}$ is not stable if the temperature T is kept constant. For example, suppose a small drop is formed with radius r, and the saturated pressure $P_\text{v}^{(r)}$ satisfies the above relation. Then if at some moment some molecules evaporate from the drop, this will decrease r, then in order that Eq. (8.44) can be satisfied, $P_\text{v}^{(r)}$ must increase.[10] This will induce more evaporation, and eventu-

[10] If T is a constant, then all other factors in Eq. (8.44), v_l, σ, and P_v^0 are constants.

ally the drop will disappear. On the other hand, if at one moment some molecules condense to the drop, this will increase r, then $P_{\rm v}^{(r)}$ must decrease in order that Eq. (8.44) can be satisfied. This will induce more condensations, until the drop becomes infinitely large (a plane surface) with $P_{\rm v}^{(r)} = P_{\rm v}^0$. From these considerations we see that no liquid drop with a finite radius is stable against thermal fluctuations. It either shrinks and disappears or grows until becomes infinitely large.

It is worth to look at this problem from another angle. We consider a one-component system for which vapor and liquid coexist at temperature T and pressure P. Suppose there are $n_{\rm v}$ kilomoles of vapor molecules and n_l kilomoles of liquid molecules; the total Gibbs free energy G of the system can be written as

$$\begin{aligned} G &= n_{\rm v}\mu_{\rm v}(T,P) + n_l\mu_l(T,P) \\ &= n\mu_{\rm v} + n_l(\mu_l - \mu_{\rm v}), \end{aligned}$$

where $n = n_{\rm v} + n_l$ is the total number of kilomoles of the system. Now if the liquid phase of the system is a spherical drop of radius r, then we have to add the surface tension contribution to G, and the total Gibbs free energy G becomes

$$G = n\mu_{\rm v} + \frac{4\pi r^3}{3\,v_l}(\mu_l - \mu_{\rm v}) + 4\pi r^2\sigma, \tag{8.45}$$

where v_l is the specific volume of the liquid phase and $n_l = \frac{4}{3}\pi r^3/v_l$. When T and P are kept constant, the stable equilibrium of the system is the state where G is minimum, which requires $dG/dr = 0$. It can be shown (see Problem 8.17) that when $\mu_{\rm v} < \mu_l$, the only stable state is $r = 0$, i.e., no liquid drop exists. This is a reasonable result as we expected, because liquid state is the unstable phase ($\mu_l > \mu_{\rm v}$). However when $\mu_{\rm v} > \mu_l$ there exists a critical radius r_c, such that the only stable state for drops with $r < r_c$ is $r = 0$; and for drops with $r > r_c$ the only stable state is $r = \infty$. Therefore we get the same result as in the above analysis; that liquid drops with finite radius are unstable and do not exist if T and P are constant. Moreover, if the saturated vapor pressure which "coexists" with the liquid drop of radius r is denoted as $P_{\rm v}^{(r)}$, the chemical potential of the vapor phase may be written as (cf. Eq. (7.51))

$$\mu_{\rm v} = \mu_{\rm v}^0 + RT\ln\left(\frac{P_{\rm v}^{(r)}}{P_{\rm v}^0}\right),$$

where μ_v^0 and P_v^0 are the chemical potential and the saturated vapor pressure for a liquid with a plane surface. Then the expression of r_c is exactly the same as Eq. (8.44), because $\mu_l = \mu_v^0$.

To have a feeling about how large the critical radius r_c is, we take water at $T = 300$ K as an example. The needed data for water are: saturated vapor pressure for a flat surface $P_v^0 \approx 3.6 \times 10^3$ N m^{-2}, $v_l \approx 1.8 \times 10^{-2}$ m^3 kilomole^{-1}, $\sigma \approx 7.0 \times 10^{-2}$ N m^{-1}, and $R = 8.31 \times 10^3$ J kilomole^{-1} K^{-1}. If we take $P_v^{(r)} = 2.71\, P_v^0$, then from Eq. (8.44) we get $r_c \approx 1.0 \times 10^{-9}$ m. This is very small; there are only about 30 molecules in the drop, and thus it should be considered as a microscopic system, therefore the applicability of Eq. (8.44) is questionable. Also a large saturated vapor pressure, $P_v^{(r)}/P_v^0 = 271\%$, is needed which requires a large and sharp drop of the temperature. Therefore non-equilibrium thermodynamic processes may be involved. If we take a more moderate saturated vapor pressure, $P_v^{(r)}/P_v^0 = 110\%$, then $r_c \approx 1.0 \times 10^{-8}$ m, which contains about 3×10^4 molecules. Although this is not large enough to be considered as a macroscopic system, it is not a small number; the probability that a drop of this size can be formed by thermal fluctuations is extremely small. A drop with a smaller size than r_c will be unstable and will disappear by evaporation according the above theory. Therefore we have the conclusion that, practically, finite size water drops do not exist under the conditions that both the temperature T and the pressure P are kept constant.

Whether the above conclusion may be applicable to the formation of water drops in the atmosphere is questionable. We have assumed that both the temperature T and the pressure P are constants in the above analysis. However our daily life experiences tell us that temperature variations due to weather condition must play an important, and possibly the key, role in the formation of water drops in the atmosphere. This may involve non-equilibrium thermodynamics processes. Moreover a drop must grow from a tiny cluster which may behave quite differently from a macroscopic drop. Therefore microscopic considerations may be also needed in the understanding the nucleation processes for larger water drops in the atmosphere.

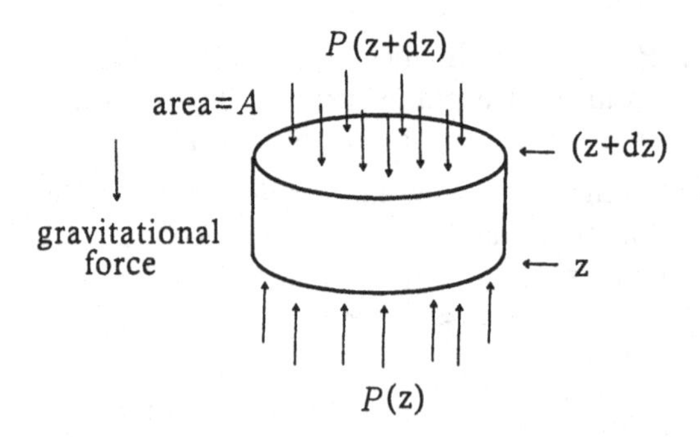

Figure 8.4: The air pressure P is a function of the elevation z.

8.6 Air Temperature as a Function of Elevation

It is well known that the temperature on top of a high mountain is lower than that at sea level. In this section we would like to understand why this happens. First we have to notice that because of the gravitational force, the density of the air is lower at higher elevations. This implies that the air pressure P is a function of the elevation z, i.e., $P = P(z)$. We consider a cylindrical air volume with a cross sectional area A which is parallel to sea level. We consider two horizontal planes with respective elevations at z and $z + dz$, as shown in Fig. 8.4, where dz is a very small height. The cylinder between z and $z + dz$ experiences three forces: (1) the force due to the air above the plane $z + dz$, which is a downward force with a magnitude $P(z + dz)A$, (2) the force due to the air below the plane z which is a upward force with a magnitude $P(z)A$, (3) the gravitational force on the air in the volume, which is a downward force with a magnitude $g\rho(z)A\,dz$ (g and $\rho(z)$ are, respectively, the gravitational acceleration and the density of the air). The balance of the forces gives

$$P(z + dz)A - P(z)A + g\rho(z)Adz = 0.$$

Expanding $P(z + dz)$ to the first order derivative, $P(z + dz) = P(z) + (dP/dz)dz$, and substituting it into the above equation, we obtain

$$\frac{dP}{dz} = -\rho(z)\,g. \tag{8.46}$$

The right hand side of the equality is negative, therefore $dP/dz < 0$, which implies that the higher the elevation the lower the pressure. However, if $T = T(z)$, then the functional form of $\rho(z)$ is unknown, therefore we cannot integrate the above equation to obtain $P = P(z)$. We define $\overline{m}$ as the mean mass per kilomole of the air, then the volume per kilomole $v(z)$ of the air is

$$v(z) = \frac{\overline{m}}{\rho(z)}, \quad \text{or} \quad \rho(z) = \frac{\overline{m}}{v(z)}. \tag{8.47}$$

Substituting the above equation into Eq. (8.46), and using the ideal gas law $P(z)v(z) = RT(z)$ for the air, we can eliminate $v(z)$ to obtain,

$$\frac{dP}{dz} = -\overline{m}\,g\frac{1}{v(z)} = -\overline{m}\,g\frac{P(z)}{RT(z)}. \tag{8.48}$$

This equation gives the relation between the temperature $T(z)$ at the elevation z and the pressure gradient $P(z)/dz$, but we still do not have enough information to integrate the above equation.

In addition to the gravitational effect, we need to find out the mechanism which keeps the air in a steady state,[11] than the term *equilibrium*even though the temperature $T(z)$ and the pressure $P(z)$ are different at different elevations z. We know that air molecules are in dynamic equilibrium rather than static equilibrium. This is because the air molecules will flow from a hotter place to a cooler place, which is known as **convection**. We are considering the case that the air may be approximated as being in a quasi-equilibrium state, and $P(z)$, $T(z)$ and $v(z)$ may be considered to be constants, even though the molecules are flowing. Since the coefficient of thermal conductivity of the air is very small, we may approximate the convection process as an adiabatic process, because the heat transferred from the flowing air molecules to the neighboring air molecules (or the reverse process) is negligible. This is a slow process, and it can also be approximated as a reversible process. Therefore this can be considered as a constant entropy process. We may write the relation between the temperature T and the elevation z as

$$\frac{dT}{dz} = \left(\frac{\partial T}{\partial P}\right)_S \frac{dP}{dz}. \tag{8.49}$$

We know that the hotter air at the lower elevation, which flows adiabatically to the higher elevation, will expand[12] and lower the temperature. The cooler air at the higher elevation, which flow adiabatically to the lower elevation, will be compressed and the temperature

[11] The system is under the action of the gravitational force. Therefore we say it in a *steady state* rather than an *equilibrium state*.

[12] This is due to the pressure of the air at the lower elevation is larger than that of the air at the higher elevation.

will be increased. The relation between the temperature change dT and the pressure change dP is given by Eq. (8.49). By using the equation Pv^γ =constant (Eq. (1.32)), which holds for an ideal gas in a reversible adiabatic process, and the ideal gas equation of state $Pv = RT$, we obtain

$$T(z)P(z)^{(1/\gamma-1)} = \text{constant (reversible adiabatic)}, \tag{8.50}$$

$$\left(\frac{\partial T}{\partial P}\right)_S = \frac{\gamma-1}{\gamma}\frac{T(z)}{P(z)}. \tag{8.51}$$

Substituting the above equation and Eq. (8.48) into Eq. (8.49), we have

$$\frac{dT}{dz} = \frac{\gamma-1}{\gamma}\frac{T(z)}{P(z)}\left[-\overline{m}\,g\frac{P(z)}{RT(z)}\right] = -\frac{(\gamma-1)\overline{m}\,g}{\gamma R}. \tag{8.52}$$

This is the equation we want to derive. We take the data for the air as: $\gamma = 1.41$ ($\approx 7/5$), $\overline{m} = 29$ kg/kilomole, g=9.8 m/sec^2, $R = 8.31 \times 10^3$ J/K·kilomole, and substitute into Eq. (8.52) to obtain

$$\begin{aligned}\frac{dT}{dz} &= -\frac{(1.41-1)\times 29\times 9.8}{1.41\times 8.31\times 10^3}\\ &= -9.9\times 10^{-3}\ \text{K/m} = -9.9\ \text{K/km}.\end{aligned} \tag{8.53}$$

The actual value for the decrease of temperature per 1000 m is about 6°C. The discrepancy between this value and the result of Eq. (8.53) is rather large in percentage, but not large in absolute value. The origin of the discrepancy is mainly due to the neglect of the vapor condensation effect. The rising air contains water vapor, which may condense to water and release heat, i.e., the latent heat of condensation, in the rising process. This will increase the temperature of the air. When this effect is taken into account the agreement is reasonable good (see Problem 8.18).

8.7 Osmotic Pressure in Dilute Solutions

The existence of *osmotic pressure* is an important phenomenon in life science. This phenomenon is associated with the behavior of a *semi-permeable membrane* in a solution. A semi-permeable membrane, such as a cell membrane, is a membrane that has no effect on the passage of some types of molecules, but for other types of molecules the passage is blocked. For example there are some semi-permeable membranes

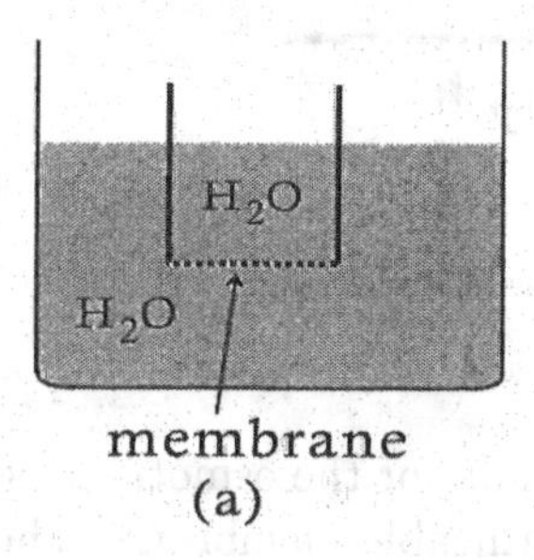

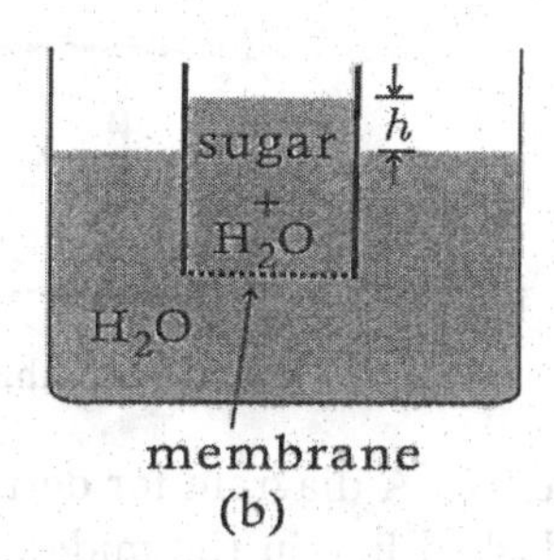

Figure 8.5: Osmotic pressure in a sugar solution. (a) A test tube with a particular type of semi-permeable membrane at its lower end is put in a glass filled with pure water. The water level inside and outside of the tube will be the same. (b) Some sugar crystals are added to the tube so that water inside the tube becomes a sugar solution, but the water outside the tube remains pure. In this case the level of the sugar solution will be higher than the water level, due to the osmotic pressure.

which stop larger molecules but have no effect on the passage of smaller molecules. Due to the presence a semi-permeable membrane, the concentration of the solutions at the different sides of the membrane will be different. This will result in a difference of pressure on the different sides of the membrane. This difference of pressure is called the **osmotic pressure**. We use Fig. 8.5 to explain the phenomenon of osmotic pressure. In Fig. 8.5(a), we have a glass of pure water. If we put in the water an empty test tube whose lower end is sealed with a suitable semi-permeable membrane, then the water molecules will flow *freely* into the tube; finally the water level both inside and outside the tube will be the same, because the semi-permeable membrane has no effect on the passage of the water molecules. Now if we add some sugar crystals to the tube, the water inside the tube becomes a sugar solution. The water outside the tube remains pure water, because the membrane blocks the passage of the sugar molecules. In order to maintain thermodynamic equilibrium, some extra water molecules will flow into the tube which results in a difference in height for the solutions inside and outside the tube, as shown in Fig. 8.5(b). The semi-permeable membrane experiences an extra down ward pressure, which is called the osmotic pressure.

Now we are going to derive the formula for the osmotic pressure. We would like to consider a general case, which will not specify the origin of the pressure, such as gravitation etc., and may be applicable to any solution. Therefore we consider Fig. 8.6 instead of Fig. 8.5(b). In Fig. 8.6,

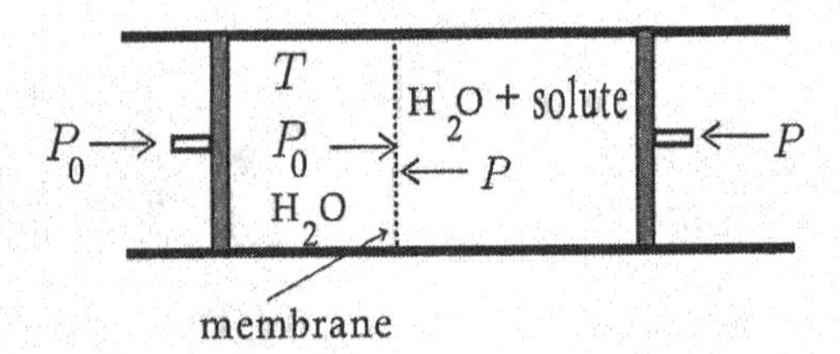

Figure 8.6: A diagram for deriving the formula for the osmotic pressure. The dashed line in the middle is a semi-permeable membrane. The left side is pure water at temperature T and pressure P_0. The right side is a solution at temperature T and pressure P. The pressure difference $P - P_0$ is the osmotic pressure.

the dashed line in the middle represents the semi-permeable membrane. On the left side is pure water, and on the right side is a solution. In thermodynamic equilibrium, both sides have the same temperature T, but the pressures are different. The pressure in water is P_0, and in the solution it is P. The pressure difference $P - P_0$ is the osmotic pressure.

Consider the solution on the right side. From Eq. (5.30),

$$dG = -SdT + VdP + \mu_{\mathrm{w}} dn_{\mathrm{w}} + \mu_{\mathrm{s}} dn_{\mathrm{s}}, \tag{8.54}$$

where μ and n are the chemical potential and the number of kilomoles. The subscripts w and s represent the quantities of water and the solute, respectively. The chemical potentials of water and the solution are functions of the temperature T, pressure P, and the mole fraction of solute x_{s}, thus $\mu_{\mathrm{w}} = \mu_{\mathrm{w}}(T, P, x_{\mathrm{s}})$ and $\mu_{\mathrm{s}} = \mu_{\mathrm{s}}(T, P, x_{\mathrm{s}})$. We consider the case of a very dilute solution, that $x_{\mathrm{s}} \ll 1$. In thermodynamic equilibrium the chemical potential of water on both sides of the membrane must be equal to each other,[13] therefore

$$\mu_{\mathrm{w}}^{(0)}(T, P_0) = \mu_{\mathrm{w}}(T, P, x_{\mathrm{s}}). \tag{8.55}$$

Here $\mu_{\mathrm{w}}^{(0)}$ is the chemical potential of pure water, and P_0 and P are, respectively, the pressure of pure water and the solution. We assume that the solution is a regular solution, and take the heat of mixing as $\Delta H = \lambda n_{\mathrm{w}} n_{\mathrm{s}}/(n_{\mathrm{w}} + n_{\mathrm{s}})$, assuming that in the mole fraction x_{s} we are

[13]Water molecules on both sides of the membrane are the same kind of molecule but in different phases.

considering, the solute and water can mix homogeneously without phase separation.[14] The total Gibbs free energy G of the solution has the form

$$\begin{aligned} G(T, P, n_{\mathrm{w}}, n_{\mathrm{s}}) &= n_{\mathrm{w}}\mu_{\mathrm{w}}^{(0)}(T,P) + n_{\mathrm{s}}\mu_{\mathrm{s}}^{(0)}(T,P) \\ &+ n_{\mathrm{w}}RT\ln x_{\mathrm{w}} + n_{\mathrm{s}}RT\ln x_{\mathrm{s}} + \lambda\frac{n_{\mathrm{w}}n_{\mathrm{s}}}{n_{\mathrm{w}}+n_{\mathrm{s}}}. \end{aligned} \tag{8.56}$$

From Eq. (5.31) we have,

$$\begin{aligned} \mu_{\mathrm{w}}(T,P,x_{\mathrm{s}}) &= \left(\frac{\partial G}{\partial n_{\mathrm{w}}}\right)_{T,P,n_{\mathrm{s}}} \\ &= \mu_{\mathrm{w}}^{(0)}(T,P) + \lambda x_{\mathrm{s}}^2 + RT\ln(1-x_{\mathrm{s}}) \\ &\approx \mu_{\mathrm{w}}^{(0)}(T,P) - x_{\mathrm{s}}RT, \quad \text{for } x_s \ll 1. \end{aligned} \tag{8.57}$$

In the last approximate form we neglected terms with order higher than x_{s}^2. Since $x_{\mathrm{s}} \ll 1$, $\ln(1-x_{\mathrm{s}}) \approx -x_{\mathrm{s}}$. We find that under the condition $x_{\mathrm{s}} \ll 1$, λ does not appear in our final equation. It seems that λ does not play any role in our result, but there is a restriction for the case $\lambda > 0$. If there is a phase separation in the solution, then our result fails, because Eq. (8.56) no longer holds.

Since $x_{\mathrm{s}} \ll 1$, we expect $P - P_0$ to be a small pressure, therefore we may take a Taylor series expansion of $\mu_{\mathrm{w}}^{(0)}(T,P)$ with respect to P around $P = P_0$,

$$\begin{aligned} \mu_{\mathrm{w}}^{(0)}(T,P) &= \mu_{\mathrm{w}}^{(0)}(T,P_0) + \left(\frac{\partial \mu_{\mathrm{w}}^{(0)}}{\partial P}\right)_T (P-P_0) \\ &+ \left(\frac{\partial^2 \mu_{\mathrm{w}}^{(0)}}{\partial P^2}\right)_T (P-P_0)^2 + O(P-P_0)^3. \end{aligned} \tag{8.58}$$

Now $\mu_{\mathrm{w}}^{(0)} = (\partial G^{(0)}/\partial n_{\mathrm{w}})_{T,P}$, therefore the coefficient of $(P-P_0)$ and $(P-P_0)^2$ in Eq. (8.58) can be written as

$$\begin{aligned} \left(\frac{\partial \mu_{\mathrm{w}}^{(0)}}{\partial P}\right)_T &= \frac{\partial}{\partial P}\left(\frac{\partial G^{(0)}}{\partial n_{\mathrm{w}}}\right)_{T,P} = \frac{\partial}{\partial n_{\mathrm{w}}}\left(\frac{\partial G^{(0)}}{\partial P}\right)_{T,n_{\mathrm{w}}} \\ &= \left(\frac{\partial V^{(0)}}{\partial n_{\mathrm{w}}}\right)_{T,P} = v_{\mathrm{w}}^{(0)}, \end{aligned} \tag{8.59}$$

[14] The result we obtained applies to both the cases $\lambda < 0$ and $\lambda > 0$, as long as there is no phase separation in the solution.

$$\left(\frac{\partial^2 \mu_W^{(0)}}{\partial P^2}\right)_T = \left(\frac{\partial v_W^{(0)}}{\partial P}\right)_T \approx 0 \text{ (incompressible)}. \tag{8.60}$$

The last approximation sign in Eq. (8.60) is obtained because the volume of water is almost pressure independent (i.e., incompressible). Substituting Eqs. (8.59)–(8.60) into Eq. (8.58) we have

$$\mu_W^{(0)}(T,P) \approx \mu_W^{(0)}(T,P_0) + v_W^{(0)}(P-P_0). \tag{8.61}$$

Substituting this into Eq. (8.57), the result is

$$\mu_W(T,P,x_s) = \mu_W^{(0)}(T,P_0) + v_W^{(0)}(P-P_0) - x_s RT. \tag{8.62}$$

From the above equation and the condition of equilibrium Eq. (8.55), we finally get the equation for the osmotic pressure $P_{osm} = P - P_0$ as

$$v_W^{(0)} P_{osm} = x_s RT. \tag{8.63}$$

Since $x_s = n_s/n$ ($n = n_W + n_s$), and $nv_W^{(0)} \approx nv_{solution} = V$, the volume of the solution, therefore Eq. (8.63) becomes

$$P_{osm} V = n_s RT. \tag{8.64}$$

This equation is known as the **van't Hoff's[15] law**. This formula can be used to calculate the osmotic pressure P_{osm} for a dilute solution. This formula has exactly the same form as the ideal gas law, except that the letters in the formula may have physical meanings different from those of the ideal gas law. In this formula P_{osm} is the osmotic pressure, V the volume of the solution, and n_s the number of kilomoles of the solute molecule (not including that of the solvent); RT has the same meaning as that in the ideal gas law.

We use an example to test the range of the applicability of Eq. (8.64). We will see how small x_s should be in order to get good agreement with the experimental results. We put n_s kilomoles of sugar ($C_{12}H_{22}O_{11}$) into 1 kg of 30°C water. Experimental data of the osmotic pressure P_{osm} and the number of kilomoles n_s are:

(1) $n_s = 0.1 \times 10^{-3}$ kilomole, $P_{osm} = 2.53 \times 10^5$ Pa,
(2) $n_s = 0.2 \times 10^{-3}$ kilomole, $P_{osm} = 5.17 \times 10^5$ Pa,
(3) $n_s = 0.3 \times 10^{-3}$ kilomole, $P_{osm} = 7.81 \times 10^5$ Pa.

[15] Jacobus H. van't Hoff, Dutch physical chemist (1852–1911), Nobel prize laureate in chemistry in 1901.

We would like to use Eq. (8.64) to compute the theoretical values of the osmotic pressure and compare them with the experimental data. How is the agreement? The solution is:

First we have to calculate the number of kilomoles n_w for 1 kg of water. A water molecule is H_2O with molecular weight $1 \times 2 + 16 = 18$. This implies that the mass of 1 kilomole of water is 18 kg. Therefore the number of kilomoles for 1 kg of water is $n_w = 1/18 = 55.6 \times 10^{-3}$ kilomole. The volume per kilomole for water is $v_w^{(0)} = 18 \times 10^{-3}$ m^3, therefore the total volume is $V = 55.6 \times 10^{-3} \times 18 \times 10^{-3} = 10^{-3}$ m^3, $R = 8.31 \times 10^3$ J kilomole^{-1} K^{-1}, and $T = 303$ K. We substitute these values into Eq. (8.64) and compute P_{osm}:

$$P_{osm} = \frac{n_s RT}{V} = \frac{(8.31 \times 10^3)(303)}{10^{-3}} n_s = 25.2 \times 10^8 \, n_s \text{ Pa}, \qquad (8.65)$$

where the unit for n_s is kilomole. Therefore from this equation we obtain the following results:

(1) $n_s = 0.1 \times 10^{-3}$ kilomole, $P_{osm} = 2.52 \times 10^5$ Pa, with $x_s = 0.1/55.7 = 0.18\%$. The percentage error for P_{osm} is $\Delta P / P_{osm} = 0.4\%$.

(2) $n_s = 0.2 \times 10^{-3}$ kilomole, $P_{osm} = 5.04 \times 10^5$ Pa, with $x_s = 0.2/55.8 = 0.36\%$. The percentage error for P_{osm} is $\Delta P / P_{osm} = 2.5\%$.

(3) $n_s = 0.3 \times 10^{-3}$ kilomole, $P_{osm} = 7.56 \times 10^5$ Pa, with $x_s = 0.3/55.9 = 0.54\%$. The percentage error for P_{osm} is $\Delta P / P_{osm} = 3.2\%$.

We see that the osmotic pressures P_{osm} calculated from Eq. (8.64) are in good agreement with experiment for very small mole fractions x_s. The agreement becomes poorer as x_s increases. In the above example for $x_s < 0.2\%$, the percentage error is less than 0.5%, but for x_s is increased to 0.3%, the percentage is more than 2%.

8.8 Blackbody Radiation: Photon Gas

Thermodynamic laws are mainly used to study the thermodynamic properties of material systems. They have also successfully been applied to study the thermodynamic properties of non-material systems, such as blackbody radiation. Blackbody radiation is composed of electromagnetic waves not atoms, and is therefore a non-material system. We know

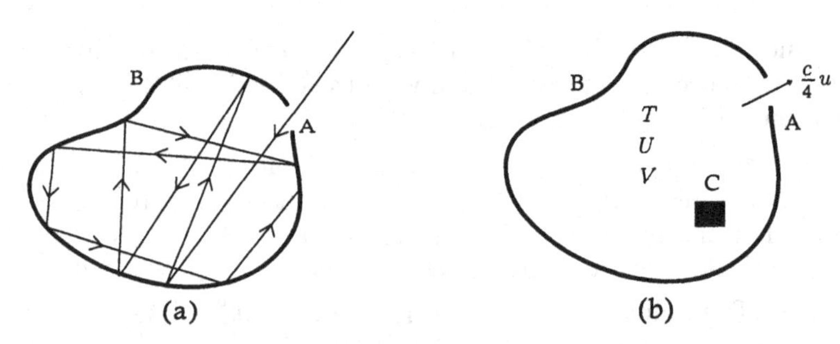

Figure 8.7: A cavity can be considered as a blackbody. (a) A cavity is enclosed by a material B with a small hole A in the wall of B. External radiation which enters the cavity by passing through A will be absorbed completely by the cavity. (b) Put a body C in the cavity described in (a). In thermal equilibrium the temperature T will be equal for both the body C and the wall of B, and the cavity is filled with the blackbody radiation with the spectral distribution of temperature T.

that any material whose temperature is not at absolute zero will radiate electromagnetic waves. This is because all material systems are composed of atoms, and each atom contains charged electrons and protons, which will radiate when the atom has an acceleration. In the cases we considered, the acceleration is due to thermal vibrations and therefore the radiation is called thermal radiation, which is different from other types of radiation, such as that due to external electric or magnetic fields. In general, the spectral distribution of thermal radiation depends not only on the temperature of the radiating system, but also on the composition of the system. There is, however, one class of systems whose spectral distribution of thermal radiation depends only on the temperature of the system, and is independent of its composition. Any system in this class is known as a *blackbody*. The thermal radiation of a blackbody is called the **blackbody radiation**. The spectral distribution of blackbody radiation depends on the temperature only, independent of the composition of its source, therefore it is a basic physical quantity.

A **blackbody** is a body whose surface will absorb all the radiation hitting it, without reflection. The body will not reflect any light shining on it, and therefore it looks black, hence the name blackbody. It is not easy to find in nature a system which possesses the characteristic of a blackbody, because it should absorb radiation of all frequencies. In the laboratory, however, we can make a system which can approximate a blackbody. We consider a cavity enclosed by a material system B with a small hole, A, in the wall of B, as shown in Fig. 8.7(a). External radiation can enter the cavity by passing through the hole A. The radiation will collide with the wall of the cavity, and it will be absorbed or reflected by the wall. The intensity of the reflected beam is reduced, because part of the radiation is absorbed. The reflected beam will repeat the collision and the absorption/reflection process. Finally all the external radiation on A will be absorbed (by the wall of the cavity) and there is no reflection.[16] Therefore the *surface* A may be considered as the surface of a blackbody, because it absorbs all the radiation on it. In the above discussion, we have assumed that the probability of a reflected beam to leak out from A to the exterior of the cavity is very small, because the area of A is assumed to be very small. In thermal equilibrium the cavity will be filled with thermal radiation from the wall of the cavity. If the temperature of the wall is T then the thermal radiation in the cavity will have the spectral distribution of blackbody radiation of temperature T, hence the radiation from the *surface* A, which comes from the cavity, is blackbody radiation of temperature T.

We introduce two physical quantities, the absorptivity a and the emissivity e. The **absorptivity** quantifies the ability for a surface to absorb radiation, and the **emissivity** quantifies the ability for a surface to emit radiation. Suppose $\dot{\varepsilon}_{\rm i}$ and $\dot{\varepsilon}_{\rm a}$ are, respectively, the amount of radiation on a surface per unit area per unit time, and the amount of radiation absorbed by a surface per unit area per unit time (the subscripts mean: i=incident, a=absorption), then a is defined as

$$a = \frac{\dot{\varepsilon}_{\rm a}}{\dot{\varepsilon}_{\rm i}}. \tag{8.66}$$

Therefore a is the percentage of the radiation on a surface that is absorbed ($0 \leq a \leq 1$). In fact the absorptivity of a surface depends on the frequency ν of the radiation. This means that different frequencies will have different absorptivity. Therefore the above definition may be considered as the mean absorptivity. Now we consider the emissivity e. Suppose the temperature of a body is T, and the amount of radiation emitted by its surface per unit area per unit time is $\dot{\varepsilon}_{\rm e}(T)$. At the same temperature T, the amount of radiation emitted by a blackbody

[16] We are interested in thermal radiation, therefore radiation with extremely high frequencies is negligible, and will be neglected. Extreme high frequency radiation like X-rays will actually penetrate the wall of the cavity and will not be absorbed.

per unit area per unit time is $\dot{\varepsilon}_{\rm b}(T)$, then the definition of e is (the subscripts mean: e=emission, b=blackbody),

$$e = \frac{\dot{\varepsilon}_{\rm e}(T)}{\dot{\varepsilon}_{\rm b}(T)}. \tag{8.67}$$

Therefore the emissivity of a body is its relative power of emitting radiation in comparison with a blackbody.

We will show that at a given temperature, a blackbody has the largest emissivity e, i.e., $0 \le e \le 1$. In fact emissivity also depends on the frequency of the radiation ν. Therefore the above definition is the mean emissivity e. We first show that $e = a$ for the same body. Among all systems, a blackbody has the largest a (equal to 1), therefore a blackbody also has the largest e. We put a body C inside the cavity as shown in Fig. 8.7(b). When C is in thermal equilibrium with the cavity (including its wall), they will have the same temperature T, and the cavity is filled with isotropic blackbody radiation of temperature T. Therefore the radiation on C per unit area per unit time is $\dot{\varepsilon}_{\rm b}(T)$, and for C

$$\dot{\varepsilon}_{\rm a} = a\dot{\varepsilon}_{\rm i} = a\dot{\varepsilon}_{\rm b}(T). \tag{8.68}$$

Since the temperature of C is T, the amount of radiation emitted by C per unit area per unit time is $\dot{\varepsilon}_{\rm e}(T)$. From Eq. (8.67), we have

$$\dot{\varepsilon}_{\rm e}(T) = e\dot{\varepsilon}_{\rm b}(T). \tag{8.69}$$

In thermal equilibrium the radiation absorbed and emitted by C must be equal to each other, i.e., $\dot{\varepsilon}_{\rm a} = \dot{\varepsilon}_{\rm e}$, therefore from Eqs. (8.68)–(8.69), we have

$$e = a, \quad \text{and} \quad e_\nu = a_\nu. \tag{8.70}$$

The above equation (8.70) is known as **Kirchhoff's[17] law**; it tells us that for a given body the emissivity is equal to the absorptivity. Therefore for a blackbody the absorptivity is the largest ($a = 1$) among all materials, and its emissivity is also the largest ($e = a = 1$) among all material systems. The second equality in the Kirchhoff's law means that for a given body not only the mean values of e and a are equal to each other, but also that for every frequency e and a are equal, i.e., $e_\nu = a_\nu$. This is because Eqs. (8.68) and (8.69) must hold for any temperature T, but the spectral functions for $\dot{\varepsilon}_{\rm b}(T)$ are uncorrelated for different T; therefore only when $e_\nu = a_\nu$ holds, can the relation $\dot{\varepsilon}_{\rm a} = \dot{\varepsilon}_{\rm e}$ be true for any given T. The frequency dependence of the absorptivity and emissivity have an

[17]Gustav R. Kirchhoff, German physicist (1824–1887).

important consequence, namely that $e_{\nu'}$ may not be equal to a_ν for the same system, if $\nu' \neq \nu$. Therefore a body that looks black (dark) may not have the strongest power to emit thermal radiation. A dark body has a strong absorption power in the frequency range of visible light, but the main frequencies of thermal radiation are in the range of the infrared for a body at around room temperature. A dark body may not have a large emissivity in that frequency range. This is the reason why wearing dark color clothes is warmer than wearing light color clothes. Although dark color clothes may absorb much more radiation from the sun than light color clothes do, their emissivity of thermal radiation may not differ too much.

In the latter half of the nineteenth century, before Planck made his postulate of the quantization of the radiation energy, physicists already knew from experiment that the energy density u of blackbody radiation is proportional to the fourth power of the temperature T, i.e.,

$$u = bT^4, \tag{8.71}$$

where b is a constant. This is an experimental law, but the T^4 dependence can be derived from thermodynamic considerations. However, we will not derive it here, but rather leave it as an exercise for the reader. We just mention the conditions which are needed to derive the T^4 law.

We know that the radiation in the cavity is blackbody radiation. The total energy U of the blackbody radiation is a function of T and is proportional to the volume V of the cavity (independent of the shape). Therefore the energy density of the blackbody u is independent of the volume V, i.e.,

$$u = \frac{U(T,V)}{V} = u(T). \tag{8.72}$$

From electricity and magnetism we know that electromagnetic waves carry momentum and exert a radiation pressure on the surface in the direction of their propagation. The blackbody radiation in the cavity is isotropic, and the radiation pressure P and the energy density u have the relation

$$P = \frac{1}{3}u. \tag{8.73}$$

From Eqs. (8.72) and (8.73) and the relation $dU = TdS - PdV$, we can derive Stefan's T^4 law, i.e., $u = bT^4$. Therefore classical thermodynamics can be used to study the thermodynamic properties of blackbody radiation.

In fact, experimentally, one does not measure u directly, but measures the amount of radiation which is emitted from a small hole A in the cavity. The amount of radiation from A per unit area per unit time is $\dot{\varepsilon}_b(T)$, therefore[18]

$$\dot{\varepsilon}_b(T) = \frac{1}{4} c\, u(T) = \sigma_B T^4, \tag{8.74}$$

where c is the speed of light. Equation (8.74) is called **Stefan's**[19] **law** or **Stefan-Boltzmann law**. This means that $\dot{\varepsilon}_b(T)$ is proportional to the fourth power of T, and the proportionality constant σ_B is called the *Stefan-Boltzmann constant.* The experimental value of σ_B is

$$\sigma_B = \frac{1}{4} cb = 5.67 \times 10^{-8}\, \mathrm{W\, m^{-2}\, K^{-4}}. \tag{8.75}$$

Therefore classical thermodynamics can be successfully used to derive the correct T^4 dependence of the Stefan-Boltzmann law, but it cannot be used to calculate the Stefan-Boltzmann constant. The constant could be correctly calculated theoretically only after Planck proposed that the energy of electromagnetic waves is quantized. Therefore blackbody radiation played an important role in the development of quantum theory in early twentieth century. According to Planck's theory, the *spectral energy density* of blackbody radiation with frequency between ν and $\nu + d\nu$ is[20]

$$u_\nu(T)\, d\nu = \frac{8\pi h}{c^3} \frac{\nu^3}{e^{h\nu/kT} - 1} d\nu, \tag{8.76}$$

where T is the temperature of the blackbody, and the constants c, h, and k are, respectively, the speed of light, the Planck constant, and the Boltzmann constant. Integrating over all the frequencies, we get the total energy density $u(T)$ of the blackbody radiation as

$$\begin{aligned} u(T) &= \int_0^\infty u_\nu(T)\, d\nu = \frac{8\pi h}{c^3} \int_0^\infty \frac{\nu^3\, d\nu}{e^{h\nu/kT} - 1} \\ &= \frac{8\pi h}{c^3} \left(\frac{kT}{h}\right)^4 \int_0^\infty \frac{x^3\, dx}{e^x - 1} = \frac{8\pi^5 k^4}{15 c^3 h^3} T^4, \end{aligned} \tag{8.77}$$

---

[18]We will derive in Chapter 9 the formula for the particle flux (the number of particles which strike a surface of unit area per unit time) $\Phi = \frac{1}{4} \overline{n}\, \overline{v}$, where $\overline{n}$ and $\overline{v}$ are, respectively, the particle density and mean speed of the particles. In a photon gas $\overline{v} = c$, and $\overline{n} = u(T)$.

[19]Josef Stefan, Austrian physicist (1835–1893).

[20]We will derive Eq. (8.76) in Sec. 11.7.

where the integral on the right hand side of the third equality can be integrated to give $\pi^4/15$ (see Appendix E). The above equation satisfies the Stefan's T^4 relation. Comparing this equation with Eq. (8.71), we have $b = 8\pi^5 k^4/15c^3h^3$. Therefore from Eq. (8.75) the theoretical expression for the Stefan-Boltzmann constant $\sigma_{\rm B}$ is

$$\sigma_{\rm B} = \frac{c}{4}\frac{8\pi^5 k^4}{15c^3h^3} = \frac{2\pi^5 k^4}{15c^2h^3}. \tag{8.78}$$

Equation (8.75) together with (8.78) is one part of the set of formulas which Planck used to determine the numerical value of the constant h. If we consider that the speed of light c is a known quantity, and the Boltzmann constant k is unknown, then we need another relation between h and k which can be used to compare with experimental data, in order that both the numerical values of h and k can be determined. We will derive such a relation at the end of this chapter.

The quantized electromagnetic wave is called a photon, and each photon carries an energy $h\nu$, where ν is the frequency of the wave. Thermal radiation is a collection of a large number of photons, and it may be called a photon gas. There is no interaction between photons and therefore a photon gas behaves like an ideal gas. However there is an important difference between an ideal gas and a photon gas, in that the total number of photons is not conserved. Now we will study the thermodynamic properties of a photon gas. From Eqs. (8.71)–(8.73) we get the internal energy U, pressure P, and the constant volume heat capacity C_V as functions of temperature T and volume V as:

$$U = uV = bVT^4, \tag{8.79}$$

$$P = \frac{1}{3}u = \frac{1}{3}bT^4, \tag{8.80}$$

$$C_V = \left(\frac{\partial U}{\partial T}\right)_V = 4bVT^3. \tag{8.81}$$

From the definitions of the entropy S, the Helmholtz free energy F, and the Gibbs free energy G, we obtain the expressions for these quantities as follows (the details of the derivation are left for the reader to complete):

$$S = \frac{4}{3}bVT^3, \tag{8.82}$$

$$F = -\frac{1}{3}bVT^4, \tag{8.83}$$

$$G = 0. \tag{8.84}$$

Therefore the chemical potential μ for the photon gas is zero, because $\mu = G/n$ and $G = 0$. The physical significance of the chemical potential μ is to control the number density of a system, which will be discussed in Chapter 12. If the chemical potential $\mu = 0$ (or a constant), this implies that the number of particles of the system is not conserved. A photon gas is a system which belongs to this class. From the consideration of classical thermodynamics we get the result $\mu = 0$ for a photon gas, which is consistent with the requirements of statistical mechanics.

Finally to conclude this chapter we introduce another formula, **Wien's**[21] **displacement law**,

$$\lambda_{\max} T = B, \quad B = 2.898 \times 10^{-3}\,\mathrm{m\,K}, \tag{8.85}$$

where $\lambda_{\max}$ is the wavelength for which the blackbody radiation of temperature T has maximum intensity, and B is a constant. This is an empirical formula and the given value of B is the experimental value. Now we will derive Eq. (8.85) by using a theoretical approach. We change the variable frequency ν in Eq. (8.76) to the wavelength λ, we obtain $(\nu\lambda = c)$

$$u_\nu(T)|d\nu| = u_\lambda(T)|d\lambda|, \tag{8.86}$$

$$u_\lambda(T) = \frac{8\pi hc}{\lambda^5}\frac{1}{e^{hc/\lambda kT} - 1}. \tag{8.87}$$

By requiring $du_\lambda/d\lambda = 0$ we obtain the equation for $\lambda = \lambda_{\max}$ as

$$\frac{hc}{\lambda_{\max} kT} = 5(1 - e^{-hc/\lambda_{\max} kT}). \tag{8.88}$$

Solving this equation, the result is

$$\frac{hc}{\lambda_{\max} kT} = 4.965, \quad \text{or} \quad \lambda_{\max} T = 0.2014\,\frac{hc}{k}. \tag{8.89}$$

Therefore Eqs. (8.85) and (8.89) provide the second set of formulas used to determine the numerical values of k and h (the first set of the formulas are Eqs. (8.75) and (8.78)). Wien's displacement law (8.85) may be used to determine the temperature of a radiator. For example, the strongest intensity of the radiation from the sun occurs at the wavelength around

[21] Wilhelm Wien, German physicist (1864–1928), Nobel prize laureate in physics in 1911.

5.1×10^{-7} m, therefore the temperature of the surface of the sun is[22] $T = 2.898 \times 10^{-3}$ m K/(5.1×10^{-7} m) ≈ 5700 K. We may also determine the main wavelength of the radiation which is radiated from a body with known temperature. For example the temperature of a human body is around 310 K, therefore the main wavelength $\lambda_{\max}$ of human radiation is around 2.898×10^{-3} m K/(310 K)=9.35×10^{-6} m. This is in the infrared range, which can not be seen by human eyes but can be detected by an infrared detector.

8.9 Problems

8.1. Show that the configuration work for a dielectric material is $đW = -E\,dP$, where E is the applied electric field, and P is the polarization of the material. You may use a parallel capacitor as an example, and calculate the work necessary to charge the capacitor.

8.2. Consider a paramagnetic salt whose equation of state is given by Curie's law; the work done by the system can be written as

$$đW = a\,dT + b\,dB.$$

Find the coefficients a and b in terms of T, B and the Curie constant C_c.

8.3. For a certain range of temperature T and magnetic field B, the magnetic free energy F_M of a magnetic system is given by

$$F_M = -a\,T - \frac{b\,B^2}{2T},$$

where a and b are constants.

[22]The temperature obtained by using the strongest intensity of the radiation from the sun is the surface temperature. The temperature at the center of the sun is about 1000 to 2000 times greater than that of the surface. This is because in the interior of the sun, continuing fission reactions produce a huge amount of heat. Radiation from the interior will be absorbed by the outer atoms and then re-emitted. The absorption and re-emission process continues until the radiation is emitted by the surface atoms. Therefore the spectral distribution of the radiation of the sun detected on the earth is due to the radiation of the atoms on the surface of the sun, and $\lambda_{\max}$ is determined by the temperature of the surface of the sun.

(a) Obtain the equation of state, $M = M(T, B)$, and sketch M vs. T at constant B.

(b) The magnetic field is increased in a reversible adiabatic process, will the temperature of the system increase or decrease?

8.4. Consider a paramagnetic material which obeys Curie's law. Calculate the work done on the system by the magnetic field B, when it is increased from B_1 to B_2 at constant T. What is your comment if the work done is not equal to the change of the magnetic potential energy of the system $-M_2B_2 - (-M_1B_1)$?

8.5. Show that the internal energy U of a paramagnetic salt, which obeys Curie's law, is a function of temperature T only (if the volume of the system is kept constant), i.e., U is independent both of B and M.

8.6. If a gas is both ideal and paramagnetic, obeying Curie's law, show that

$$TdS = C_{V,M}dT - PdV + BdM,$$

where $C_{V,M}$ is the heat capacity at constant volume and constant magnetization. If $C_{V,M}$ is a constant, show that the above equation can be integrated to obtain

$$S = C_{V,M} \ln T + nR \ln V - \frac{M^2}{2C_c} + \text{constant},$$

where C_c is the Curie constant.

8.7. A paramagnetic salt obeys Curie's law and has a constant M heat capacity given by $C_M = a/T^2$ (a is a constant). Show that:

(a)

$$S = -\frac{a}{2T^2} - \frac{M^2}{2C_c} + \text{constant}.$$

(b)

$$C_B - C_M = C_c \frac{B^2}{T^2}.$$

(c) In a reversible adiabatic process, M and B have the relation

$$M = \frac{bB}{\sqrt{1 + (C_c/a)B^2}}, \quad b = \text{constant}.$$

8.8. Consider a magnetic material which obeys the Curie-Weiss law

$$M = a\frac{B}{T-\theta},$$

where a and θ are constants, and B is the magnetic field.

(a) Show that $\left(\frac{\partial U}{\partial M}\right)_T = -\frac{\theta M}{a}$, where U is the internal energy.

(b) Show that $\left(\frac{\partial C_M}{\partial M}\right)_T = 0$, where C_M is the constant M heat capacity.

(c) Show that the internal energy U is given by

$$U(T, M) = \int_{T_0}^{T} C_M dT - \frac{\theta}{2a}M^2 + \text{constant},$$

where T_0 is a reference temperature.

(d) Show that the entropy S is given by

$$S(T, M) = \int_{T_0}^{T} C_M \frac{dT}{T} - \frac{M^2}{2a} + \text{constant},$$

where T_0 is a reference temperature.

8.9. The isothermal susceptibility χ of a magnetic substance is defined as $\chi = \mu_0(\partial M/\partial B)_T$, where M is the total magnetic moment of the system, μ_0 is the permeability of the vacuum, and B is the applied magnetic field.

(a) Show that $(\partial\chi/\partial T)_B \to 0$ as $T \to 0$.

(b) Show that Curie's law, $M = C_c B/T$, for a paramagnetic substance cannot be valid near 0 K.

8.10. Consider a paramagnetic material whose constant M and constant B heat capacities are C_M and C_B, respectively. Show that the following relation holds:

$$\frac{\chi_S}{\chi} = \frac{C_M}{C_B},$$

where χ_S and χ are, respectively, the adiabatic and isothermal susceptibilities, which are defined as

$$\chi_S = \mu_0\left(\frac{\partial M}{\partial B}\right)_S \quad \text{and} \quad \chi = \mu_0\left(\frac{\partial M}{\partial B}\right)_T.$$

8.11. Consider a paramagnetic material which obeys Curie's law (with the Curie constant C_c), and has a heat capacity $C_0 = A/T^2$ for zero magnetization (A is a constant).

(a) Show that the Helmholtz free energy $F = U - TS$ of the system is given by

$$F(T, M) = a - bT - \frac{A}{2T} + \frac{TM^2}{2C_c},$$

where a and b are constants.

(b) The entropy S is given by

$$S(T, M) = b - \frac{A}{2T^2} - \frac{M^2}{2C_c}.$$

(c) The constant B heat capacity C_B is given by

$$C_B = \frac{A + C_c B^2}{T^2}.$$

(d) The adiabatic magnetic susceptibility at a constant field $B = B_0$ is given by

$$\chi_S = \mu_0 \left(\frac{\partial M}{\partial B} \right)_S \Bigg|_{B=B_0} = \mu_0 \frac{AC_c}{T(A + C_c B_0^2)},$$

which reduces to the isothermal magnetic susceptibility $\chi = \mu_0 (\partial M/\partial B)_T = \mu_0 C_c/T$, if $B_0 = 0$.

8.12. For temperature T in the range $0 < T < T_c$ the Helmholtz free energy of a film can be expressed as

$$F = bA \left(1 - \frac{T}{T_c} \right)^n,$$

where A is the area of the film, and b, T_c, and n are constants depending on the film properties.

(a) What experiments do we have to do in order to determine the values of b, T_c, and n?

(b) If it is experimentally found that $1 < n < 2$, what are your comments on the surface entropy s and the surface heat capacity c_A?

8.13. If the surface tension $\sigma(T)$ is known, show that the heat required for an isothermal expansion of the surface area from A_1 to A_2 is given by

$$Q = -T\frac{d\sigma}{dT}(A_2 - A_1).$$

Next show that the reversible adiabatic expansion is given by the condition

$$A\frac{d\sigma}{dT} = \text{constant}.$$

8.14. Calculate the work necessary to slowly increase the radius of a spherical rubber balloon by 5 percent. The initial radius of the balloon is 20 cm and the surface tension of a thin rubber sheet is taken to be 2×10^4 N m^{-1}.

8.15. The equation of state for a surface film can be written as $\sigma = \sigma_0(1 - T/T_c)^n$, where $n = 1.25$, $\sigma_0 = 0.2$ J m^{-2}, and $T_c = 600$ K.

(a) Calculate the entropy s and the specific heat c_A and at $T = 300$ K.

(b) Calculate the temperature change as the area of the film is increased from 0 to 2×10^{-3} m^2 reversibly and adiabatically.

8.16. (a) The vapor pressure of water at 20°C, when no other gas molecules are present, is 17.5 torr. Find the change of vapor pressure when the water is open to the atmosphere. Neglect any effect of the dissolved air.

(b) Find the pressure required to increase the vapor pressure of water by 1 torr.

8.17. Consider a spherical liquid drop with radius r, which coexists with its saturated vapor at temperature T and pressure P. The total Gibbs free energy G of the system, including the contribution of the surface tension σ, is given by Eq. (8.45). Suppose that both T and P are kept constant in the following discussions.

(a) Show that when $\mu_v < \mu_l$, the minimum of G occurs for $r = 0$ only. Note that this implies that no liquid drop with $r \neq 0$ is stable.

(b) Show that when $\mu_v > \mu_l$, the minimum of G occurs for $r = 0$ and $r = \infty$. Note that this also implies that no liquid drop with $r \neq 0$ is stable (a drop with $r = \infty$ is no longer a drop, it has a flat surface).

(c) Consider the case (b), i.e., $\mu_v > \mu_l$. Show that there exists a critical r_c, such that for drops with $r < r_c$, it will evaporate until $r = 0$; and for drops with $r > r_c$, it will grow until r becomes infinity.

(d) Show that the expression of r_c of (c) is exactly the same as Eq. (8.44), if we replace r by r_c and $P_v^{(r)}$ by P.

8.18. In deriving Eq. (8.52) we neglected the effect of condensation of vapor in the process of the adiabatic expansion of air. In this problem we will consider this effect.

(a) Show that when water vapor condensation forms in the adiabatic expansion of air, the temperature change of an air mass is

$$dT = \frac{(\gamma - 1)}{\gamma} \frac{T}{P} dP - \frac{(\gamma - 1)}{\gamma} \frac{\ell}{nR} dn_w,$$

where $\gamma = c_p/c_v$, ℓ is the latent heat of vaporization per kilomole of water, and n and n_w are, respectively, the number of kilomoles of air and water vapor present; assume $n_w \ll n$. In deriving the above relation the air is approximated as an ideal gas.

(b) Assuming the vapor is always saturated in the process and can also be approximated as an ideal gas, show that

$$\frac{dn_w}{dz} = \frac{nP_v\ell}{RPT^2} \frac{dT}{dz} - \frac{nP_v}{P^2} \frac{dP}{dz},$$

where P_v is the saturated vapor pressure at temperature T.

(c) Show that the final result is

$$\frac{dT}{dz} = -\left[\frac{(\gamma - 1)}{\gamma} \frac{\overline{m}g}{R}\right] \frac{1 + \dfrac{P_v}{P} \dfrac{\ell}{RT}}{1 + \dfrac{(\gamma - 1)}{\gamma} \dfrac{P_v}{P} \left(\dfrac{\ell}{RT}\right)^2}.$$

(d) Use the following data to calculate dT/dz at 0°C and $P = 1$ atm: $P_v/P \approx 0.0061$, $\ell \approx 44.9 \times 10^6$ J/kilomole, $\overline{m} \approx 29$ kg/kilomole, and $\gamma \approx 1.40$.

8.19. (a) Show that for a liquid containing a nonvolatile solute in equilibrium with its vapor at a given temperature T and pressure P,

$$g^{(3)} = \mu^{(2)} = g^{(2)} + RT \ln(1 - x),$$

where the superscripts (2) and (3) denote the quantities in the liquid and the vapor phase, respectively; $g^{(2)}$ and $g^{(3)}$ are the specific Gibbs free energy for the pure solvent; and x is the mole fraction of the solute in the solution. We have assumed that the liquid solution is an ideal solution.

(b) For a pure substance show that at constant pressure

$$d\left(\frac{g}{T}\right) = h\,d\left(\frac{1}{T}\right),$$

where g and h are the specific Gibbs free energy and enthalpy of the substance, respectively.

(c) Use part (b) to show that for a small change in x at constant pressure, part (a) reduces to

$$(h^{(3)} - h^{(2)})d\left(\frac{1}{T}\right) = R\,d\,\ln(1-x).$$

(d) In the limit of small x

$$dT = \frac{RT^2}{\ell_{32}}dx,$$

where ℓ_{32} is the latent heat of vaporization. This shows that the boiling temperature is elevated if a solute is added to a liquid.

8.20. One kilogram of sea water contains 35 grams of NaCl, which dissolves into Na^+ and Cl^- ions in water. Calculate the osmotic pressure difference between sea water and fresh water at $T = 300$ K. The gram molecular weight of NaCl is 58.44.

8.21. In an ordinary biological cell, the ratio of the number of water molecules to other kinds of molecules is about 200:1. Consider this as a dilute solution to calculate the osmotic pressure at $T = 300$ K when a cell is put into pure water. One kilomole of water has a mass 18 kg, and a volume of 0.018 m^3.

8.22. Derive Eqs. (8.79)–(8.84) by using the following two known equilibrium properties for the blackbody radiation: (a) the internal energy U is proportional to the volume of the cavity V, i.e., $U = uV$, where $u = u(T)$ is a function of T only; and (b) the radiation pressure $P = \frac{1}{3}u$.

8.23. Consider reversible adiabatic expansion or compression for blackbody radiation in thermal equilibrium inside a cavity of volume V at temperature T. Show that during the process the following equation holds: $TV^{1/3} = \text{constant}$.

8.24. Consider a blackbody radiation in thermal equilibrium inside a cavity of volume V and at temperature T. The system is expanded isothermally until the volume becomes $2V$. Find the heat absorbed by the system in this process.

8.25. Let a photon gas in a cylinder be carried through a Carnot cycle $a \to b \to c \to d \to a$, in which $a \to b$ and $c \to d$ are reversible isothermal processes, and $b \to c$ and $d \to a$ are reversible adiabatic processes. Suppose $T_a = T_b = T$, $T_c = T_d = T - dT$, $V_a < V_b$, and $V_c > V_d$. The pressure P of the photon gas is related to its internal energy u by $P = u/3$, where u is a function of T only.

(a) Plot the Carnot cycle in the P-V plane.

(b) Calculate the work done by the system during the cycle.

(c) Calculate the heat flowing to the system during the cycle.

(d) Show that u is proportional to T^4 by considering the efficiency of the cycle.

Chapter 9

The Kinetic Theory of Gases

9.1 Introduction

In the previous chapters, except in Chapter 3, we used macroscopic empirical results to establish the thermodynamic laws, and to derive many useful thermodynamic relations. In all these studies and discussions we did not consider the roles played by the microscopic constituents of a thermodynamic system. Therefore most of the numerical values of the physical quantities in thermodynamics can only be measured experimentally; they cannot be evaluated theoretically. These physical quantities include the specific heat, the equation of state of a system, etc.[1] Theoretical calculations of the numerical values of these quantities can be done only when a microscopic theory for the foundation of thermodynamics has been established. Moreover physical insight into some thermodynamic quantities, such as heat and entropy, and many irreversible processes such as heat conduction and diffusion, can only be understood in terms of microscopic descriptions. In this chapter we will use the **kinetic theory of gases** to study the simplest system, the classical ideal gas. In this theory a gas is considered as composed of many non-interacting microscopic particles (molecules), except for the *hard collisions.* The theory can successfully derive many equilibrium thermodynamic properties of a classical ideal gas, including the heat capacity, average molecular speed, equation of state, etc. Moreover the theory can also be applied to

[1]If the equation of state of a system is known, then several quantities can be calculated, such as the expansivity, the compressibility and, for a magnetic system, the magnetic susceptibility, etc.

calculate several coefficients of transport processes of a dilute gas, such as diffusion, thermal conduction, etc. These phenomena belong to the category of non-equilibrium thermodynamics, because in these cases the system is under the action of external fields.

The kinetic theory of gases, however, cannot be extended to condensed matter, such as liquids and solids, in which the inter-atomic (or intermolecular) interactions are much stronger and more complicated than those in a gaseous system. Therefore another approach is required to study these more condensed systems. The main point is that it is impractical to describe the details of the atomic (or molecular) motions of the system. The new approach uses the statistical method to study a system with a very large number of particles. We will study this approach in Chapter 10.

9.2 Basic Assumptions

We are considering an idealized classical gaseous system of N molecules in a volume V at temperature T, which has the following features: (a) the molecules obey the classical equations of motion; (b) there is no intermolecular interaction, except for binary collisions; (c) the number density of the molecules is independent of time and space (assuming no external fields); (d) the distribution of molecular speed is also independent of time and space, and the velocity distribution is isotropic. In order to satisfy these features, the following requirements must be met:

(1) Thermal equilibrium (uniform distribution) is established by collisions, therefore in order to satisfy the conditions of (c) and (d), the system must have a very large N/V, so that a molecule can make many collisions before an experimental measurement is made. We will see that the mean collision time τ for a molecule is proportional to $\sqrt{T}/P$ at temperature T and pressure P. At room temperatures and $P = 1$ atm, τ is of the order of 10^{-10} to 10^{-11} seconds. Therefore at room temperatures, this may set a lower limit for the pressure of around 10^{-6} atm or N/V around 10^{13} per cm^3. This implies that a molecule can make about $10^4 - 10^5$ collisions per second.

(2) The condition (b) requires that N/V not be too large. If we use a sphere of diameter d_0 to approximate a molecule, then we require that $\bar{d} \gg d_0$, where $\bar{d} = (V/N)^{1/3}$ is the mean distance between

two neighboring molecules. Under this condition the interaction between molecules can be neglected, because the mean distance between molecules is large enough for the interactions to be neglected. At STP ($T = 0°\text{C}, P = 1$ atm), the volume of 1 mole of molecules is 22.4 liters, therefore $N/V = (6.02 \times 10^{23}/22.4 \times 10^{-3})$ $\text{m}^{-3} = 2.69 \times 10^{25}$ m^{-3}, and $\overline{d} \approx 3.3 \times 10^{-9}$ m. For most molecules d_0 is around $2.5 \sim 5 \times 10^{-10}$ m, and the condition $\overline{d} \gg d_0$ can be satisfied at STP. We may roughly say that the STP value of N/V sets the upper limit of N/V for (b) to be satisfied.

(3) Condition (a) is the requirement of the classical limit, that quantum effects can be neglected. The condition for the classical approximation to be valid is[2]

$$\lambda_{\text{th}} \ll \overline{d}, \quad \lambda_{\text{th}} = \frac{h}{\sqrt{2\pi mkT}}, \tag{9.1}$$

where m is the mass of each particle and h and k are, respectively, the Planck constant and the Boltzmann constant. The symbol λ_{th} denotes the *thermal wavelength* of a particle. In quantum mechanics a particle can be considered as a wave with the wavelength $\lambda = h/p$, where p is the momentum, and $p = \sqrt{2m\varepsilon}$ (ε is the kinetic energy). In Eq. (9.1) πkT may be considered as the thermal energy, thus λ_{th} is called the thermal wavelength. The condition of Eq. (9.1) requires low pressure (large $\overline{d}$) and high temperature (small λ_{th}). Now we estimate the values of $\overline{d}$ and λ_{th}. At STP, $\overline{d} \approx 3.3 \times 10^{-9}$ m, and $\lambda_{\text{th}} < 7.0 \times 10^{-11}$ m (take m to be the mass of H_2). Therefore the condition of Eq. (9.1) is satisfied. Since $\overline{d} \propto (T/P)^{1/3}$, $\lambda_{\text{th}} \propto T^{-1/2}$, therefore, if we exclude the extreme case of very high pressure or very low temperature, or very low pressure and temperature,[3] we may say that most of the gases can be considered as classical gases. The condition of Eq. (9.1) is in general satisfied.

Finally, we have to make the assumption that the collisions between molecules are elastic. This condition can be easily satisfied. Inelastic

[2]This condition will be derived in Chapter 10.

[3]At normal pressure and a low enough temperature all gases will be condensed to liquids. However if the pressure is low enough, a gas may still be stable in the gaseous phase even at very low temperatures, thus the condition of Eq. (9.1) may no longer hold. For example, in Sec. 6.5 we gave an example of a quantum gas at extremely low T and P.

collisions involves energy level changes of the orbital electrons during the collisions. This in general requires an exchange of energy of the order of 1 eV or more. The average kinetic energy of a molecular at temperature T is around kT, which is about 0.025 eV for $T = 300$ K. Therefore the probability of an inelastic collision is extremely small. However the collisions between the molecules and the wall of the container may be inelastic in general. Since the energy levels of the wall are continuous, a molecule may lose or gain energy from the wall during its collision with the wall. However there is no difficulty in handling the inelastic collisions with the wall, as long as the molecules and the wall are in thermal equilibrium, i.e., they have the same temperature. In this case, *on the average*, there is no exchange of energy between the molecules and the wall, and the final result will be the same if we treat all the collisions as elastic.

9.3 The Distribution Function and Molecular Flux

If we want to understand the state of a classical particle, we need to know the position vector $\boldsymbol{r}$ and the velocity vector $\boldsymbol{v}$ of the particle as a function of time t. However when we consider a system with very large number of particles (molecules), there is no need for us to know all the $\boldsymbol{r}$'s and $\boldsymbol{v}$'s of the molecules as functions of t. We introduce a quantity $\mathcal{F}(\boldsymbol{r}, \boldsymbol{v}, t)$, known as the distribution function. The physical meaning is that $\mathcal{F}(\boldsymbol{r}, \boldsymbol{v}, t) d^3r\, d^3v$ represents, at time t, the number of molecules with position vectors between $\boldsymbol{r}$ and $\boldsymbol{r} + d\boldsymbol{r}$, and velocity vectors between $\boldsymbol{v}$ and $\boldsymbol{v} + d\boldsymbol{v}$. Since we are considering a spatially homogeneous system, the distribution function should be independent of $\boldsymbol{r}$ and t. Also the velocity distribution is isotropic, so the distribution function is a function of the speed v only, independent of the direction of $\boldsymbol{v}$. Therefore we may define a speed distribution function $F(v)$,

$$F(v)\, dv = \text{number of molecules with speeds between } v \text{ and } v + dv, \tag{9.2}$$

and

$$\int_0^\infty f(v) dv = 1, \quad f(v) \equiv \frac{F(v)}{N}, \tag{9.3}$$

where N is the total number of molecules, and $f(v) = F(v)/N$ is called the **probability density function.** This means that $f(v)dv$ is the probability of finding a molecule with a speed between v and $v + dv$.

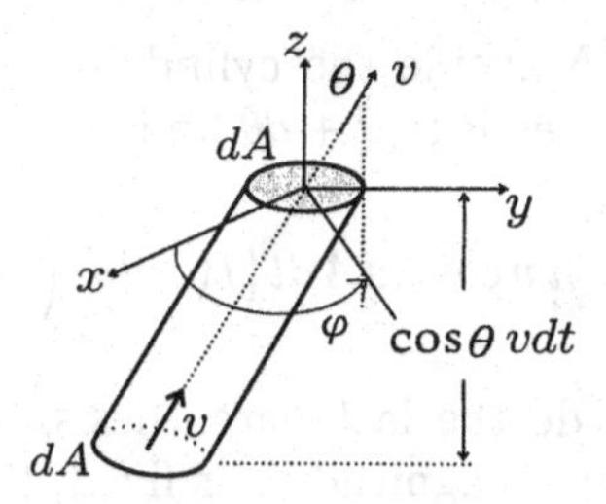

Figure 9.1: A graph to calculate the molecular flux.

From $f(v)$ we can calculate the average value of any function of v. The most common example being the mean value of v^n, i.e., $\overline{v^n}$:

$$\overline{v^n} = \int_0^\infty v^n f(v)dv. \tag{9.4}$$

We know that the mean values of the speed of a system, such as $\overline{v}$ and $\overline{v^2}$ are both dependent on the temperature T, therefore the probability density function $f(v)$ not only depends on v, it also depends on T.

We consider a system with N molecules in a volume V, which satisfies the basic assumptions described in the previous section. We use $\overline{n} = N/V$ to denote the average number of molecules per unit volume. Now we will calculate how many molecules strike the wall of the container per unit area per unit time. We call this quantity the **molecular flux** Φ. We use spherical coordinates to describe the velocity of a molecule $\boldsymbol{v} = (v, \theta, \phi)$, as shown in Fig. 9.1. We take one of the walls of the container as the xy plane, i.e., $z = 0$, and the container is in the space of $z < 0$. We consider a small area dA on the wall which is centered at the origin. We are going to calculate how many molecules will strike dA per unit time. For the convenience of calculation, we classify the molecules according to their velocity (v, θ, ϕ). We consider molecules with speed within the range v to $v + dv$, with their direction being between θ to $\theta + d\theta$ and ϕ to $\phi + d\phi$. Any molecule with velocity in the above range, and whose position at time t is inside the cylinder shown in Fig. 9.1, will strike dA in the time interval from t to $t + dt$. The reason is that in time dt the molecule will travel a distance $v\,dt$ and with its direction aiming at dA (we take dt to be so small that the probability of suffering a collision during dt is negligible). The volume of the cylinder is $v \cos\theta\, dt\, dA$, and the total number of molecules in it is $\overline{n}\, v \cos\theta\, dA\, dt$. The probability that a molecule has a speed between v to $v + dv$ is $f(v)dv$, therefore the number of molecules dN inside the cylinder with a speed within v to $v+dv$, a direction between θ to $\theta+d\theta$ and ϕ to $\phi+d\phi$, is

$$dN(v,\theta,\phi) = \overline{n}\, v \cos\theta \, dA\, dt\, [f(v)dv] \left(\frac{\sin\theta\, d\theta\, d\phi}{4\pi} \right), \tag{9.5}$$

where the quantity inside the last parenthesis, $\sin\theta\, d\theta\, d\phi/(4\pi)$, is the probability that when the magnitude v is fixed, the direction is within θ to $\theta+d\theta$ and ϕ to $\phi+d\phi$. This can be shown as follows. On the surface of a sphere of radius v, an infinitesimal area da within the boundaries $\theta \to \theta+d\theta$ and $\phi \to \phi+d\phi$ can be written as

$$da = v^2 \sin\theta\, d\theta\, d\phi. \tag{9.6}$$

Therefore the probability of finding the value of θ in the range θ to $\theta+d\theta$, and the value of ϕ within ϕ to $\phi+d\phi$ is $(0 \le \theta \le \pi,\ 0 \le \phi \le 2\pi)$

$$\frac{v^2 \sin\theta\, d\theta\, d\phi}{v^2 \int_0^\pi \sin\theta\, d\theta \int_0^{2\pi} d\phi} = \frac{1}{4\pi} \sin\theta\, d\theta\, d\phi. \tag{9.7}$$

The molecular flux Φ is therefore

$$\begin{aligned} \Phi &= \int \frac{dN}{dA\, dt} = \frac{\overline{n}}{4\pi} \int_0^\infty v f(v) dv \int_0^{\pi/2} \sin\theta \cos\theta\, d\theta \int_0^{2\pi} d\phi \\ &= \frac{1}{4} \overline{n}\, \overline{v}. \end{aligned} \tag{9.8}$$

It should be noted that the integration of θ is from 0 to $\pi/2$ (i.e., from $0°$ to $90°$), not to π, because for $\theta > 90°$ the direction of $\boldsymbol{v}$ is in the negative z direction. A molecule with this $\boldsymbol{v}$ will move away from the wall and will not strike dA.

9.4 Gas Pressure and the Ideal Gas Law

We know that the pressure on the wall of the container is due to the collisions of the molecules on the wall. In this section we will discuss the problem of collisions between the molecules and the wall. Most of the books consider the collisions as elastic collisions with mirror reflections, i.e., the reflected speed is the same as the incident speed, and the reflected angle is equal to the incident angle. Although the results obtained are correct, these assumptions are not general enough, so that they can

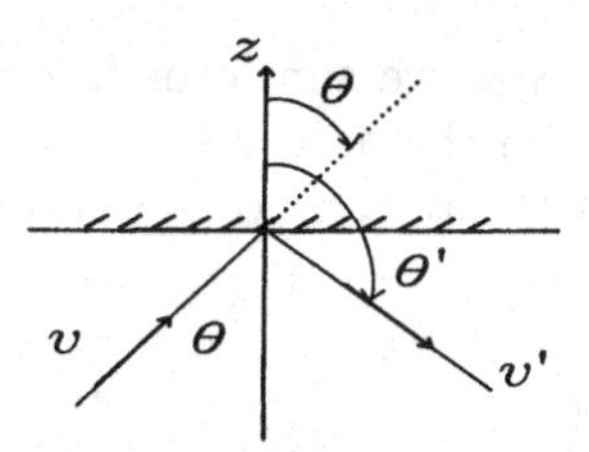

Figure 9.2: The diagram for a collision between a molecule and the wall of a container, which is in the xy plane. The incident and the reflected velocities of the molecule are respectively (v, θ, ϕ) and (v', θ', ϕ'). To simplify the diagram, the angles ϕ and ϕ' are not shown.

be applied only to very smooth and flat surfaces. In fact the results obtained are valid for any surface, and apparently these assumptions are not necessary. Therefore we will not make any assumption about the nature of the collisions, but only require that the molecules and the wall are in thermal equilibrium (i.e., have the same temperature), because this is what happens in a real situation. Therefore we may say that we do not make any assumptions about the collisions between the molecules and the wall. One of the important consequences of thermal equilibrium between the molecules and the wall is that the probability density function before collision $f(v)$ and after collision $f(v')$ are the same, although the speed before collision v and after collision v' may not be the same. Now we will use this condition to discuss the problem of collisions between the molecules and the wall.

We consider the collision between the ith molecule and the wall whose normal is in the z-direction, as shown in Fig. 9.2. The velocities of the molecule before and after the collision are respectively denoted as $\boldsymbol{v}_i$ and $\boldsymbol{v}'_i$. At first we consider the momentum change in the directions parallel to the wall. Since there is no difference between the x- and the y-direction, we discuss only the x-direction.[4] The momentum change of the x-component of the ith molecule is

$$\Delta p_{ix} = mv'_{ix} - mv_{ix} = mv' \sin\theta' \cos\phi' - mv \sin\theta \cos\phi, \tag{9.9}$$

where m is the mass of a molecule, v_{ix} and v'_{ix} are the x-components of $\boldsymbol{v}_i$ and $\boldsymbol{v}'_i$, respectively. In the second equality we used spherical coordinates

[4] If elastic collisions and mirror reflections are assumed, then the momentum change parallel to the wall is easily seen to be zero.

to express v_{ix} and v'_{ix}. Now we sum over the x-components of all the molecules which will strike the area dA in the time interval from t to $t + dt$. Using Eq. (9.5) we may convert $\sum_i$ into an integral

$$\begin{aligned}
\Delta p_{\text{tot},x} &= \sum_i \Delta p_{ix} \\
&= \int mv' \sin\theta' \cos\phi' \, dN(v', \theta', \phi') - \int mv \sin\theta \cos\phi \, dN(v, \theta, \phi) \\
&= m \int v' \sin\theta' \cos\phi' \, \overline{n} \, v' f(v') \, dv' |\cos\theta'| dA \, dt \, \frac{\sin\theta' \, d\theta' \, d\phi'}{4\pi} \\
&\quad -m \int v \sin\theta \cos\phi \, \overline{n} \, v f(v) \, dv \cos\theta \, dA \, dt \, \frac{\sin\theta \, d\theta \, d\phi}{4\pi} \\
&= 0,
\end{aligned} \tag{9.10}$$

where the final result is 0, because $\int_0^{2\pi} \cos\phi' \, d\phi' = \int_0^{2\pi} \cos\phi \, d\phi = 0$. By using the same procedure we may also obtain the result $\Delta p_{\text{tot},y} = 0$. Therefore, when we sum over all molecules, the net momentum change parallel to the wall is zero, but an individual molecule may have a momentum change parallel to the wall. The result tells us that the wall does not suffer any force parallel to the surface, even though *we have not made any assumptions about the collisions.* Therefore this result is quite general and should hold for any wall, which may be rough and not flat. Moreover it is consistent with the known fact that the pressure is always normal to the surface; the component parallel to the surface is zero. It is worth mentioning the following points:

(1) Although we have imposed the condition that the probability density functions $f(v')$ and $f(v)$ are the same, but even if they are not the same function we still obtain the result $\Delta p_{\text{tot},x} = \Delta p_{\text{tot},y} = 0$. This tells us that *even when the molecules and the wall are not in thermal equilibrium, the pressure still has no components parallel to the wall.*

(2) For an individual molecule, the momentum change $\Delta p_{i,x}$ or $\Delta p_{i,y}$ may not be equal to zero. Only for the sum over all molecules does one obtain the result $\Delta p_{\text{tot},x} = \Delta p_{\text{tot},y} = 0$.

(3) In the third equality we took the absolute value of the expression $\cos\theta'$. The number dN must be positive, but for the reflected molecules the angle θ' has a value between $\pi/2$ and π, which implies that $\cos\theta' \leq 0$. Therefore we need to take the absolute value to make $dN > 0$.

Now we consider the momentum change in the z-direction. First we consider the momentum change of the ith molecule,

$$\Delta p_{iz} = mv'_{iz} - mv_{iz} = mv' \cos\theta' - mv \cos\theta. \tag{9.11}$$

Referring to Eq. (9.10), we obtain the total momentum change of the molecules which strike the area dA in the time interval from t to $t + dt$,

$$\begin{aligned}
\Delta p_{\text{tot},z} &= m \int v' \cos\theta' \, \overline{n}\, v' f(v')\, dv' |\cos\theta'| dA\, dt \, \frac{\sin\theta'\, d\theta'\, d\phi'}{4\pi} \\
&\quad - m \int v \cos\theta \, \overline{n}\, v f(v)\, dv \cos\theta \, dA\, dt \, \frac{\sin\theta\, d\theta\, d\phi}{4\pi} \\
&= \frac{\overline{n}\, m\, dA\, dt}{2} \int_0^\infty v'^2 f(v') dv' \int_{\pi/2}^{\pi} \sin\theta' \cos\theta' |\cos\theta'| d\theta' \\
&\quad - \frac{\overline{n}\, m\, dA\, dt}{2} \int_0^\infty v^2 f(v) dv \int_0^{\pi/2} \sin\theta \cos^2\theta \, d\theta \\
&= -\frac{1}{3} \overline{n}\, m\, \overline{v^2}\, dA\, dt.
\end{aligned} \tag{9.12}$$

Note that in the second equality, the integration over ϕ gives a factor 2π, and so does the integration over ϕ'. In this derivation we have used the condition that $f(v')$ and $f(v)$ are the same function, which implies $\overline{v'^2} = \overline{v^2}$, and the two integrated terms can be added to become one term. We also note that the integration over the angle θ' is from $\pi/2$ to π, and that of the angle θ is from 0 to $\pi/2$. This is because θ' is the angle for the reflected velocity and θ is the angle of the incident velocity.

From mechanics we know that $\Delta p_{\text{tot},z}/dt$ is the force exerted on the molecules by the area dA of the wall. This is a force normal to the wall, and the reaction force acting on the wall (due to the molecules) per unit area is the pressure P on the wall,

$$P = -\left(\frac{\Delta p_{\text{tot},z}}{dt}\right) \frac{1}{dA} = \frac{1}{3} \overline{n}\, m\, \overline{v^2}. \tag{9.13}$$

Using $\overline{n} = N/V$, the above equation can be written as

$$PV = \frac{1}{3} N m \overline{v^2} = NkT. \tag{9.14}$$

The last equality is due to the ideal gas law, $PV = NkT$, where k is the Boltzmann constant. From the above equation we obtain

$$\frac{1}{2} m \overline{v^2} = \frac{3}{2} kT. \tag{9.15}$$

This means that the average kinetic energy per molecule is equal to $\frac{3}{2}kT$ when the temperature of the system is T. The energy kT is usually referred to as the thermal energy of a particle which is in an environment of temperature T. This is because kT is roughly equal the average kinetic energy per molecule, by Eq. (9.15).

This result is quite general for a classical particle. We will show in Chapter 11 that this result also holds for molecules in liquids and solids. It is worth noting that we call this average kinetic energy **thermal energy** because this is the mean energy of *random motions*. An important characteristic of the random motion is that the velocity of the center of mass of the system is zero. If the velocity of the center of mass of the system is non-zero, then this is a coherent motion of the system. The energy associated with a coherent motion is a **mechanical energy** not a thermal energy.[5] Therefore if a system has a center of mass motion, then we have to exclude this part of the energy in order to get the correct thermal energy. The requirements that a system has no center of mass motion are $\overline{v_x} = \overline{v_y} = \overline{v_z} = 0$.

9.5 The Equipartition Theorem

The system we considered is isotropic, and therefore $\overline{v_x^2} = \overline{v_y^2} = \overline{v_z^2}$, thus

$$\overline{v^2} = \overline{v_x^2} + \overline{v_y^2} + +\overline{v_z^2} = 3\overline{v_x^2}. \tag{9.16}$$

From Eq. (9.15) we have

$$\frac{1}{2}m\overline{v_x^2} = \frac{1}{2}m\overline{v_y^2} = \frac{1}{2}m\overline{v_z^2} = \frac{1}{2}kT. \tag{9.17}$$

This means that the mean thermal energy is the same for each degree of freedom which is equal to $\frac{1}{2}kT$. This is known as the **equipartition theorem** in statistical mechanics. We will show in Chapter 11 that this theorem holds not only for the kinetic energy terms, which have the form bp^2 (with b a constant and p a generalized momentum), but also holds for the potential energy terms, which have the form aq^2 (with a a constant and q a generalized coordinate). This theorem tells us that the mean value of each term of the above form is equal to $\frac{1}{2}kT$, independent of the

[5] A mechanical energy can be completely (i.e., 100%) converted to work, but a thermal energy (heat) can not be completely converted to work, because of the second law.

values of a and b. But it should be noted that this theorem applies only to classical systems, it does not hold for quantum systems.

Now we consider a classical ideal gas consisting of N molecules in thermal equilibrium at temperature T. The Hamiltonian of each molecule contains f terms of the form aq^2 or bp^2, therefore the number of degrees of freedom is f. Then the internal energy U of the system can be written as

$$U = N\bar{\varepsilon} = \frac{f}{2}NkT = \frac{f}{2}nRT, \tag{9.18}$$

where n is the number of kilomoles, and $\bar{\varepsilon} = fkT/2$ is the mean energy per molecule. Therefore the constant volume heat capacity per kilomole c_v, the constant pressure heat capacity per kilomole c_p, and γ are respectively (note that $c_p = c_v + R$)

$$c_v = \frac{f}{2}R, \quad c_p = \left(\frac{f}{2}+1\right)R, \quad \gamma = \frac{c_p}{c_v} = \frac{f+2}{f}. \tag{9.19}$$

Now we use Eq. (9.19) to discuss the values of c_v, c_p and γ for monatomic gases and diatomic gases.

1. Monatomic gases:
   Molecules such as He, Ne, Ar, Kr, etc. are moantomic. The number of degrees of freedom is three, i.e., $f = 3$ (each molecule has 3 kinetic energy terms p_x^2, p_y^2 and p_z^2), therefore

$$c_v = \frac{3}{2}R, \quad c_p = \frac{5}{2}R, \quad \gamma = \frac{5}{3} \approx 1.67. \tag{9.20}$$

2. Diatomic gases:
   Molecules such as H_2, O_2, N_2, CO and Cl_2 are diatomic. There are 7 energy terms in the Hamiltonian of each molecule, which includes the 3 terms of the translational motion of the center of mass, 2 rotational motion terms,[6] and 2 vibrational terms.[7] Therefore one

[6]In general there are three rotational energy terms of the form $I_i\omega_i^2/2$, where I_i and ω_i are respectively the moment of inertia and the angular velocity of the ith component. For a diatomic molecule, the moment of inertia along the direction connecting the two atoms can be approximated as zero. Therefore there are only two terms left.

[7]The vibrational energy has the form $\frac{1}{2}\alpha r^2 + \frac{1}{2}\mu\dot{r}^2$, where r is the distance between the two atoms, α and μ are, respectively, the spring constant and the reduced mass.

would expect $f = 7$, and $\gamma = (f+2)/f = 9/7 \approx 1.29$. However *experimental values of γ are close to 1.40=7/5 rather than 1.29 at room temperatures for almost all diatomic gases.* Experimental value of γ for a gas may be obtained from measurement of the velocity of sound v_s, using the relation $v_s = \sqrt{\gamma/\rho\kappa}$ where ρ and κ are, respectively, the mass density and the isothermal compressibility. The experimental values of γ for some of the diatomic gases are listed in Table 9.1. Therefore the value of f should be 5 rather than 7. This is one of the examples where classical theory fails. In order to understand why f is 5 instead 7, we have to interpret this result in terms of quantum mechanics.

In classical mechanics all the energies are continuous quantities, but in quantum mechanics energy may be continuous or quantized. The translational kinetic energy may be considered as continuous in quantum mechanics,[8] but the rotational and vibrational energies are quantized. In order to excite the rotational or the vibrational motion of a diatomic molecule, the molecule must absorb an energy which is equal to the energy difference $\Delta\varepsilon$ between the ground state and the first excited state. If $\Delta\varepsilon \ll kT$, then the molecules can easily be excited and the quantization of the energy levels may be neglected, so classical theory works well. However if $\Delta\varepsilon \geq kT$, then the quantum effects become important, and classical theory may fail. This is because the molecules do not have enough energy to be excited to the first excited state, and the associated degree of freedom may be considered as non-existent.

For diatomic molecules, the characteristic temperature $\theta_{\rm rot}$ for the rotational motion[9] are mostly smaller than 10 K, and the rotational energy levels can be considered as continuous at room temperatures. However the characteristic temperature $\theta_{\rm vib}$ for the vibrational motion[10] are mostly larger than 2000 K (see Table 9.1), therefore the vibrational degrees of motion can be considered as non-existent at room temperatures. Thus, at room temperatures, the number of degrees of freedom f for a diatomic molecule is 5 rather than 7, and

$$c_v = \frac{5}{2}R, \quad c_p = \frac{7}{2}R, \quad \gamma = \frac{7}{5} = 1.40. \tag{9.21}$$

[8]This is why classical statistical mechanics works well for monatomic gases.
[9]The characteristic temperature for rotation is defined as $\theta_{\rm rot} = \Delta\varepsilon_{\rm rot}/k$.
[10]The characteristic temperature for vibration is defined as $\theta_{\rm vib} = \Delta\varepsilon_{\rm vib}/k$.

Table 9.1 Some Experimental Values of γ, θ_{rot} and θ_{vib}

Monatomic	γ	Diatomic	γ	θ_{rot}(K)	θ_{vib}(K)
He	1.66	H_2	1.40	85.4	6140
Ne	1.64	O_2	1.40	2.1	2239
Ar	1.67	N_2	1.40	2.9	3352
Kr	1.69	CO	1.42	2.8	3080
Xe	1.67	Cl_2	1.36	0.36	810

9.6 The Distribution of Molecular Speeds

Interactions between particles is the most important mechanism for a system to reach thermal equilibrium. From the macroscopic point of view, thermal equilibrium requires that the temperature must be equal everywhere in the system. From the microscopic point of view, thermal equilibrium requires that the energy distribution among the particles must satisfy the distribution function at the given temperature. The only interaction between ideal gas molecules is the hard core collision between the molecules. In this section we will prove that from the kinetics of the hard core collisions between the molecules we can derive the speed distribution function of the ideal gas molecules.

We consider a system of N molecules each with mass m. For simplicity we assume that each molecule is a sphere with a diameter d_0. There is no interaction between the molecules except for the hard core collisions. We are considering a dilute gas such that the mean distance between the molecules $\overline{d}$ is much greater than the diameter of a molecule d_0, i.e., $\overline{d} \gg d_0$. Therefore we may just consider binary collisions; three-body collisions may be neglected. We consider a binary collision where the velocities before and after the collision are, respectively, $(\boldsymbol{v}_1, \boldsymbol{v}_2)$ and $(\boldsymbol{v}_1', \boldsymbol{v}_2')$, as shown in Fig. 9.3(a). We have discussed in Sec. 9.2 that the collisions between molecules are elastic collisions. Therefore both the conservation of energy and momentum hold in each collision,

$$\boldsymbol{v}_1 + \boldsymbol{v}_2 = \boldsymbol{v}_1' + \boldsymbol{v}_2', \tag{9.22}$$

$$v_1^2 + v_2^2 = v_1'^2 + v_2'^2. \tag{9.23}$$

We use the notation $(\boldsymbol{v}_1, \boldsymbol{v}_2) \to (\boldsymbol{v}_1', \boldsymbol{v}_2')$ to denote the collision with the specified incoming and outgoing velocities. To simplify our discussions,

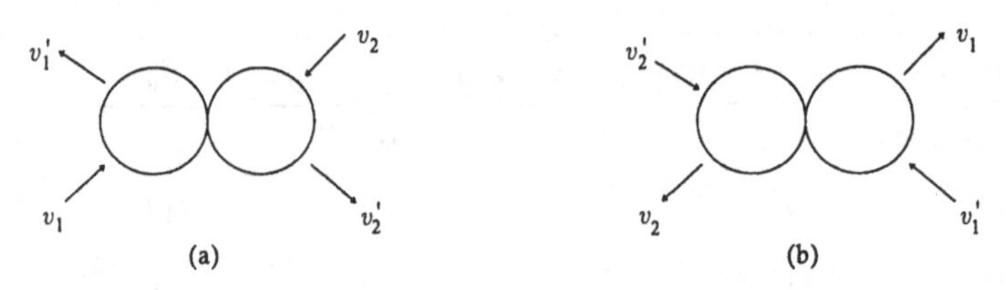

Figure 9.3: A collision between two molecules. (a) A collision with initial velocities $(\boldsymbol{v}_1, \boldsymbol{v}_2)$ and final velocities $(\boldsymbol{v}_1', \boldsymbol{v}_2')$. (b) The inverse collision of (a): $(\boldsymbol{v}_1', \boldsymbol{v}_2') \to (\boldsymbol{v}_1, \boldsymbol{v}_2)$.

we call this collision the *direct* collision. We note that for each direct collision there is always a corresponding *inverse* collision, which interchanges the incoming and the outgoing velocities. The direct and its inverse collision obey the same conservation laws (Eqs. (9.22)–(9.23)), therefore for each collision its inverse collision can also happen. The inverse collision is denoted as $(\boldsymbol{v}_1', \boldsymbol{v}_2') \to (\boldsymbol{v}_1, \boldsymbol{v}_2)$, as shown in Fig. 9.3(b). Now we use the notation $\mathcal{F}(\boldsymbol{v})d^3v$ to denote the number of molecules with the velocity between $\boldsymbol{v}$ to $\boldsymbol{v} + d\boldsymbol{v}$, then the number of the direct collisions per unit time is $a\mathcal{F}(\boldsymbol{v}_1)\mathcal{F}(\boldsymbol{v}_2)$, where a is a constant.[11] The number of the corresponding inverse collisions per unit time is $a'\mathcal{F}(\boldsymbol{v}_1')\mathcal{F}(\boldsymbol{v}_2')$, where a' is another constant. The direct collisions will reduce the number of molecules with velocities $\boldsymbol{v}_1$ and $\boldsymbol{v}_2$, and increase the number of molecules with velocities $\boldsymbol{v}_1'$ and $\boldsymbol{v}_2'$. On the other hand the inverse collisions have the reverse effect, they will increase the number of molecules with $\boldsymbol{v}_1$ and $\boldsymbol{v}_2$ and decrease the number of molecules with $\boldsymbol{v}_1'$ and $\boldsymbol{v}_2'$. Sincere we are considering an equilibrium situation, $\mathcal{F}(\boldsymbol{v})$ will not change with time. This requires that $a\mathcal{F}(\boldsymbol{v}_1)\mathcal{F}(\boldsymbol{v}_2) = a'\mathcal{F}(\boldsymbol{v}_1')\mathcal{F}(\boldsymbol{v}_2')$. Also we are considering an isotropic system, which implies that every direction is equivalent to each other. Therefore we must have $a = a'$, and

$$\mathcal{F}(\boldsymbol{v}_1)\mathcal{F}(\boldsymbol{v}_2) = \mathcal{F}(\boldsymbol{v}_1')\mathcal{F}(\boldsymbol{v}_2'), \tag{9.24}$$

$$\ln \mathcal{F}(\boldsymbol{v}_1) + \ln \mathcal{F}(\boldsymbol{v}_2) = \ln \mathcal{F}(\boldsymbol{v}_1') + \ln \mathcal{F}(\boldsymbol{v}_2'). \tag{9.25}$$

Since $\mathcal{F}(\boldsymbol{v})$ is independent of the direction of[12] $\boldsymbol{v}$, $\mathcal{F}(\boldsymbol{v}) = \mathcal{F}(|\boldsymbol{v}|) = \mathcal{F}(v)$.

[11] Except for the restrictions of the range of the velocities, the distance between the centers of the two colliding molecules must come closer than d_0 in order that a collision can happen. Therefore there is a proportionality constant a.

[12] See the condition (d) in Sec. 9.2.

Comparing Eqs. (9.23) and (9.25) we have the solution

$$\ln \mathcal{F}(\mathbf{v}) = \ln \mathcal{F}(v) = -\alpha v^2, \tag{9.26}$$
$$\mathcal{F}(\mathbf{v}) = \mathcal{F}(v) = A\,\exp(-\alpha v^2), \tag{9.27}$$

where A and α are constants to be determined later. Comparing $\mathcal{F}(v)$ with the distribution function $F(v)$ defined in Sec. 9.3, we get

$$\begin{aligned} F(v)dv &= \int_{\text{angles}} \mathcal{F}(v)d^3v \\ &= \int_0^{2\pi} d\phi \int_0^{\pi} \sin\theta\, d\theta\, v^2 \mathcal{F}(v)dv \\ &= 4\pi\, v^2 \mathcal{F}(v)dv. \end{aligned} \tag{9.28}$$

The constants A and α can be determined from the conditions that the total number of molecules is N and the internal energy of the system at temperature T is[13] $U = N\bar{\varepsilon} = \frac{3}{2}NkT$.

$$N = \int_0^\infty F(v)dv, \tag{9.29}$$

$$U = \frac{3}{2}NkT = \int_0^\infty (\frac{1}{2}mv^2)F(v)dv. \tag{9.30}$$

From the above two equations and Eqs. (9.27)–(9.28),

$$\begin{aligned} \frac{3kT}{m} &= \frac{\int_0^\infty v^2 F(v)dv}{\int F(v)dv} = \frac{\int_0^\infty v^4 e^{-\alpha v^2} dv}{\int v^2 e^{-\alpha v^2} dv} \\ &= \frac{3\sqrt{\pi}/(8\alpha^{5/2})}{\sqrt{\pi}/(4\alpha^{3/2})} = \frac{3}{2\alpha}, \end{aligned} \tag{9.31}$$

where we have used the following formulas for integration (see Appendix A):

$$\int_0^\infty e^{-ax^2} dx = \frac{1}{2}\int_{-\infty}^\infty e^{-ax^2} dx = \frac{1}{2}\sqrt{\frac{\pi}{a}}, \tag{9.32}$$

$$\int_0^\infty x^{2n} e^{-ax^2} dx = (-1)^n \frac{1}{2}\frac{d^n}{da^n}\sqrt{\frac{\pi}{a}}, \quad n = 1, 2, 3, \cdots . \tag{9.33}$$

[13] For simplicity we consider a monatomic gas.

From Eq. (9.31) we have

$$\alpha = \frac{m}{2kT}. \tag{9.34}$$

From Eq. (9.29), by using Eqs. (9.27)–(9.28) and (9.33), one obtains

$$N = 4\pi A \int_0^\infty v^2 e^{-\alpha v^2} dv = 4\pi A \frac{\sqrt{\pi}}{4\alpha^{3/2}},$$

$$A = \left(\frac{\alpha}{\pi}\right)^{3/2} N = \left(\frac{m}{2\pi kT}\right)^{3/2} N, \tag{9.35}$$

$$F(v)dv = 4\pi N \left(\frac{m}{2\pi kT}\right)^{3/2} v^2\, e^{-mv^2/2kT} dv. \tag{9.36}$$

This equation is known as the **Maxwell-Boltzmann distribution function** for molecular speed. Since $\frac{1}{2}mv^2 = \varepsilon$ is the energy of a molecule of a monatomic gas, the last factor in Eq. (9.36) can be written as $\exp(-\varepsilon/kT)$, which is known as the **Boltzmann factor**.[14] This factor not only appears in the distribution function of an ideal gas, but will also appear in other classical systems and localized quantum particles. The physical meaning of this factor is that it is the probability of finding a particle with energy ε in a classical system at temperature T. We will discuss this factor again in Chapter 10. From Eq. (9.36) we see that the function $F(v)$ is not only a function of the molecular speed v, but also depends on the temperature T of the system. At lower T, the maximum of $F(v)$ occurs at smaller v, which implies that most of the molecules have smaller v. And when T is increased, the maximum of $F(v)$ shifts to a larger value of v with a lower peak and a wider width, as shown in Fig. 9.4. This implies that the mean speed will increase but the speed distribution is wider.

From Eqs. (9.3) and (9.36) we obtain the probability density function $f(v)$,

$$f(v) = \frac{F(v)}{N} = 4\pi \left(\frac{m}{2\pi kT}\right)^{3/2} v^2\, e^{-mv^2/2kT}. \tag{9.37}$$

We may use Eq. (9.4) to calculate some mean values of the powers of the speed v, i.e., $\overline{v^n}$.

---

[14] There is a factor v^2 in Eq. (9.36) which comes from the fact that d^3v is a three-dimensional differential, therefore it is not included in the Boltzmann factor.

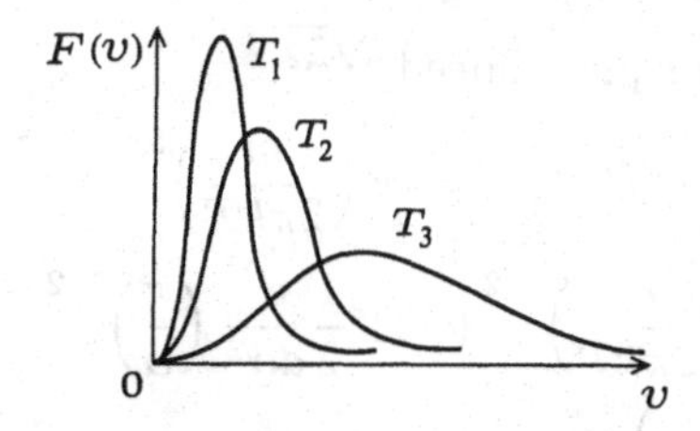

Figure 9.4: The Maxwell-Boltzmann distribution function $F(v)$ vs. v diagram for three different temperatures, $T_1 < T_2 < T_3$.

1. The mean speed $\overline{v}$:

$$\begin{aligned}\overline{v} &= \int_0^\infty v f(v)\, dv = 4\pi \left(\frac{m}{2\pi kT}\right)^{3/2} \int_0^\infty v^3\, e^{-mv^2/2kT} dv \\ &= 4\pi \left(\frac{m}{2\pi kT}\right)^{3/2} \frac{1}{2} \int_0^\infty v^2\, e^{-mv^2/2kT} dv^2 \\ &= \left(\frac{8kT}{\pi m}\right)^{1/2} = 1.596 \left(\frac{kT}{m}\right)^{1/2},\end{aligned} \tag{9.38}$$

where the integration in the third equality may be obtained by an integration by parts, and use of the formula in Eq. (9.32). Note that even though $\overline{v} \neq 0$, the average values of each component of the velocity $\boldsymbol{v}$ is zero, i.e., $\overline{v_x} = \overline{v_y} = \overline{v_z} = 0$. In order to verify these results we use the distribution function given in Eq. (9.27) to obtain the mean values, by noting that $v^2 = v_x^2 + v_y^2 + v_z^2$:

$$\begin{aligned}\overline{v_x} &= \frac{1}{N} \int v_x \mathcal{F}(\boldsymbol{v})\, d^3v = \frac{A}{N} \int v_x\, e^{-\alpha v^2}\, dv_x\, dv_y\, dv_z \\ &= \frac{A}{N} \int_{-\infty}^{\infty} v_x\, e^{-\alpha v_x^2}\, dv_x \int_{-\infty}^{\infty} e^{-\alpha v_y^2}\, dv_y \int_{-\infty}^{\infty} e^{-\alpha v_z^2}\, dv_z \\ &= 0.\end{aligned} \tag{9.39}$$

Since the integrand in the integral of v_x is an odd function of v_x, the integration over v_x will be 0. Similarly, $\overline{v_y} = \overline{v_z} = 0$.

2. The root mean square speed $\sqrt{\overline{v^2}}$:

$$\begin{aligned}\overline{v^2} &= \int_0^\infty v^2 f(v)\, dv = 4\pi \left(\frac{m}{2\pi kT}\right)^{3/2} \int_0^\infty v^4\, e^{-mv^2/2kT} dv \\ &= 4\pi \left(\frac{m}{2\pi kT}\right)^{3/2} (-1)^2 \frac{1}{2} \frac{d^2}{d\alpha^2} \left(\frac{\pi}{\alpha}\right)^{1/2} \\ &= \left(\frac{3kT}{m}\right),\end{aligned} \tag{9.40}$$

where in the third equality we have used the formula in Eq. (9.33). The result of the above equation is exactly the same as we have obtained in Eq. (9.15), i.e., $\frac{1}{2}mv^2 = \frac{3}{2}kT$. We define $v_{\rm rms} \equiv \sqrt{\overline{v^2}}$ (rms= root mean square),

$$v_{\rm rms} = \sqrt{\overline{v^2}} = \left(\frac{3kT}{m}\right)^{1/2} = 1.732 \left(\frac{kT}{m}\right)^{1/2} > \overline{v}. \tag{9.41}$$

3. The most probable speed $v_{\rm m}$:

The most probable speed $v_{\rm m}$ is the speed that the probability density function $f(v)$ has the maximum. This means that the probability of finding a molecule with speed $v_{\rm m}$ is larger than any other speed v. Therefore the condition of $v_{\rm m}$ is that it makes $f(v)$ a maximum, i.e.,

$$\begin{aligned}\frac{df}{dv} &= 4\pi \left(\frac{m}{2\pi kT}\right)^{3/2} \frac{d}{dv} \left(v^2\, e^{-mv^2/2kT}\right) \\ &= 0, \quad \text{when } v = v_{\rm m}.\end{aligned} \tag{9.42}$$

Solving the above equation,

$$2v\, e^{-mv^2/2kT} + v^2 \left(-\frac{mv}{kT}\right) e^{-mv^2/2kT} = 0, \quad \text{when } v = v_{\rm m}$$

$$v_{\rm m} = \left(\frac{2kT}{m}\right)^{1/2} = 1.414 \left(\frac{kT}{m}\right)^{1/2} < \overline{v}. \tag{9.43}$$

Therefore the most probable speed $v_{\rm m}$ is smaller than the mean speed $\overline{v}$, which implies that there is a larger percentage of the molecules whose speed is smaller than the average speed $\overline{v}$.

Comparing the above three speeds at the same temperature, we find that $v_{\rm rms}$ is the largest, $\overline{v}$ is the next, and $v_{\rm m}$ is the smallest,

$$\begin{aligned}v_{\rm m} : \overline{v} : v_{\rm rms} &= 1.414 : 1.596 : 1.732 \\ &= 1 : 1.128 : 1.225.\end{aligned} \tag{9.44}$$

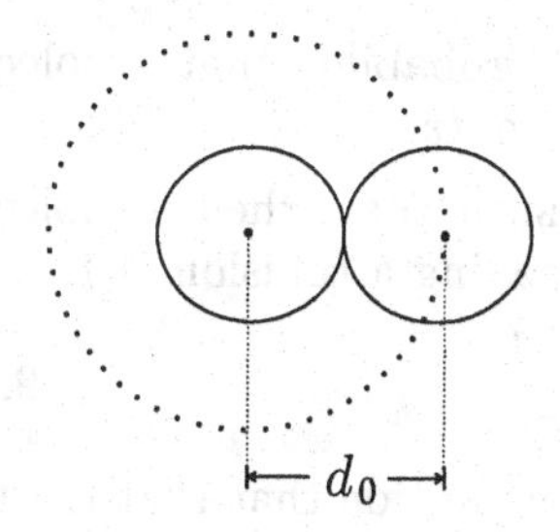

Figure 9.5: The collision cross section area for a hard core collision. Each target molecule spans an obstacle area, which is a circle centered at the center of the target molecule with an area πd_0^2.

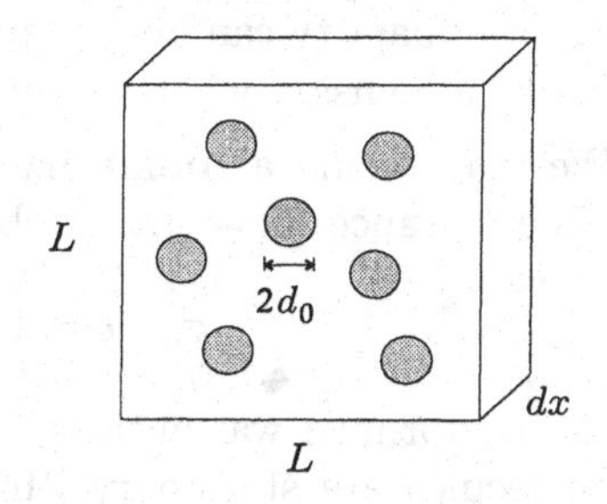

Figure 9.6: The probability that a bullet molecule will strike an obstacle area. A dark circle represents a cross section area of a target molecule.

9.7 The Mean Free Path

In collision theory, we usually consider that one of the molecules is stationary, a *target*, and the other molecule is moving toward the target, a *bullet*. We consider the collisions as hard core collisions. Then the collision will happen only when the distance between the centers of the bullet and the target molecules is smaller than the diameter of a molecule d_0. Therefore every target molecule will span an *obstacle area* which is a circle centered at the center of the target molecule with an area πd_0^2. Only when the center of a bullet molecule passes through the obstacle area, will a collision happen, as shown in Fig. 9.5. In collision theory the obstacle area is called the collision cross section σ_0. We consider the velocity $\boldsymbol{v}$ of a bullet molecule moving in the x-direction. We imagine that there is a target zone on the yz plane with an area $L \times L$ and a thickness dx. Within this target zone, the number of molecules is $\overline{n}\,(L \times L)\,dx$. We take dx to be so small that within the zone there is no overlap of the obstacle areas (viewed from the direction of x). Therefore the total obstacle area is $\pi d_0^2\,\overline{n}\,L^2\,dx$, as shown in Fig. 9.6. The bullet molecules which are moving toward the target zone are spatially uniform; the percentage of the bullet molecules that strike the obstacle area (i.e., the percentage of the bullet molecules that suffer a collision) is

$$\frac{\sigma_0\,\overline{n}\,L^2\,dx}{L^2} = \sigma_0\,\overline{n}\,dx \quad (\sigma_0 = \pi d_0^2). \tag{9.45}$$

The above quantity can be interpreted as the probability that a molecule will make a collision when it travels a distance dx.

We may define a **mean free path** ℓ, such that when a molecule travels a distance $dx = \ell$ its probability of making a collision is 1:

$$\sigma_0 \overline{n} \ell = 1, \quad \text{or} \quad \ell = \frac{1}{\sigma_0 \overline{n}}. \tag{9.46}$$

The above formula was obtained with the assumption that that the target molecules are stationary. In fact all the molecules are moving at speeds obeying the Maxwell-Boltzmann distribution. If this is taken into account then Eq. (9.46) becomes

$$\ell = \frac{1}{\sqrt{2}\sigma_0 \overline{n}}. \tag{9.47}$$

Since d_0 is a constant, ℓ is proportional to the inverse of the density of the molecules $\overline{n}$. However $\overline{n} = N/V = P/kT \propto P/T$, therefore $\ell \propto T/P$, which implies that at higher temperature T and lower pressure P, the mean free path ℓ is longer. We use the molecules in the atmosphere as an example to calculate the mean free path. At STP, $\overline{n} \approx 2.69 \times 10^{25}$ m^{-3} and $d_0 \approx 3.6 \times 10^{-10}$ m, therefore $\sigma_0 \approx \pi(3.6 \times 10^{-10})^2 \approx 4.07 \times 10^{-19}$ m^2 and $\ell \approx (\sqrt{2} \times 2.69 \times 10^{25} \times 4.07 \times 10^{-20})^{-1} \approx 6.46 \times 10^{-8}$ m $\approx 18\, d_0$. This means that $\ell \gg d_0$, and thus the air satisfies the condition of a dilute gas.

In the following we use the statistical method to define a function called the **survival function** $p(x)$. By using this method we cannot derive the mean free path in terms of microscopic quantities such as we have derived in Eq. (9.46), but it gives us another point of view for understanding the physical meaning of the mean free path ℓ. We consider a molecule which moves in a straight line (assumed to be the x-axis) if it makes no collisions. We use the function $p(x)$ to denote the probability that the molecule does not make any collision in the interval between 0 and x. The probability that the molecule makes a collision in the interval dx is denoted as $a\,dx$, where dx is very small such that $dx \ll \ell$ and therefore $a\,dx \ll 1$. The probability that the molecule will not make a collision in dx is $1 - a\,dx$, therefore

$$p(x + dx) = p(x)(1 - a\,dx). \tag{9.48}$$

Expanding $p(x + dx) = p(x) + (dp/dx)dx$ and substituting it into the above equation, we have

$$\frac{dp}{dx} = -ap(x). \tag{9.49}$$

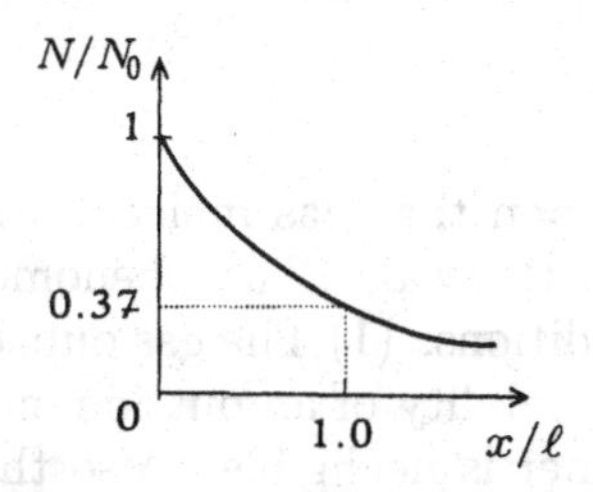

Figure 9.7: The graph for the survival function $p(x)$ vs. x.

The above equation can be integrated with the condition $p(0) = 1$, which implies that the molecule starts from $x = 0$, and the result is

$$p(x) = e^{-ax}. \tag{9.50}$$

The above expression of $p(x)$ is the probability that a molecule does not make any collision in the interval from 0 to x. We use $g(x)dx$ to denote the probability that a molecule will make a collision in the interval from x to $x + dx$, then

$$g(x)dx = p(x) - p(x + dx) = -\frac{dp}{dx}dx = ae^{-ax}dx. \tag{9.51}$$

From the definition of the mean free path we have

$$\ell = \overline{x} = \int_0^\infty xg(x)dx = a\int_0^\infty xe^{-ax}dx = \frac{1}{a}. \tag{9.52}$$

The above definition of the mean free path ℓ is exactly the same as that defined in Eqs. (9.45)–(9.46). Therefore from Eq. (9.50)

$$p(x) = e^{-x/\ell}. \tag{9.53}$$

We plot $p(x)$ vs. x in Fig. 9.7. Note that $e^{-1} \approx 0.37$, this means that a little over 1/3 (37%) of the molecules do not make a collision after traveling a distance ℓ. The other 2/3 (63%) of the molecules make a collision when traveling a distance equal to or less than ℓ. Therefore the *mean free path* is a *mean distance*, it does not mean that every molecule will make a collision after traveling a distance ℓ.

9.8 Effusion

Effusion is the phenomenon that gas molecules leak out of a container through a small hole in the wall. The phenomenon of **effusion** must satisfy the following conditions: (1) The gas outside the container is very dilute, such that the probability of an outside molecule passing through the hole into the container is negligible. Also the outside molecules do not interfere with the effusion process. (2) The hole must be very small so that the molecules inside the container remain in thermal equilibrium at any instant of time and at any place therein, just as if the hole did not exist. Point (2) implies that the number of molecules striking the wall near the hole per unit time per unit area will not be affected by the existence of the hole. Now we will estimate how small the hole must be in order that Point (2) can be satisfied. Suppose the hole is a circle with a diameter D, then we may require that $D \ll \ell$, where ℓ is the mean free path. The reason is that thermal equilibrium is maintained by collisions, therefore if $D \ll \ell$ then most of the molecules will not make any collisions near the hole, and thermal equilibrium will not be affected by the existence of the hole. We have calculated the mean free path at STP as $\ell \sim 5 \times 10^{-8}$ m, which is very small. Therefore at STP the phenomenon of effusion can hardly be observed experimentally. However because the mean free path is inversely proportional to the pressure, i.e., $\ell \propto P^{-1}$, therefore at a low enough pressure D may become large enough so that effusion may be observed in the laboratory.

From the formula for molecular flux Eq. (9.8), we have

$$\begin{aligned} \Phi &= \frac{1}{4}\overline{n}\,\overline{v} = \frac{1}{4}\overline{n}\int_0^\infty v f(v) dv \equiv \int_0^\infty \Phi_v\, dv, \\ \Phi_v\, dv &= \frac{1}{4}\overline{n}\, v f(v)\, dv \equiv \frac{1}{4} v\, \overline{n}_v\, dv. \end{aligned} \tag{9.54}$$

Here $\overline{n}_v dv = \overline{n} f(v) dv$ is the number of molecules per unit volume in the container with speed in the range between v to $v + dv$, and $\Phi_v dv$ is the number of molecules with speed in the range between v to $v + dv$ which will pass through the hole per unit time per unit area. Therefore the mean speed of the molecules which escape from the hole is

$$\overline{v}_{\mathrm{e}} = \frac{\displaystyle\int_0^\infty v\Phi_v dv}{\displaystyle\int_0^\infty \Phi_v dv} = \frac{\displaystyle\int_0^\infty v^4 e^{-mv^2/2kT} dv}{\displaystyle\int_0^\infty v^3 e^{-mv^2/2kT} dv} = \frac{3}{4}\left(\frac{2\pi kT}{m}\right)^{1/2}, \tag{9.55}$$

where we have used Eq. (9.33) to get

$$\int_0^\infty v^4 e^{-mv^2/2kT} dv = \frac{3}{2\cdot 4}\left(\frac{2kT}{m}\right)^{5/2} \pi^{1/2},$$

and the method of an integration by parts to obtain

$$\int_0^\infty v^3 e^{-mv^2/2kT} dv = \frac{1}{2}\int_0^\infty x\, e^{-mx/2kT} dx = \frac{1}{2}\left(\frac{2kT}{m}\right)^2 .$$

We compare the mean speed $\overline{v}_e$ for those molecules which escape from the hole with the mean speed $\overline{v}$ of the molecules inside the container:

$$\frac{\overline{v}_e}{\overline{v}} = \frac{\frac{3}{4}\left(\frac{2\pi kT}{m}\right)^{1/2}}{\left(\frac{8kT}{\pi m}\right)^{1/2}} = \frac{3\pi}{8} = 1.18 > 1. \tag{9.56}$$

Therefore the mean speed of the escaped molecules is larger than that of the molecules in the container. This implies that molecules with higher speed have a larger probability of escaping from the hole.

From the results $\overline{n} = N/V = P/kT$ and $\overline{v} = (8kT/\pi m)^{1/2}$, we obtain

$$\Phi = \frac{1}{4}\overline{n}\,\overline{v} = \frac{P}{\sqrt{2\pi mkT}}. \tag{9.57}$$

We see that Φ is proportional to $P/\sqrt{T}$. Therefore Φ is larger for higher pressure and lower temperature. Finally it is worth mentioning that effusion has a useful application in isotope separation. From Eq. (9.57) we see that at the same T and P, Φ is proportional to the inverse of the square root of the molecular mass. Therefore, if the container contains two different kinds of molecules with different masses, the effusion rate will be different for the two different molecules. The effusion rate will be larger for the molecules with smaller mass; this process can be used in isotope separation.

9.9 Transport Processes

A transport process occurs when there is an external field acting on a system. The molecules of the system will transport the message of the field from one end (or place) to the other. This is known as the **transport phenomenon**. One of the most well known transport phenomena in a gaseous system is thermal conduction. Thermal conduction occurs when the temperatures of the two ends of the system are different, i.e., there is a *temperature gradient* across the system. The molecules will transport thermal energy from the end with higher T to the end with lower T. Since the temperature is not everywhere the same in the system, the system is not in a state of thermal equilibrium. However if the temperature gradient is small enough, every small region of the system may be considered to be in *local thermal equilibrium*. Every small region has a definite local temperature $T(z)$, where z is the direction of the gradient. Under this condition, all the laws of equilibrium thermodynamics may be considered to be valid, at least approximately. However now the temperature is no longer uniform but is position dependent, i.e., $T = T(z)$. Since collision is the main mechanism for reaching thermal equilibrium, it is reasonable to assume that when a molecule makes a collision, it will reach *local thermal equilibrium* at the place of collision. When a molecule makes a collision it will get the *local property* of the place of collision.

For example if it makes the first collision at z_1, then its equilibrium temperature will become $T(z_1)$. Then it goes on to make the second collision at z_2, where it loses its characteristic of $T(z_1)$ and picks up the local temperature of $T(z_2)$. In the process, the molecule loses thermal energy proportional to $T(z_1) - T(z_2)$. The process goes on until the molecule reaches the other end of the system. Apparently, if the temperature gradient is imposed by an external source, so that $T(z)$ is independent of time, then there is a heat flow from the end of higher T to the end of lower T. This is known as thermal conduction. However, if there is no external source to maintain the temperature gradient, then the end with higher T will lose energy (thus the temperature will be lowered), while the end with lower T will gain energy (thus the temperature will be increased), and eventually the system will reach thermal equilibrium, so that the temperature will be uniform across the system.

As described in the above we may use the properties of collisions and the kinetic theory to understand some of the transport processes of a classical gas. In addition to thermal conduction, which we have just mentioned, we will also discuss the phenomena of viscosity and diffusion. Viscosity is the phenomenon that the opposite walls of the container have relative motion. We may assume that one wall is motionless, and the other wall is moving. The molecules will transport momentum from

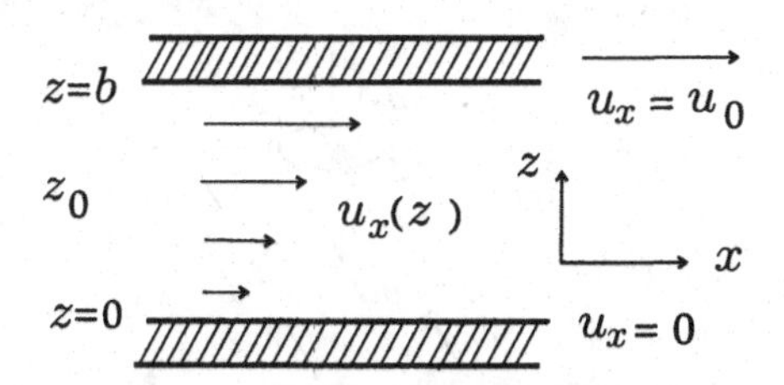

Figure 9.8: A relative motion of the opposite walls of the container of a system. The wall at $z = 0$ is motionless, and the wall at $z = b$ moves with a velocity u_0 in the x-direction.

the moving wall to the motionless wall. The phenomenon of diffusion occurs when the density of the gas is non-uniform. The molecules in the higher density place will transport more mass (i.e., more molecules) to the place with lower density, than the mass transported from the lower density place to the higher density place.

Now we proceed to discuss the respective details of the phenomena of viscosity, thermal conduction, and diffusion.

1. Viscosity

This is the phenomenon that a system has a velocity gradient. We consider a system of gas molecules which are confined between a pair of parallel walls, whose locations are at $z = 0$ and $z = b$. The wall at $z = 0$ is motionless, and the wall at $z = b$ is moving in the x-direction with a speed u_0, as shown in Fig. 9.8. When the molecules are in thermodynamic equilibrium with the walls, the mean velocity of the molecules at the wall will be equal to that of the wall. Therefore at $z = 0$ and $z = b$ the average x-component of the velocity of the molecules are, respectively, $\overline{v_x} = 0$ and $\overline{v_x} = u_0$. Thus the average x-component of the velocity of the molecules $\overline{v_x}$ is no longer 0 (as in Eq. (9.39)), but will be a function of z. If u_0 is small, we may approximate the relation between $\overline{v_x}$ and z as linear, i.e.,

$$\overline{v_x} = u_x(z) = z\,\frac{du_x}{dz} = \frac{z}{b}\,u_0. \tag{9.58}$$

Here we take $u_x(z)$ to be a linear function with $du_x/dz = u_0/b$; the above equation satisfies the boundary conditions, $\overline{v_x} = 0$ at $z = 0$ and $\overline{v_x} = u_0$ at $z = b$. Viscosity is the phenomenon associated with the existence of

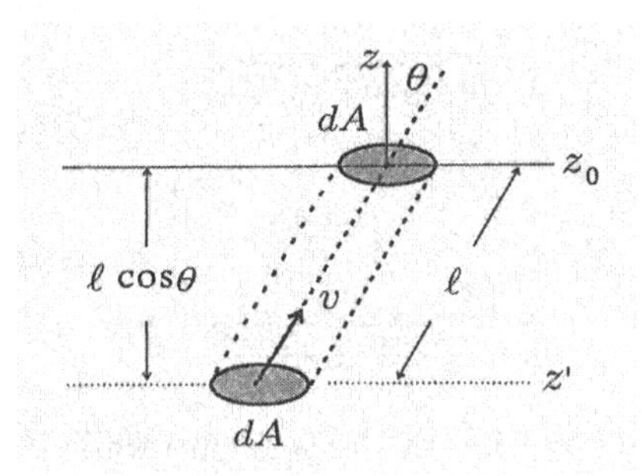

Figure 9.9: A molecule passes through an imaginary plane at $z = z_0$, where θ is the angle between $\boldsymbol{v}$ and the z-axis, thus $v \cos\theta = v_z$.

the velocity gradient $du_x/dz \neq 0$. Since the average x-component of the momentum $\overline{p_x}$ of the molecules is a function of the position z, we have

$$\overline{p_x}(z) = m\,\overline{v_x} = mz\,\frac{du}{dz}, \quad 0 \leq z \leq b. \tag{9.59}$$

In a collision, the molecules with larger (smaller) $\overline{p_x}$ will give this information to the molecules with smaller (larger) $\overline{p_x}$. Thus there is momentum transfer in the collisions; this is the phenomenon of viscosity.

We consider an imaginary plane at $z = z_0$ $(0 < z_0 < b)$. A molecule may pass through the z_0 plane from above $(z > z_0)$ or from below $(z < z_0)$. We consider molecules with velocity $\boldsymbol{v} = (v, \theta, \phi)$ when they pass through the z_0 plane. From the definition of the mean free path, we may assume that a molecule making the last collision will *on the average*, travel a distance ℓ before it reaches the plane at z_0. Therefore the mean position where these molecules made the last collision before passing through z_0 is at $z' = z_0 - \ell\cos\theta$, where ℓ is the mean free path and θ is the angle between $\boldsymbol{v}$ and the z-axis, as shown in Fig. 9.9. The molecules which made the last collision at z' will pick up the mean momentum $\overline{p_x}(z')$. Since ℓ is a small distance we may expand $\overline{p_x}(z')$ around z_0,

$$\overline{p_x}(z') = \overline{p_x}(z_0 - \ell\cos\theta) = \overline{p_x}(z_0) - \ell\cos\theta\frac{d\overline{p_x}}{dz} + \cdots. \tag{9.60}$$

The values of $\overline{p_x}(z_0)$ and $d\overline{p_x}/dz$ are equal for all molecules, but different molecules have different values of $\ell\cos\theta$. This is because θ depends on the direction of $\boldsymbol{v}$, thus we have to calculate the average value of $\ell\cos\theta$.

We consider separately the molecules coming from $z < z_0$ and those coming from $z > z_0$. From Eq. (9.5) we get the differential molecular flux, $d\Phi(v,\theta,\phi)$, for molecules with speed $v \to v + dv$ and direction $\theta \to \theta + d\theta$, $\phi \to \phi + d\phi$, which pass through a unit area per unit time:

$$\begin{aligned} d\Phi(v,\theta,\phi) &= \frac{dN(v,\theta,\phi)}{dA\,dt} \\ &= \frac{\overline{n}}{4\pi} v f(v)\, dv\, \sin\theta\, \cos\theta\, d\theta\, d\phi. \end{aligned} \tag{9.61}$$

By using the above equation, we obtain the mean value of $\ell\cos\theta$ for molecules from $z < z_0$ (the molecules with $v_z > 0$, thus $0 \le \theta \le \pi/2$),

$$\begin{aligned} \overline{z}_\ell^{(+)} \equiv \overline{\ell\cos\theta}^{(+)} &= \frac{\displaystyle\int \ell\cos\theta\, d\Phi(v,\theta,\phi)}{\displaystyle\int d\Phi(v,\theta,\phi)} \\ &= \frac{\displaystyle\frac{\overline{n}}{4\pi}\int_0^\infty v f(v)\, dv \int_0^{\pi/2} \ell\cos^2\theta\, \sin\theta\, d\theta \int_0^{2\pi} d\phi}{\displaystyle\frac{1}{4}\overline{n}\,\overline{v}} \\ &= \frac{2}{3}\ell \quad \text{(molecules with } v_z > 0), \end{aligned} \tag{9.62}$$

where the denominator is the molecular flux $\Phi = \frac{1}{4}\overline{n}\,\overline{v}$, which has been evaluated in Eq. (9.8). Equation (9.62) tells us that before passing through the $z = z_0$ plane an average molecule makes its final collision at a position whose vertical distance from the $z = z_0$ plane is $2/3$ of the mean free path, i.e., $\overline{z}_\ell^{(+)} = \frac{2}{3}\ell$.

Therefore, from Eq. (9.60), we see that the momentum transferred per unit time per unit area across the z_0 plane by the molecules coming from $z < z_0$ is

$$\begin{aligned} G_x^{(+)} &= \Phi\left[\overline{p_x}(z_0) - \overline{z}_\ell^{(+)}\frac{d\overline{p_x}}{dz}\right] \\ &= \frac{1}{4}\overline{n}\,\overline{v}\left[\overline{p_x}(z_0) - \frac{2}{3}\ell\frac{d\overline{p_x}}{dz}\right], \end{aligned} \tag{9.63}$$

where the subscript x in $G_x^{(+)}$ denotes that the transferred momentum is in the x-direction, and the superscript $(+)$ denotes that the direction of transport of the momentum is in the $+z$ direction.

By using the same procedure, we obtain the mean value of $\ell\cos\theta$ and the transferred momentum $G_x^{(-)}$ per unit time per unit area by the molecules coming from $z > z_0$ (the molecules with $v_z < 0$, thus $\pi/2 \leq \theta \leq \pi$) as

$$\overline{z}_\ell^{(-)} = -\frac{2}{3}\ell \ \text{(molecules with } v_z < 0), \tag{9.64}$$

$$G_x^{(-)} = \frac{1}{4}\overline{n}\,\overline{v}\left[\overline{p_x}(z_0) + \frac{2}{3}\ell\frac{d\overline{p_x}}{dz}\right]. \tag{9.65}$$

The superscript $(-)$ in $G_x^{(-)}$ denotes that the direction of the transport of momentum is in the $-z$ direction.

Finally we obtain the *net momentum transfer* per unit time per unit area across the plane at $z = z_0$ as

$$\begin{aligned} G_x^{\text{tot}} &= G_x^{(+)} - G_x^{(-)} = -\frac{1}{3}\overline{n}\,\overline{v}\,\ell\frac{d\overline{p_x}}{dz} \\ &= -\frac{1}{3}\overline{n}\,m\,\ell\,\overline{v}\frac{du_x}{dz}. \end{aligned} \tag{9.66}$$

In the above equation, there is a minus sign in the last equality, which implies that the net momentum transfer is in the $-z$ direction, in the opposite direction of du_x/dz (which is in the $+z$ direction). Momentum divided by time is force, so we may rewrite the above equation in terms of more familiar quantities, defining a quantity called the **coefficient of viscosity** η,

$$|G_x^{\text{tot}}| = \frac{F_x}{A_z} = \eta\frac{du_x}{dz} \quad \text{or} \quad \eta = \frac{F_x/A_z}{du_x/dz}, \tag{9.67}$$

where F_x is the viscous force between different layers of molecules (x is the direction of the force), A_z is the area for each layer with the normal in the z-direction, and du_x/dz is the velocity gradient. Comparing Eqs. (9.66) and (9.67), we obtain

$$\eta = \frac{1}{3}\overline{n}\,m\,\ell\,\overline{v}. \tag{9.68}$$

Since ℓ is inversely proportional to $\overline{n}$ (Eq. (9.47)), η is proportional to the molecular mass m and the mean speed $\overline{v}$, and does not explicitly depend on the number density $\overline{n}$. Since $\overline{v}$ is proportional to $\sqrt{T/m}$, η is proportional to $\sqrt{mT}$. For different materials at the same temperature, the material with larger mass m will have a larger η. For the same material, a higher temperature will give a larger η, which implies that for the same velocity gradient, the viscous force is larger for higher temperature.

This result is not consistent with our experiences about liquids. For liquids, the higher the temperature the smaller the viscous force. Therefore Eq. (9.68) holds for gaseous systems only. This gives us a hint that the origin of the viscous force for liquids and for gases are different. The gas molecules exchange momentum only when they collide with each other. Between collisions the molecules are in free motion, and are independent of each other. At higher temperatures the molecules have larger momentum and shorter mean collision time. Therefore the higher the temperature, the more effective the momentum transfer, and thus the coefficient of viscosity η is larger. For liquids, however, the neighboring molecules interact strongly *at all times*. The stronger the interaction the larger the momentum exchange that can occur. At a higher temperature the average momentum for each molecule is larger, and the interaction between the molecules become relatively smaller. The efficiency of momentum exchange will also become smaller. Therefore the higher the temperature the smaller will be the coefficient of viscosity η. Finally we see that the coefficient of viscosity η has the units N m^{-2} s. The pressure has the units N m^{-2}, the units of η may also be written as Pa s.

2. Thermal conductivity

Thermal conduction occurs when there is a temperature gradient in a system. We consider a system whose temperature T at $z = 0$ is T_0 and at $z = b$, $T = T_0 + \Delta T$. Therefore the system has a temperature gradient $\Delta T/b$ in the $+z$ direction ($b > 0$). The system is then not in a state of thermal equilibrium. However when the gradient $\Delta T/b$ is very small, we may assume that for every small region the system is in a state of *local thermal equilibrium* with a well-defined temperature T as a function of z,

$$T(z) = T_0 + \frac{z}{b}\,\Delta T, \quad 0 \le z \le b, \tag{9.69}$$

which satisfies the boundary conditions $T(0) = T_0$ and $T(b) = T_0 + \Delta T$. Therefore the mean energy per molecule will be a function of the position z,

$$\bar{\varepsilon} = \bar{\varepsilon}(z) = \frac{f}{2}\,kT(z), \tag{9.70}$$

where f is the number of *classical degrees of freedom*.[15] Since the mean energy is different for molecules at different positions, molecules will

[15] This means that f excludes the degrees of freedom which cannot be excited due to large energy gaps.

transport energy by collisions. Consider the energy transport across the imaginary surface at $z = z_0$. Using Eq. (9.62), molecules with velocity $v = (v, \theta, \phi)$ will, on the average, make their last collision at $z' = z_0 - \ell \cos\theta$ before crossing the surface at $z = z_0$. These molecules have the following mean energy when crossing the surface at $z = z_0$,

$$\overline{\varepsilon}(z') = \overline{\varepsilon}(z_0 - \ell\cos\theta) = \overline{\varepsilon}(z_0) - \ell\cos\theta \frac{d\overline{\varepsilon}}{dz} + \cdots . \tag{9.71}$$

Since the mean energy for a molecule depends on the position z, there is a *net* transport of energy at the surface $z = z_0$. By using the same argument used in deriving Eqs. (9.63) and (9.65), we obtain the respective energy transport in the $+z$ direction $j_Q^{(+)}$, and in the $-z$ direction $j_Q^{(-)}$, across the surface at $z = z_0$ per unit area per unit time as

$$j_Q^{(+)} = \frac{1}{4}\overline{n}\,\overline{v}\left[\overline{\varepsilon}(z_0) - \frac{2}{3}\ell\frac{d\overline{\varepsilon}}{dz}\right], \tag{9.72}$$

$$j_Q^{(-)} = \frac{1}{4}\overline{n}\,\overline{v}\left[\overline{\varepsilon}(z_0) + \frac{2}{3}\ell\frac{d\overline{\varepsilon}}{dz}\right]. \tag{9.73}$$

The physical meaning of $j_Q^{(+)}$ ($j_Q^{(-)}$) is the transport of energy per unit area per unit time in the $+z$ ($-z$) direction. The transported energy is thermal energy (heat), so we call it the **thermal current density** or **heat current density**. The *net thermal current density* is

$$j_Q = j_Q^{(+)} - j_Q^{(-)} = -\frac{1}{3}\overline{n}\,\ell\,\overline{v}\frac{d\overline{\varepsilon}}{dz} = -\frac{1}{3}\overline{n}\,\ell\,\overline{v}\,c_v^*\frac{dT}{dz}, \tag{9.74}$$

where we have used the definition of the heat capacity *per molecule*, $c_v^* = d\overline{\varepsilon}/dT$,

$$\frac{d\overline{\varepsilon}}{dz} = \frac{d\overline{\varepsilon}}{dT}\frac{dT}{dz} = c_v^*\frac{dT}{dz}. \tag{9.75}$$

There is a minus sign on the right hand side of Eq. (9.74), this implies that the direction of the heat current density $\boldsymbol{j}_Q$ is in the $-z$ direction, which is opposite to the direction of the temperature gradient dT/dz (which is in the $+z$ direction). We define a quantity called the **coefficient of thermal conductivity** λ:

$$\boldsymbol{j}_Q = -\lambda\,\boldsymbol{\nabla} T, \quad \text{or} \quad j_Q\hat{\boldsymbol{z}} = -\lambda\frac{dT}{dz}\hat{\boldsymbol{z}}. \tag{9.76}$$

Comparing Eqs. (9.76) and (9.74), we obtain

$$\lambda = \frac{1}{3}\overline{n}\,\ell\,\overline{v}\,c_v^*. \tag{9.77}$$

From Eqs. (9.77) and (9.68),

$$\lambda = \frac{c_v^*}{m}\eta. \tag{9.78}$$

Therefore the coefficient of thermal conductivity λ is a measure of the energy transport by molecules, and it is proportional to the heat capacity of a molecule c_v^*. The coefficient of viscosity η is a measure of momentum transport by molecules, and it is proportional to the mass of a molecule m. Finally we see that the coefficient of thermal conductivity λ has the units J m^{-1} s^{-1} K^{-1}.

3. Diffusion

We consider a system consisting of two kinds of molecules, we shall refer to them as type 1 molecules and type 2 molecules.[16] For simplicity we assume that the respective number densities of type 1 and type 2, n_1 and n_2, are both functions of z, but the total number density n of the system is uniform, i.e.,

$$n_1(z) + n_2(z) = n = \text{constant}. \tag{9.79}$$

Since n is uniform, both the temperature and pressure of the system are well defined. We also assume that both type 1 and type 2 molecules obey the Maxwell-Boltzmann distribution function of speed, and therefore Eqs. (9.62) and (9.64) we derived for the phenomenon of viscosity still hold.

Now we consider the diffusion phenomenon of type 1 molecules. The existence of type 2 molecules does not affect our analysis, and we can just consider type 1 molecules. For simplicity, we use *molecules* to denote type 1 molecules in the following. As we have discussed in the case of viscosity, *molecules* with velocity $\boldsymbol{v} = (v, \theta, \phi)$ will, on the average, make their last collision at $z' = z_0 - \ell\cos\theta$ before crossing the surface at $z = z_0$. The number density of *molecules* at that position is

$$n_1(z') = n_1(z_0 - \ell\cos\theta) = n_1(z_0) - \ell\cos\theta\,\frac{dn_1}{dz} + \cdots. \tag{9.80}$$

The number density n_1 has a gradient, i.e., it is not uniform, therefore there is a *net* transport of the number of *molecules* for the *molecules*

[16] Type 1 and type 2 molecules may also be of the same kind of molecule but of different isotopes.

crossing the surface at $z = z_0$. By using Eqs. (9.62), (9.64) and (9.8), we obtain the respective particle transport in the $+z$ direction $j_p^{(+)}$, and in the $-z$ direction $j_p^{(-)}$, across the surface at $z = z_0$ per unit area per unit time as

$$j_p^{(+)} = \frac{1}{4}\,\overline{n}^{(+)}\,\overline{v} = \frac{1}{4}\,\overline{v}\left[n_1(z_0) - \frac{2}{3}\,\ell\,\frac{dn_1}{dz}\right], \tag{9.81}$$

$$j_p^{(-)} = \frac{1}{4}\,\overline{n}^{(-)}\,\overline{v} = \frac{1}{4}\,\overline{v}\left[n_1(z_0) + \frac{2}{3}\,\ell\,\frac{dn_1}{dz}\right], \tag{9.82}$$

where $\overline{n}^{(\pm)} = n_1(z_0 - \overline{z}_\ell^{(\pm)}\,dn_1/dz)$ are, respectively, the number density of *molecules* crossing the surface at z_0, with a "+" sign for the *molecules* moving in the $+z$ direction, and a "−" sign for the *molecules* moving in the $-z$ direction. Therefore the *net particle current density* j_p is

$$j_p = j_p^{(+)} - j_p^{(-)} = -\frac{1}{3}\,\overline{v}\,\ell\,\frac{dn_1}{dz}. \tag{9.83}$$

There is a minus sign on the right hand side of the equality, which implies that the particle current density $\boldsymbol{j}_p$ is in the $-z$ direction, in the opposite direction of the density gradient dn_1/dz, which is in the $+z$ direction. We define a quantity called the **coefficient of diffusion** D:

$$\boldsymbol{j}_p = -D\,\boldsymbol{\nabla} n_1, \quad \text{or} \quad j_p\hat{\boldsymbol{z}} = -D\,\frac{dn_1}{dz}\,\hat{\boldsymbol{z}}. \tag{9.84}$$

Comparing with Eq. (9.83), we have

$$D = \frac{1}{3}\,\overline{v}\,\ell. \tag{9.85}$$

Therefore the coefficient of diffusion D has the units m^2 s^{-1}.

9.10 Problems

9.1. The speed distribution function of a group of N particles is given by $\Delta N_v = av\,\Delta v$ for $v_0 \geq v > 0$, and $\Delta N_v = 0$ for $v > v_0$, where a is a constant. (a) Draw the graph of the distribution function. (b) Show that the constant $a = 2N/v_0^2$. (c) Compute the average speed of the particles. (d) Compute the root-mean-square speed of the particles.

9.2. Compute $v_{\rm m}$, $\overline{v}$, and $v_{\rm rms}$ for a nitrogen molecule at 300 K.

9.3. A cubic box 0.1 m on a side contains 3×10^{22} molecules of O_2 at 300 K, which is considered as a classical ideal gas. (a) On the average, how many collisions does each molecule make with the walls of the box in one second? (b) What pressure does the oxygen exert on the wall of the box?

9.4. (a) Show that the Maxwell-Boltzmann speed distribution function for N molecules can be written as

$$\mathcal{N}(x)\,dx = \frac{4N}{\sqrt{\pi}}\,x^2\,e^{-x^2}\,dx,$$

where $x = v/v_{\rm m}$ ($v_{\rm m}$ is the most probable speed), and $\mathcal{N}(x)\,dx$ is the number of molecules with speeds in the range x to $x+dx$.

(b) Show that the number of molecules with speeds less than some specified speed v_0 is given by

$$N_{0\to x_0} = N\left[\operatorname{erf}(x_0) - \frac{2}{\sqrt{\pi}}x_0\,e^{-x_0^2}\right],$$

where $x_0 = v_0/v_{\rm m}$ and $\operatorname{erf}(x)$ is the error function, defined by

$$\operatorname{erf}(x) \equiv \frac{2}{\sqrt{\pi}}\int_0^x e^{-y^2}\,dy.$$

9.5. Consider a two-dimensional ideal monatomic gas of N molecules at temperature T, constrained to move only in the xy plane with area A. The mass of each molecule is m. The pressure P of the system is then defined as the force per unit length.

(a) What is the probability density function $f(v)$ for the two-dimensional gas?

(b) Find the mean values of v and v^2.

(c) Find the equation of state of the system.

(d) Derive the expression for the number of molecules striking a unit length of the "wall" per unit time. Express your answer in terms of $\overline{n} \equiv N/A$, and the mean speed $\overline{v}$.

9.6. Consider a container of volume V, which is filled with a photon gas. The energy of each photon is given by $\varepsilon = pc$, where p is the momentum and c the speed of light. In thermal equilibrium, the number of photons can be considered as a constant and the walls of the container do not absorb or emit photons. The collision between a photon and the wall can be considered as elastic and a mirror reflection for the photon. Show that the equation of state of the photon gas is $PV = \frac{1}{3}U$, where P is the pressure and U the total energy of the photons.

9.7. Oxygen molecules fill a cubic container with a side of 10 cm at a temperature of 100°C. At what pressure will the mean free path be equal to half of the linear size of the container? What will be the mean free time (time interval between two collisions) at this pressure? The diameter of an oxygen molecule may be taken as 3×10^{-10} m.

9.8. Find the pressure dependence at constant temperature of the mean free path and the collision frequency.

9.9. A group of oxygen molecules start their free path at the same instant. The pressure is such that the mean free path is 3 cm. After how long a time will half of the group still remain unscattered? Assume all molecules have a speed equal to the *rms* speed. The temperature is 300 K.

9.10. The mean free path of a certain gas is 2 cm. Consider 10,000 mean free paths. How many are longer than (a) 2 cm? (b) 5 cm? (c) How many are longer than 1 cm but shorter than 2 cm? (d) How many are between 1.95 cm and 2.05 cm long?

9.11. Find the *rms* free path in terms of the mean free path ℓ. What is the most probable free path?

9.12. Use the simple argument for calculating the mean free path to obtain the mean free path in a mixture of gases of densities n_1 and n_2 and molecular radii a_1 and a_2, respectively.

9.13. A thin-walled vessel of volume V contains N particles which slowly leak out of a small hole of area A. No particles enter the volume through the hole. Find the time required for the number of particles to decrease to $N/2$, assuming that the temperature inside the the volume is kept as a constant in the process. Express your answer in terms of A, V, and $\overline{v}$.

9.14. A container is divided into two parts by a partition containing a small hole of diameter D. Both parts are filled with an ideal gas. The temperatures of the two parts are held at $T = 150$ K and $T = 300$ K, respectively, through heating of the wall. The diameter of the hole D is small enough that the system can reach a steady state, and the two parts can be maintained in separate quasi-equilibrium states.

(a) What is the ratio of the mean free paths ℓ_1/ℓ_2 between the two parts when $D \ll \ell_1$ and $D \ll \ell_2$?

(b) What is the ratio ℓ_1/ℓ_2 when $D \gg \ell_1$ and $D \gg \ell_2$?

9.15. A box of volume V is divided into two parts of equal volume by a thin partition. The left side initially contains an ideal gas at pressure P_0, and the right side is initially a vacuum. A small hole of area A is punched in the partition. Derive an expression for the pressure P_1 on the left side as a function of time. Assume that the temperature remains a constant and is the same on both sides of the partition. Express your answer in terms of P_0, A, V and the mean speed $\overline{v}$.

9.16. A container is divided into two parts of equal volume by means of a plane partition, in the middle of which is a very small hole. Initially both parts contain identical ideal gas at a temperature of 300 K and a low pressure P. The temperature of one-half of the container is then raised to 600 K, while the temperature of the other half remains at 300 K. Determine the pressure difference between the two parts of the container when steady state is achieved. Express your answer in terms of the initial pressure P.

9.17. A container of volume V contains N molecules of mass m at temperature T and pressure P. A small hole of area A is punched in the wall of the container. The area A is so small that the leakage of the molecules from the hole can be considered as effusion. The intensity of the molecular beam from the hole I is defined as the number of molecules coming out of the hole per unit area per unit time. Find the change of I if (all cases are considered as effusion):

(a) A is increased by a factor of 4, but T and P remain unchanged.

(b) T is increased by a factor of 4, but P remains unchanged.

(c) P is increased by a factor of 4, but T remains unchanged.

(d) m is increased by a factor of 4, but T and P remain unchanged.

9.18. A container has porous walls containing many tiny holes. Gas molecules can pass through these holes by effusion and then be pumped off to a collecting chamber. The container is filled with a dilute gas consisting of two types of molecules that have different masses, m_1 and m_2. The concentration of these molecules are c_1 and c_2, respectively (c_i is the ratio of the number of molecules of type i to the total number of molecules). Suppose that c_1 and c_2 in the container can be kept constant by providing fresh gas to the container. If c_1' and c_2' denote the concentrations of the two types of molecules in the collecting chamber, what is the ratio c_2'/c_1'?

9.19. (a) How will the coefficient of viscosity change with the mass of gas molecule at a constant temperature?

(b) Show that at a constant temperature the coefficient of viscosity is independent of the pressure.

(c) How does the coefficient of viscosity change with temperature?

9.20. (a) Derive an expression for the temperature dependence of the thermal conductivity of an ideal gas.

(b) Calculate the thermal conductivity of helium (considered as an ideal gas) at 300 K. For helium, take the diameter of a molecule to be 2.0×10^{-10} m and the mass to be $4\times1.67\times10^{-27}$ kg.

9.21. Show that the ratio of the pressure to the coefficient of viscosity is proportional to $\overline{v}/\ell$, where ℓ is the mean free path. Find the proportionality constant. What is the physical meaning of $\overline{v}/\ell$?

9.22. The radius of an air molecule is approximately 1.8×10^{-10} m; its mass is about 4.8×10^{-26} kg. The mass density of air is about 1.29 kg m^{-3} at standard conditions. Assume that air is a diatomic ideal gas. Estimate the values of the coefficient of viscosity, the thermal conductivity, and the diffusion coefficient.

9.23. There is a small uniform pressure gradient in an ideal gas at constant temperature so that there is a mass flow in the direction of the gradient. Using the mean free path approach, show that the rate of flow of mass in the direction of the pressure gradient per unit area and per unit pressure gradient is $m\overline{v}\ell/3kT$.

Chapter 10

Statistical Thermodynamics

10.1 Introduction

In Chapter 9 we used the microscopic kinetic theory of gases (with hard core collisions) to study the thermodynamic properties of classical dilute gases. We successfully derived several thermodynamic properties of the classical ideal gas, which could only be obtained by experimental measurements in macroscopic thermodynamics. Included among the properties we have derived were the equation of state, the magnitude of the constant volume heat capacity, several transport coefficients, and the Maxwell-Boltzmann speed distribution. Unfortunately this approach cannot be extended to study properties of condensed matter systems, such as liquids and solids. In these systems the interactions between particles are much more complicated; they cannot be simplified as hard core collisions. There are complicated relative motions between the particles such as vibrations, and the contributions from the orbital electrons may also be important. Apparently the relatively simple kinetic theory cannot be used to treat these complicated systems. Therefore we need to use another approach to develop a microscopic theory of thermodynamics. The approach, now known as **statistical thermodynamics** or **statistical mechanics**, will be introduced and discussed in this chapter. In this approach we need not treat the problems of collisions nor the details of the motions of the particles. We only need to know all the possible energy states of the system and use the method of statistics to calculate the mean values of various physical quantities. We then take the mean values as the *physical values*. The statistical fluctuations are, in general, negligibly small for macroscopic systems whose number of particles are macroscopically large. Therefore statistical mechanics is a sophisticated and reliable theory. The connection between microscopic statistical thermodynamics and macroscopic thermodynamics is based on the *Boltzmann relation* $S = k \ln \Omega$, which we introduced in Chapter 3 (see Eq. (3.4)). In $S = k \ln \Omega$, S is the entropy of the system, k the Boltzmann constant, and Ω is the microscopic *multiplicity* (or the *total number of accessible states*). Since any thermodynamic system is composed of atoms or molecules, which obey quantum mechanical equations of motion, we need to use quantum mechanics to calculate the microscopic number of states.

An important and basic element in statistical mechanics is the use of an *ensemble*, which is a collection of a very large number of (hypothetical) systems. The principle of maximum entropy is used in the *microcanonical ensemble*, in which all the (hypothetical) systems are *isolated* and have the same internal energy, volume, and number of particles, but different microstates. However an isolated system is often not the most suitable description, from the perspective of both theory and experiment. The ensemble theory is thus extended to treat systems with the temperature fixed, instead of the internal energy fixed (the volume and number of particles are still kept fixed). The ensemble for treating a system which holds the temperature constant is called the *canonical ensemble*. For this ensemble a function called the *partition function* is defined. The main task of statistical mechanics is to calculate the partition function of a given system, because, once it is known, most of the thermodynamics properties of the system can be obtained from the partition function. However the canonical ensemble is unable to treat the quantum gas systems. For this class of systems, the *grand canonical ensemble* is introduced, and the associated *grand partition function* is defined. The definitions and properties of these ensembles and partition functions will be given in the latter part of this chapter.

Before discussing the more sophisticated theory of ensembles and partition functions, it seems appropriate for us to first study the properties and statistics of the simplest thermodynamic system, that of *free particles*, which does not require a knowledge of an ensemble and partition function. This will give us more insight into the nature and statistics of free particles.

Quantum mechanical particles are divided into two classes according to the symmetry property of their wave function. One class of particles, called *Fermi particles* are those which obey *Fermi-Dirac*[1] *statistics.* The other class of particles, called *Bose particles* are those which obey *Bose-Einstein statistics.* Therefore in quantum statistical mechanics, we have two different ideal gases, the *ideal Fermi gas* and the *ideal Bose gas*, which behave quite differently. However in the classical limit (high temperature and low pressure) these two gases will approach the *classical ideal gas*, which we have studied and discussed in the previous chapters. The particles of the classical ideal gas obey another statistics called *Maxwell-Boltzmann statistics*, which is the classical limit of the above two quantum statistics. We have already learned some of the properties of Maxwell-Boltzmann statistics in Chapter 9. The main differences between these three statistics are characterized by their *distribution functions.* The physical meaning of the distribution function is, for a given temperature, how the system distributes its energy to its constituting particles. We will derive the distribution functions for *the free particles* of the three statistics in the first part of this chapter.

10.2 Three Kinds of Statistics

We consider a system of an ideal gas which consists of N particles (atoms or molecules). All these N particles are identical and can not be distinguished. These N particles should obey one of the three kinds of statistics which we have mentioned in the introduction. Particles obeying to different statistics behave quite differently. We describe in the following the main features of the three different statistics, and how we can choose the correct statistics for the particles under consideration.

1. Fermi-Dirac statistics

This is also known as **Fermi statistics**, or for brevity, **FD statistics. Particles obeying FD statistics are identical and cannot be distinguished. The orbit of each particle can not be traced,** and **each one-particle quantum state can at most accommodate one particle.** The last statement is known as the *Pauli*[2] *exclusion principle.* Particles obeying these conditions are called **Fermi particles**

[1] Paul A. M. Dirac, British physicist (1902–1984), Nobel prize laureate in physics in 1933.

[2] Wolfgang Pauli, Austrian physicist (1900–1958), Nobel prize laureate in physics in 1945.

(for brevity, **fermions**). Fermions are particles with half-integer spin, $\frac{1}{2}\hbar, \frac{3}{2}\hbar, \frac{5}{2}\hbar, \cdots$ ($\hbar = h/2\pi$, where h is the Planck constant). Examples are electrons, protons, neutrons, and ^{3}He atoms.

2. Bose-Einstein statistics

This is also known as **Bose statistics**, or for brevity, **BE statistics. Particles obeying BE statistics are identical and cannot be distinguished. The orbit of each particle can not be traced,** and **there is no restriction on the number particles in each one-particle quantum state**, i.e., the *Pauli exclusion principle does not apply to particles obeying BE statistics.* Particles obeying these conditions are called **Bose particles** (for brevity, **bosons**). Bosons are particles with integer spin, $0, \hbar, 2\hbar, \cdots$. Examples are photons, hydrogen molecules, ^{4}He atoms.

3. Maxwell-Boltzmann statistics

This is also known as **classical statistics**, or **Boltzmann statistics**, or for brevity, **MB statistics. Traditionally particles obeying MB statistics are considered to be distinguishable, and there is no restriction on the number of particles in each one-particle quantum state. However at the final stage, the total number of microscopic states must be divided by** $N!$, in order to get the correct result. If the total number of microstates (i.e., microscopic states) is not divided by $N!$, then there will be a Gibbs paradox, which we have discussed in Chapter 3. The origin of the Gibbs paradox is due to the indistinguishability of the particles, hence the more precise way to describe the **particles obeying MB statistics is traceable but indistinguishable**. In nature particles which obey MB statistics for all temperatures and pressures do not actually exist. In nature there are only fermions and bosons, whose behavior will be approximately identical to each other under the classical limit (high temperature and low pressure);[3], in this limit we may say that the particles behave like classical particles which obey Maxwell-Boltzmann statistics, or classical statistics. In the classical limit the occupation number for each one-particle quantum state will be much smaller than 1, the Pauli exclusion principle does not play any role in this limit, and therefore there is no distinction between fermions and bosons.

[3]The criterion for the classical limit to be valid will be derived in Sec. 10.9.

10.3 Quantum States and the Density of States

In statistical mechanics, we frequently need to know the density of states, i.e., the distribution of quantum states, to calculate various physical quantities. In this section we discuss the quantum state of a free particle (with no internal structure) and its associated density of states. We are considering non-interacting particles, therefore it is sufficient for us to consider the one-particle Hamiltonian h_1. When there is no external field h_1 contains only the kinetic energy term,

$$h_1 = \frac{p^2}{2m} = -\frac{\hbar^2}{2m}\nabla^2, \tag{10.1}$$

where m is the mass of the particle. For simplicity, we consider motion in the x-direction only, and assume that the particle is restricted to move in the range $0 \le x \le L$. The corresponding Schrödinger[4] equation is

$$h_{1x}\psi(x) = -\frac{\hbar^2}{2m}\frac{\partial^2}{\partial x^2}\psi(x) = \varepsilon_x\psi(x), \tag{10.2}$$

where ε_x is the kinetic energy for motion in the x-direction. Apparently the solution of the wave function $\psi(x)$ may have the form

$$\psi(x) = A\sin k_x' x + B\cos k_x' x, \quad 0 \le x \le L, \tag{10.3}$$

where A, B and k_x' are constants. If we require $\psi(0) = \psi(L) = 0$, then we have $B = 0$, and k_x' must satisfy

$$k_x' = n_x\frac{\pi}{L}, \quad n_x = 1, 2, 3, \cdots. \tag{10.4}$$

The condition that the wave function must vanish at the boundaries, $\psi(0) = \psi(L) = 0$, is known as the *rigid wall boundary condition.*

For macroscopic systems, physical properties should be independent of the existence of the surface, and we may use another more convenient boundary condition, known as the *periodic boundary condition.* The real system is restricted to the region $0 \le x \le L$, but we imagine that the system can be extended to infinity for both $x > L$ and $x < 0$, by repeating the system itself, then the resulting imaginary system is a

[4]Erwin Schrödinger, Austrian physicist (1887–1961), Nobel prize laureate in physics in 1933.

periodic system with a period L, i.e., the system satisfies the periodic condition:

$$\psi(x+L) = \psi(x). \tag{10.5}$$

Therefore the wave function can be written as

$$\psi(x) = A\, e^{ik_x x}, \tag{10.6}$$

where A and k_x are constants. In order that the boundary condition in Eq. (10.5) is satisfied, we must choose

$$k_x = n_x \frac{2\pi}{L}, \quad n_x = 0, \pm 1, \pm 2, \pm 3, \cdots . \tag{10.7}$$

Comparing Eqs. (10.7) and (10.4), we see that the difference of two neighboring k_x's (Δk_x) is twice that of k'_x ($\Delta k'_x$), i.e., $\Delta k_x = 2\Delta k'_x$. However k_x may be positive or negative, while k'_x can be only positive (or only negative),[5] therefore the total number of k_x which satisfies Eq. (10.7) is equal to the total number of k'_x which satisfies Eq. (10.4). For macroscopic systems, both Δk_x and $\Delta k'_x$ approach zero, therefore there will be no difference in using Eq. (10.4) or Eq. (10.7) to calculate the total number of the microstates. In the following we use the periodic boundary condition Eq. (10.7) for the discussion. Substituting Eq. (10.7) into Eq. (10.2), we get

$$\varepsilon_x = \frac{p_x^2}{2m} = \frac{\hbar^2 k_x^2}{2m} = \frac{2\pi^2\hbar^2}{mL^2} n_x^2.$$

Now we consider a particle moving in the x, y and z directions; let the particle be confined in the volume, $0 \le x, y, z \le L$, then the total energy is

$$\begin{aligned} \varepsilon &= \varepsilon_x + \varepsilon_y + \varepsilon_z = \frac{1}{2m}(\hbar^2 k_x^2 + \hbar^2 k_x^2 + \hbar^2 k_x^2) \\ &= \frac{2\pi^2\hbar^2}{mL^2}(n_x^2 + n_y^2 + n_z^2), \quad n_x, n_y, n_z = 0, \pm 1, \pm 2, \cdots . \end{aligned} \tag{10.8}$$

Therefore a set of integers (n_x, n_y, n_z) determines a quantum state (not including the spin quantum number) of the particle, where $n_x, n_y, n_z = 0, \pm 1, \pm 2, \cdots$. The energy of this quantum state is determined by Eq. (10.8).

[5]Take the negative value of k'_x will change the wave function from $\psi(x)$ to $-\psi(x)$, therefore $-k'_x$ and k'_x represent the same state.

Since n_x, n_y and n_z are discrete numbers, the energy is not actually continuous; we call the values energy levels. For a macroscopic L, the energy difference between neighboring levels is very small, and ε can be considered as quasi-continuous. To simplify the notation, Eq. (10.8) can be written as

$$\varepsilon = \varepsilon_q = \frac{2\pi^2\hbar^2}{mV^{2/3}}\underline{n}_q^2, \quad \underline{n}_q^2 \equiv n_x^2 + n_y^2 + n_z^2, \tag{10.9}$$

where $V = L^3$ is the volume of the system, and the subscript q is the abbreviation for the set of the quantum numbers (n_x, n_y, n_z). It should be noted that each set of (n_x, n_y, n_z) corresponds to a quantum number q; a different set of (n_x, n_y, n_z) corresponds to a different q', i.e., $q \neq q'$. It is possible that $\underline{n}_q^2 = \underline{n}_{q'}^2$, in this case $\varepsilon_q = \varepsilon_{q'}$ $(q \neq q')$; this is called a degenerate level in quantum mechanics.

Now we would like to know the order of magnitude of $\underline{n}_q^2$ for a macroscopic system. If we take ^{4}He atoms as an example with the following data, $T = 100$ K, $V = (10^{-2}\text{ m})^3$, $m = 6.65 \times 10^{-27}$ kg, $\hbar = 1.05 \times 10^{-34}$ J s, $k = 1.38 \times 10^{-23}$ J K^{-1}, and approximate ε_q as $\frac{3}{2}kT$, then we have

$$\begin{aligned}\underline{n}_q^2 &= \varepsilon_q \frac{mV^{2/3}}{2\pi^2\hbar^2} \\ &\approx \frac{3}{2}(1.38 \times 10^{-23})(100)\frac{(6.65 \times 10^{-27})(10^{-6})^{2/3}}{2(3.14)^2(1.05 \times 10^{-34})^2} \\ &\approx 6.3 \times 10^{15}.\end{aligned}$$

This is a very large number. Therefore we may consider the energy ε_q as continuous, because the difference of the neighboring values of $\underline{n}_q^2$ is very small, so the energy difference between the neighboring levels $|\Delta\varepsilon_q|$ may be considered as zero in comparison with the energy ε_q,

$$\frac{|\Delta\varepsilon_q|}{\varepsilon_q} = \frac{|\Delta\underline{n}_q^2|}{\underline{n}_q^2} \sim \frac{1}{\underline{n}_q^2} \sim 10^{-15}.$$

Since the energy levels can be considered as continuous, in many calculations, the summation may be replaced by an integral, which makes the calculations easier. We may frequently encounter the conversion of a summation to an integral like

$$\sum_{\mathbf{p}} F(\mathbf{p}) \to \frac{V}{h^3}\int F(\mathbf{p})\, dp_x\, dp_y\, dp_z, \tag{10.10}$$

where $F(\boldsymbol{p})$ is a function of the momentum $\boldsymbol{p}$. Now we explain why there should be a factor V/h^3 in front of the integral sign in order to get the correct result. From Eq. (10.7), the difference between two neighboring p_x is $\Delta p_x = \hbar \Delta k_x = h/L$, therefore

$$\Delta p_x \, \Delta p_y \, \Delta p_z = \left(\frac{h}{L}\right)^3 = \frac{h^3}{V}. \tag{10.11}$$

This means that the quantum state determined by each set of integers (n_x, n_y, n_z) occupies a volume of h^3/V in momentum space. Therefore in front of the integral sign on the right hand side of Eq. (10.10) we have the factor V/h^3, in order to get the correct counting of the number of quantum states. In many cases, physical quantities are functions of the energy ε and do not depend on the momentum $\boldsymbol{p}$ explicitly, then Eq. (10.10) may be rewritten as[6]

$$\frac{V}{h^3} \int F(\varepsilon) \, dp_x \, dp_y \, dp_z = \int F(\varepsilon) \, g(\varepsilon) \, d\varepsilon, \tag{10.12}$$

where $g(\varepsilon)$ is called the **density of states**. The physical meaning of $g(\varepsilon)\, d\varepsilon$ is the number of quantum states for the energy of a particle in the range from ε to $\varepsilon + d\varepsilon$. Since $F(\varepsilon)$ is independent of the direction of the momentum $\boldsymbol{p}$, by using spherical coordinates $dp_x \, dp_y \, dp_z = 4\pi p^2 \, dp$ and $\varepsilon = p^2/2m$, we easily obtain the density of states $g(\varepsilon)$ (not including the spin states) from Eq. (10.12) as

$$g(\varepsilon) = \frac{4\sqrt{2}\pi \, V}{h^3} m^{3/2} \varepsilon^{1/2}.$$

Therefore $g(\varepsilon)$ is proportional to $\varepsilon^{1/2}$. The dependence of $g(\varepsilon)$ on ε depends on the dimensionality of the system. For a two-dimensional systems, $g(\varepsilon)$ depends on ε^0 (i.e., independent of ε), and for one-dimensional systems, $g(\varepsilon)$ depends on $\varepsilon^{-1/2}$. If we include the number of the spin states $g_s = 2s + 1$, then the above formula for $g(\varepsilon)$ must be multiplied by the factor g_s, therefore the density of states $g(\varepsilon)$ including the spin states is

$$g(\varepsilon) = g_s \frac{4\sqrt{2}\pi \, V}{h^3} m^{3/2} \varepsilon^{1/2}. \tag{10.13}$$

For example the spin for an electron is $\frac{1}{2}\hbar \equiv s\hbar$, therefore $g_s = 2$.

[6] For simplicity we use the same F to replace $F(\boldsymbol{p})$ by $F(\varepsilon)$.

10.4 Entropy and the Most Probable Distribution

We consider a system of an ideal gas of N particles in a volume V, with total energy U. We would like to know in thermal equilibrium, how the total energy U is distributed among its N particles. This is known as the distribution function. In the previous section, we noticed that each quantum state of a free particle can be denoted by a set of quantum numbers q, and the energy levels can be considered as quasi-continuous. For the convenience of calculation, we divide the quasi-continuous energy levels into discrete levels and denote them as ε_j, $j = 1, 2, 3, \cdots$. From the macroscopic point of view, $\varepsilon_{j+1} - \varepsilon_j$ is a very small positive energy ($\varepsilon_{j+1} > \varepsilon_j$). We use the notation g_j to denote the number of quantum states q (including spin states), whose energy satisfies the condition $\varepsilon_{j+1} > \varepsilon_q \geq \varepsilon_j$ (see Eq. (10.9)). Then we let $\varepsilon_q = \varepsilon_j$ and neglect the difference between ε_q and ε_j, which is negligibly small. The energy level ε_j is therefore g_j-fold degenerate. Since ε is quasi-continuous, the condition $g_j \gg 1$ can be satisfied for proper divisions of the energy levels ε_j.

Now we distribute the N particles into the energy levels, such that there are n_1 particles in the level ε_1, n_2 particles in ε_2, and n_3 particles in ε_3, etc. We choose large enough[7] g_j such that the condition $n_j \gg 1$ can be satisfied, which will simplify our calculations. We use the notation $\{n_i\}$ to denote the set of numbers, $n_1, n_2, n_3, \cdots$, i.e., $\{n_i\} \equiv (n_1, n_2, n_3, \cdots)$, which must satisfy the conditions that the total number of particles is N, and the total internal energy of the system is U,

$$\sum_{j=1} n_j = N, \tag{10.14}$$

$$\sum_{j=1} n_j \, \varepsilon_j = U. \tag{10.15}$$

The set of numbers $\{n_i\}$ denotes a *distribution*. Since the set of numbers $\{n_i\}$ only specifies the number of particles n_j in the level j, but does not specify how the n_j particles are distributed among the g_j states, each distribution $\{n_i\}$ contains a very large number of microstates. We use the notation $W(\{n_i\})$ to denote the number of microstates for the

[7]Although g_j is large, the condition that $\varepsilon_{j+1} - \varepsilon_j$ is small should still hold.

distribution $\{n_i\}$. From the Boltzmann relation (3.4), the entropy S of the system is

$$S = k \ln \Omega(N, V, U) = k \ln \left[\sum_{\{n_i\}}' W(\{n_i\}) \right], \tag{10.16}$$

where Ω is the multiplicity for the macrostate (i.e., the macroscopic state), and primed sum, $\sum'$, indicates that only those $\{n_i\}$'s which satisfy the conditions Eqs. (10.14)–(10.15) are included in the sum.

Despite of these restrictions, there are still a very large number of distributions $\{n_i\}$ which are included in the sum. The calculation is therefore not an easy one. For a system with very large N, the number $W(\{n_i\})$ will vary significantly for a slight change of $\{n_i\}$. There is a sharp maximum for $W(\{n_i\})$ which occurs at a particular distribution $\{\overline{n}_i\}$, which is called the **most probable distribution** ($n_j = \overline{n}_j$ for all j). The distribution $\{\overline{n}_i\}$ makes $W(\{n_i\})$ achieve a sharp maximum, which is a characteristic of a macroscopic system. A sharp maximum implies that

$$W(\{\overline{n}_i\}) \gg W(\{n_i\}), \quad \text{if } \{n_i\} \neq \{\overline{n}_i\}.$$

Therefore we may rewrite Eq. (10.16) as

$$S = k \ln W(\{\overline{n}_i\}), \tag{10.17}$$

which is the result that the contributions to Ω from the distributions $\{n_i\} \neq \{\overline{n}_i\}$ can be neglected.[8] This equation also implies that the most probable distribution $\{\overline{n}_i\}$ is also the equilibrium distribution, which is the distribution we are looking for.

Now our problem is to look for the distribution $\{\overline{n}_i\}$ which will make W a maximum. Since the same distribution will also make $\ln W$ a maximum, we will use the condition that $\ln W$ is a maximum to determine the distribution $\{\overline{n}_i\}$, which simplifies the calculations greatly. All the n_j are numbers much greater than 1, so we may approximate the n_j's as quasi-continuous variables, and use the methods of calculus to determine the condition that $\ln W$ is a maximum, i.e., the first order derivative is zero,

$$\delta \ln W = \sum_{j=1} \frac{\partial \ln W}{\partial n_j} \delta n_j = 0, \quad \text{when } n_j = \overline{n}_j. \tag{10.18}$$

[8]We may write Eq. (10.16) as $S = k \ln[c\, W(\{\overline{n}_i\})] = k \ln c + k \ln W(\{\overline{n}_i\})$. For very large N, $\ln c / \ln W(\{\overline{n}_i\}) \to 0$, therefore Eq. (10.17) is satisfied.

However the numbers n_j must satisfy the two conditions Eqs. (10.14)–(10.15), therefore δn_j must satisfy the following two equations,

$$\delta N = \sum_{j=1} \delta n_j = 0, \tag{10.19}$$

$$\delta U = \sum_{j=1} \varepsilon_j \, \delta n_j = 0. \tag{10.20}$$

We may solve the above problem in two steps. First, from Eqs. (10.19)–(10.20) we solve for δn_1 and δn_2 in terms of δn_3, δn_4, δn_5, $\cdots$. Second, we substitute the solved δn_1 and δn_2 into Eq. (10.18) to eliminate δn_1 and δn_2, then Eq. (10.18) can be written as

$$\sum_{j=3} \mathcal{W}_j \, \delta n_j = 0, \quad \text{when } n_j = \overline{n}_j. \tag{10.21}$$

It should be noted that the sum in the above equation does not include the terms of $j = 1$ and $j = 2$. Since the δn_j ($j \geq 3$) are independent variables, which may be varied arbitrarily, the coefficient for each term must equal to zero, thus

$$\mathcal{W}_j = 0, \quad j \geq 3, \quad \text{when } n_j = \overline{n}_j. \tag{10.22}$$

Solving the above coupled equations, we may obtain the distribution $\overline{n}_j$ that we are looking for. This can be a very complicated problem, because the number of independent variables are very large. Therefore it is not practical to solve the problem this way. There is an easier way to solve this problem known as the method of *Lagrange*[9] *multipliers*. We multiply Eqs. (10.19) and (10.20), respectively, by $-\alpha$ and $-\beta$, which are constants independent of n_j. The resulting two equations are added to Eq. (10.18) to obtain

$$\sum_{j=1} \left(\frac{\partial \ln W}{\partial n_j} - \alpha - \beta \, \varepsilon_j \right) \delta n_j = 0, \quad \text{when } n_j = \overline{n}_j. \tag{10.23}$$

In the above equation, the undetermined constants α and β, which are independent of n_j, are called the Lagrange multipliers. We may choose α and β such that

$$\frac{\partial \ln W}{\partial n_j} - \alpha - \beta \, \varepsilon_j = 0 \text{ for } j = 1, 2, \text{ when } n_j = \overline{n}_j. \tag{10.24}$$

[9] Joseph L. Lagrange, French mathematician (1736–1813).

When the above two equations are satisfied ($j = 1, 2$), Eq. (10.23) takes the form of Eq. (10.21), and therefore, from Eq. (10.22),

$$\mathcal{W}_j = \frac{\partial \ln W}{\partial n_j} - \alpha - \beta\,\varepsilon_j = 0, \quad \text{when } n_j = \overline{n}_j,\ j \geq 1. \tag{10.25}$$

The above equation holds also for $j = 1$ and $j = 2$, because they are just Eq. (10.24). The above result tells us that the method of Lagrange multipliers treats all of the δn_j's as independent variables, but it does not look for the maximum of $\ln W$, instead it looks for the maximum of $\ln W - \alpha \sum n_j - \beta \sum n_j\,\varepsilon_j$. In this way Eq. (10.25) is obtained. When we solve for $\overline{n}_j$ from Eq. (10.25), the solution we obtained will contain the unknown constants α and β. We have to solve for the constants α and β in terms of known physical quantities, otherwise the solution for $\overline{n}_j$ is not complete. We may say that the extra needed effort to solve for the constants α and β is the price we have to pay for using the method of the Lagrange multipliers. We note that $\{\overline{n}_i\}$ must satisfy Eqs. (10.14)–(10.15), and therefore we have two equations to solve for the two unknowns α and β.

Before closing this section, we introduce a very basic assumption in statistical mechanics. *In an isolated system, all the microstates which satisfy the conditions* Eqs. (10.14)–(10.15) *have equal probability to appear in the macrostate with the total number of particles N, total internal energy U and volume V.* This is known as the **postulate of equal probability**. Therefore in the equation for the entropy S, Eq. (10.16), the contribution due to each distribution $\{n_i\}$ is equal to the total number of microstates $W(\{n_i\})$ in the distribution. This implies that each microstate has an equal contribution to the entropy. Therefore the total number of microstates $W(\{n_i\})$ in the distribution $\{n_i\}$ is called the **thermodynamic probability** for the distribution $\{n_i\}$. Of course this is not a normalized probability, because $W(\{n_i\})$ is a number much larger than 1. In classical statistical mechanics, a microstate is determined by the position vector $\boldsymbol{r}_i$ and the momentum vector $\boldsymbol{p}_i$ of all the particles ($i = 1, 2, 3 \cdots, N$). Since all the position vectors $\boldsymbol{r}_i$ depend on time, every microstate will change with time. The postulate of equal probability requires that the system stays in each microstate for equal time. In quantum statistical mechanics, one solves the time independent Schrödinger equation and the state of the system is represented by a time independent stationary state wave function. However the system is constantly interacting with its surroundings (at least the measuring equipment), so the wave function of the system should be a mixed state, which is composed of many stationary states with time dependent coefficients. Therefore the assumption of equal probability requires that each stationary state which satisfies the conditions Eqs. (10.14)–(10.15) has equal probability to appear (including time duration) in the macrostate with the number of particles N, total energy U, and volume V.

10.5 The Maxwell-Boltzmann Distribution

In this section we will use Eq. (10.25) to derive the most probable distribution $\{\overline{n}_j\}$ for an ideal gas of N particles which obeys *Maxwell-Boltzmann statistics* (MB statistics). We may consider the particles which obey MB statistics as distinguishable and divide the final total number of microstates by $N!$ in order to get the correct result. We use the method of Lagrange multipliers to obtain the distribution $\{\overline{n}_j\}$, therefore in evaluating the total number of the microstates $W(\{n_j\})$ there is no need to consider the restrictions of Eqs. (10.14)–(10.15). At first we count the number of ways to divide N into groups of numbers $n_1, n_2, n_3, \cdots$; the number of ways is

$$\frac{N!}{n_1!\,n_2!\,n_3!\cdots}. \tag{10.26}$$

Next, we count the number of ways to distribute the n_j particles in the g_j states for the energy level ε_j. For example, for the ε_1 level there are g_1 states and the number of particles in the level is n_1. Since the particles are considered as distinguishable, for each particle there are g_1 possible microstates, and in total $g_1^{n_1}$ possible microstates for n_1 particles. By considering all levels and the factor in Eq. (10.26), we have the total number of possible microstates:

$$W'(\{n_i\}) = N! \prod_j \frac{g_j^{n_j}}{n_j!}. \tag{10.27}$$

The number $W'(\{n_i\})$ is not the correct total number of microstates, because the particles are actually indistinguishable and we need to divide it by $N!$ in order to get the correct number. Therefore the correct result is

$$W_{\mathrm{MB}}(\{n_i\}) = \prod_j \frac{g_j^{n_j}}{n_j!}. \tag{10.28}$$

In MB statistics, we require $g_j \gg n_j \gg 1$, therefore we may use the Stirling approximation (see Appendix C),

$$\ln N! \approx N \ln N - N \quad (\text{large } N), \tag{10.29}$$

to calculate the value of $\ln W_{\rm MB}$. We obtain

$$\ln W_{\rm MB} = \sum_j (n_j \ln g_j - n_j \ln n_j + n_j). \tag{10.30}$$

Substituting the above equation into Eq. (10.25), we find

$$\ln g_j - \ln \overline{n}_j - \alpha - \beta\,\varepsilon_j = 0,$$

$$\frac{\overline{n}_j}{g_j} = e^{-\alpha-\beta\,\varepsilon_j}. \tag{10.31}$$

This is the *Maxwell-Boltzmann distribution*, but the distribution function contains two unknown constants α and β, which have to be solved for in terms of known physical quantities. Otherwise we cannot say that the distribution function is known. From the above equation and Eqs. (10.14), (10.10), (9.32) and $\varepsilon = (p_x^2 + p_y^2 + p_z^2)/2m$, we have

$$\begin{aligned} N &= \sum_j \overline{n}_j = \sum_j g_j\, e^{-\alpha-\beta\,\varepsilon_j} \\ &= \frac{V}{h^3}\, e^{-\alpha} \left(\int_{-\infty}^{\infty} e^{-(\beta/2m)p_x^2}\, dp_x \right)^3 \\ &= e^{-\alpha}\, V \left(\frac{2\pi m}{h^2 \beta} \right)^{3/2}. \end{aligned} \tag{10.32}$$

By using the above equation and Eqs. (10.15) and (10.31), we obtain

$$\begin{aligned} U &= \sum_j \overline{n}_j\, \varepsilon_j = \sum_j g_j\, \varepsilon_j\, e^{-\alpha-\beta\,\varepsilon_j} \\ &= -\frac{\partial N}{\partial \beta} = \frac{3}{2} N \frac{1}{\beta}. \end{aligned} \tag{10.33}$$

Comparing this equation with Eq. (9.30), the constant β is obtained as

$$\beta = \frac{1}{kT}. \tag{10.34}$$

From Eqs. (10.32) and (10.34) the expression for α can be written as

$$e^{-\alpha} = \frac{N}{V}\left(\frac{h^2}{2\pi mkT}\right)^{3/2}. \tag{10.35}$$

Substituting the above equation and Eq. (10.34) into Eq. (10.31), and considering j as the absolute value of the momentum $|\boldsymbol{p}|$, we have $g_j = (V/h^3)4\pi p^2\,dp$ and

$$\overline{n}_j \equiv \overline{n}(p)\,dp = N\,4\pi\left(\frac{1}{2\pi mkT}\right)^{3/2} p^2\,e^{-p^2/2mkT}dp. \tag{10.36}$$

This equation is exactly the same as Eq. (9.37), the Maxwell distribution for speed $F(v)\,dv$, where the independent variable is the speed $v = p/m$ instead of the momentum p.

10.6 The Physical Meaning of α

In Eq. (10.35) we obtained the expression for $e^{-\alpha}$, but its physical meaning is not clear. Therefore it is worthwhile studying this expression in more detail. From the expression for entropy S Eq. (10.17), Eq. (10.30) and the MB distribution (10.31), we have

$$\begin{aligned} S &= k\sum_j(\overline{n}_j \ln g_j - \overline{n}_j \ln \overline{n}_j + \overline{n}_j) \\ &= k\left(N - \sum_j \overline{n}_j \ln\frac{\overline{n}_j}{g_j}\right) = k\left[N + \sum_j \overline{n}_j(\alpha + \beta\varepsilon_j)\right] \\ &= k\,N(1+\alpha) + k\,\beta\sum_j \overline{n}_j\,\varepsilon_j \\ &= k\,N(1+\alpha) + \frac{U}{T}. \end{aligned} \tag{10.37}$$

The Helmholtz free energy F and $(\partial F/\partial N)_{T,V}$ are, respectively,

$$F = U - TS = -NkT(1+\alpha),$$

$$\begin{aligned} \mu &= \left(\frac{\partial F}{\partial N}\right)_{T,V} = -kT(1+\alpha) - NkT\left(\frac{\partial \alpha}{\partial N}\right)_{T,V} \\ &= -\alpha\,kT, \end{aligned} \tag{10.38}$$

where we used Eq. (10.35) and $(\partial\alpha/\partial N)_{T,V} = -1/N$. Therefore we obtain the result

$$\alpha = -\frac{\mu}{kT}. \tag{10.39}$$

We have to note that in Eq. (10.38) the chemical potential μ is defined as the chemical potential *per particle*. The chemical potential μ discussed in the previous chapters is the chemical potential *per kilomole*. Therefore the latter μ is N_A times the former μ, where N_A is the number of molecules per kilomole, i.e., the Avogadro's number.

Combining Eqs. (10.31), (10.34), and (10.39), we get the final form of the **Maxwell-Boltzmann distribution** (MB distribution) or the **Boltzmann distribution** as

$$\frac{\overline{n}_j}{g_j} = e^{\mu/kT}e^{-\varepsilon_j/kT}. \tag{10.40}$$

10.7 The Fermi-Dirac Distribution

Particles which obey *Fermi-Dirac statistics* (FD statistics) are indistinguishable, therefore there is only one way to divide the N particles into groups n_1, n_2, n_3,$\cdots$, because any interchange of particles does not give a new microstate. Therefore the factor (10.26), which applies in MB statistics, is equal to 1 in FD statistics. Hence the total number of microstates $W(\{n_i\})$ is due to the ways of distributing the particles in each energy level. We look at the jth level, in which there are g_j quantum states and the number of particles to be distributed is n_j. Each state can at most accommodate one particle, hence the way to distribute the particles is to choose n_j of the g_j states and put one particle in each of the chosen n_j states. This means that the chosen n_j states are filled and the remaining $g_j - n_j$ states are empty. The total number of microstates is therefore equal to $g_j!/n_j!\,(g_j - n_j)!$, which requires $g_j \geq n_j$. By considering all the energy levels, we have the total number of microstates

$$W_{\rm FD}(\{n_i\}) = \prod_j \frac{g_j!}{n_j!\,(g_j - n_j)!}. \tag{10.41}$$

We may use the same procedures as in MB statistics, the method of Lagrange multipliers and the Stirling approximation (10.29), to derive the most probable distribution from the condition of the maximum of $\ln W_{\rm FD}$. By using this method we will get the correct distribution. However there is a difficulty in FD statistics when using this method. At very low temperatures, due to the Pauli exclusion principle, $g_j - n_j$ may become a very small number, therefore the application of the Stirling approximation is no longer appropriate. Here we will use another approach to derive the FD distribution, without using the Stirling approximation.

We consider two ideal Fermi gases A and A′ with exactly the same energy states, but with a slightly different number of particles and total energy. The number of particles of A and A′ are, respectively, N and N', and the total energies are, respectively, U and U'. The relations between N and N' and between U and U' are

$$N - N' = 1, \quad U - U' = \varepsilon_{j_\nu}, \tag{10.42}$$

where ε_{j_ν} is the energy of the j_νth energy level of the system (for both A and A′). Therefore for each distribution $\{n_i\}$ in A, there is a corresponding distribution $\{n'_i\}$ in A′, which satisfies

$$n'_j = n_j \ (j \neq j_\nu), \quad n'_{j_\nu} = n_{j_\nu} - 1. \tag{10.43}$$

Note that for the distribution $n_{j_\nu} = 0$ in A, there will be no corresponding distribution in A′. The total number of microstates for the distribution $\{n_i\}$ in A is

$$W_{\mathrm{A}}(\{n_i\}) = \prod_j \frac{g_j!}{n_j!\,(g_j - n_j)!}. \tag{10.44}$$

The total number of microstates for the corresponding distribution $\{n'_i\}$ in A′ is

$$W_{\mathrm{A'}}(\{n'_i\}) = \prod_j \frac{g_j!}{n'_j!\,(g_j - n'_j)!}. \tag{10.45}$$

The ratio of these two numbers is

$$\frac{W_{\mathrm{A'}}}{W_{\mathrm{A}}} = \prod_j \frac{n_j!\,(g_j - n_j)!}{n'_j!\,(g_j - n'_j)!} = \frac{n_{j_\nu}}{g_{j_\nu} - n'_{j_\nu}},$$

so

$$n_{j_\nu} W_{\mathrm{A}}(\{n_i\}) = (g_{j_\nu} - n'_{j_\nu}) W_{\mathrm{A'}}(\{n'_i\}).$$

This equation holds for any distribution $\{n_i\}$, including the distribution for which $n_{j_\nu} = 0$. In this case A′ has no corresponding distribution, i.e., $W_{\mathrm{A'}}(\{n'_i\}) = 0$, and both sides of the above equation are 0. Summing over all the distributions, we get

$$\sum_{\{n_i\}} n_{j_\nu} W_{\mathrm{A}}(\{n_i\}) = \sum_{\{n'_i\}} (g_{j_\nu} - n'_{j_\nu})\, W_{\mathrm{A'}}(\{n'_i\}). \tag{10.46}$$

We have mentioned that the total number of the microstates $W(\{n_i\})$ for a given distribution $\{n_i\}$ may be considered as a thermodynamic probability, therefore we may define the mean value $\overline{f}$ for any function f which is a function of the n_j's, $f(n_1, n_2, \cdots)$,

$$\overline{f} = \frac{1}{\Omega}\sum_{\{n_i\}} f(n_1, n_2, \cdots)\, W(\{n_i\}), \quad \Omega = \sum_{\{n_i\}} W(\{n_i\}). \tag{10.47}$$

Therefore Eq. (10.46) may be written as

$$\frac{\overline{n}_{j_\nu}}{g_{j_\nu} - \overline{n'}_{j_\nu}} = \frac{\Omega'}{\Omega}. \tag{10.48}$$

As in MB statistics, we consider the case that $g_j \gg 1$, $n_j \gg 1$ for all j, we may therefore make the approximation that $\overline{n'}_{j_\nu} = \overline{n}_{j_\nu}$. By using the Boltzmann relation for A and A$'$, $S = k\ln\Omega$ and $S' = k\ln\Omega'$, where S and S' are respectively the entropy of A and A$'$, Eq. (10.48) becomes

$$\ln\frac{\overline{n}_{j_\nu}}{g_{j_\nu} - \overline{n}_{j_\nu}} = \frac{S' - S}{k} = \frac{\Delta S}{k}. \tag{10.49}$$

For a one-component open system, when the volume V is kept constant, the change of the internal energy U can be obtained from Eq. (5.35):

$$\Delta U = T\Delta S + \mu\Delta N.$$

If we take $\Delta N = -1$ (the difference of the number of particles between A$'$ and A), then μ is the chemical potential per particle, and $\Delta U = -\varepsilon_{j_\nu}$ is the internal energy difference between A$'$ and A, therefore

$$\Delta S = \frac{\mu - \varepsilon_{j_\nu}}{T}.$$

Substituting the above equation into Eq. (10.49) and re-labeling the subscript j_ν by j (because j_ν is an arbitrary energy level), we get the **Fermi-Dirac distribution** (FD distribution) or the **Fermi distribution**:

$$\frac{\overline{n}_j}{g_j} = \frac{1}{e^{(\varepsilon_j - \mu)/kT} + 1}. \tag{10.50}$$

We see that $\overline{n}_j/g_j \leq 1$, which implies that the average number of particles in each state is less than 1. This is required by the Pauli exclusion principle.

If we use the method of Lagrange multipliers and the Stirling approximation to calculate the distribution function $\{\overline{n}_i\}$ for the FD statistics, we will get

$$\frac{\overline{n}_j}{g_j} = \frac{1}{e^{\alpha+\beta\,\varepsilon_j}+1}. \tag{10.51}$$

In this method, in addition to the difficulty that $g_j - n_j$ may not be large in comparison with 1, we still have to determine the constants α and β. Therefore it is not really a good method. The constants α and β can be shown, as in MB statistics, to be

$$\alpha = -\frac{\mu}{kT}, \quad \beta = \frac{1}{kT}. \tag{10.52}$$

Therefore the distributions (10.51) (the most probable distribution) and (10.50) (the mean distribution) are exactly the same.

10.8 The Bose-Einstein Distribution

Now we consider particles which obey *Bose-Einstein statistics* (BE statistics). The main difference between BE statistics and FD statistics is that the Pauli exclusion principle does not apply in BE statistics. This implies that particles obeying BE statistics are indistinguishable, and there is no restriction on the number of particles in each state. As in FD statistics, the factor (10.26) is 1 in the calculation of the total number of the microstates. We consider the jth energy level. In this level there are g_j quantum states which accommodate n_j particles. We may imagine that there are $g_j - 1$ partitions which divide the *state space* into g_j states. We then put the n_j particles into these states, and there is no restriction on how we put the particles in the states. We may consider this problem as the number of ways to arrange $n_j + g_j - 1$ objects, in which the $g_j - 1$ partitions are identical objects and the n_j particles are also identical. The situation is shown in Fig. 10.1, in which each partition is represented by a vertical line and each particle is represented by a black dot. Therefore the number of microstates for the jth level is $(n_j + g_j - 1)!/n_j!\,(g_j - 1)!$. By considering all the energy levels, the total number of microstates $W_{\rm BE}(\{n_i\})$ for the distribution $\{n_i\}$ is

$$W_{\rm BE}(\{n_i\}) = \prod_j \frac{(n_j + g_j - 1)!}{n_j!\,(g_j - 1)!}. \tag{10.53}$$

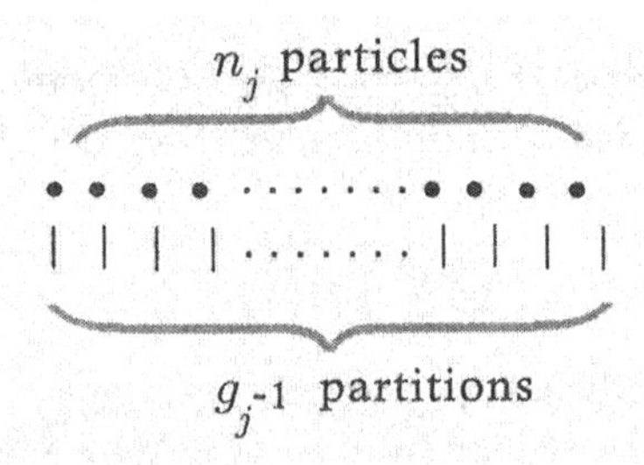

Figure 10.1: Calculation of the number of microstates in BE statistics. For the jth level, there are $g_j - 1$ identical partitions (vertical lines) and n_j identical particles (black dots).

We may use the same method as in FD statistics to derive the Bose-Einstein distribution. However in the following we use the method of Lagrange multipliers and the Stirling approximation (10.29) to derive the distribution, which may be compared with the result of Eq. (10.51).

$$\begin{aligned}\ln W_{\rm BE}(\{n_i\}) &= \sum_j [(n_j + g_j - 1)\ln(n_j + g_j - 1) \\ &\quad - n_j \ln n_j - (g_j - 1)\ln(g_j - 1)], \\ \frac{\partial \ln W_{\rm BE}}{\partial n_j} &= \ln(n_j + g_j - 1) - \ln n_j.\end{aligned}$$

Substituting the above equation into Eq. (10.25), we obtain

$$\ln(\overline{n}_j + g_j - 1) - \ln \overline{n}_j - \alpha - \beta\,\varepsilon_j = 0.$$

Solving the above equation, using the conditions $g_j \gg 1$ and $g_j - 1 \approx g_j$, we get

$$\frac{\overline{n}_j}{g_j} = \frac{1}{e^{\alpha + \beta\,\varepsilon_j} - 1}. \tag{10.54}$$

This is the **Bose-Einstein distribution** (BE distribution) or the **Bose distribution**. By comparing with Eq. (10.50), or by substituting the constants α and β in Eq. (10.52) into the above equation, we obtain the BE distribution as a function of the temperature T:

$$\frac{\overline{n}_j}{g_j} = \frac{1}{e^{(\varepsilon_j - \mu)/kT} - 1}. \tag{10.55}$$

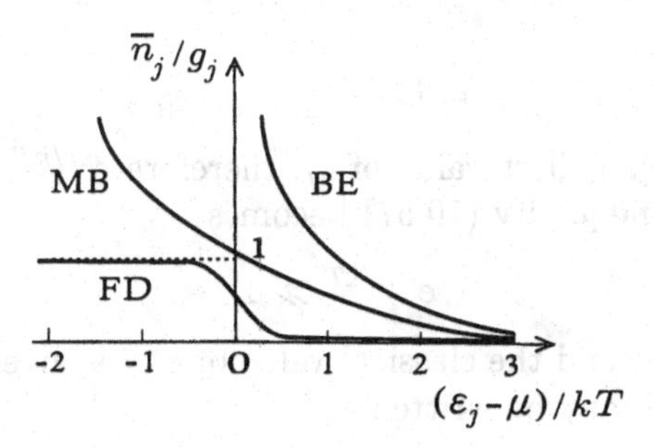

Figure 10.2: The distributions for the MB, FD and BE statistics.

10.9 The Classical Limit

We have derived the respective distributions for particles obeying the three different statistics, Maxwell-Boltzmann, Fermi-Dirac and Bose-Einstein. The results are given in Eqs. (10.40), (10.50) and (10.55). These three distributions can be combined into a single formula

$$\frac{\overline{n}_j}{g_j} = \frac{1}{e^{(\varepsilon_j-\mu)/kT} + a}, \tag{10.56}$$

where

$$a = \begin{cases} +1 & \text{for FD statistics,} \\ -1 & \text{for BE statistics,} \\ 0 & \text{for MB statistics.} \end{cases}$$

We plot the diagram for the average number of particles per state $\overline{n}_j/g_j$ vs. $(\varepsilon_j - \mu)/kT$ in Fig. 10.2. The quantity $\overline{n}_j/g_j$ must be greater than 0 in order to have a physical meaning, therefore in BE statistics μ must be equal to or less than all ε_j, i.e., $\varepsilon_j - \mu \geq 0$ for all j. In FD statistics $\overline{n}_j/g_j$ is always less than 1, which satisfies the requirement of the Pauli exclusion principle. We see that when $(\varepsilon_j - \mu)/kT$ is small or negative, the three distributions are quite different. However when $(\varepsilon_j - \mu)/kT$ is positive and large, both the FD and the BE distributions will approach the MB distribution, and there is little difference between FD and BE statistics. If this happens we say that the system reaches the *classical limit.*

In the quantum (FD, BE) statistics $|a| = 1$, and in the classical statistics $a = 0$, therefore from Eq. (10.56) we obtain the condition for the classical limit

$$e^{(\varepsilon_j-\mu)/kT} \gg 1, \quad \text{for all } j. \tag{10.57}$$

We take 0 as the smallest value of ε_j, therefore $e^{\varepsilon_j/kT} \geq 1$ for all j, so the condition of inequality (10.57) becomes

$$e^{-\mu/kT} \gg 1. \tag{10.58}$$

Since $-\mu/kT = \alpha$, and the classical value of $e^{-\alpha}$ is given in Eq. (10.35), inequality (10.58) may be written as

$$\frac{V}{N}\left(\frac{2\pi mkT}{h^2}\right)^{3/2} \gg 1. \tag{10.59}$$

This is **the condition of the classical limit**, which can be rewritten in terms of the average distance between two particles in the system $\overline{d}$, $\overline{d} \equiv (V/N)^{1/3}$,

$$\overline{d} \gg \lambda_{\text{th}}, \quad \lambda_{\text{th}} \equiv \frac{h}{\sqrt{2\pi mkT}}, \tag{10.60}$$

where λ_{th} is called the **thermal wavelength**. In quantum mechanics, a particle may be considered as a wave with wavelength $\lambda = h/p = h/\sqrt{2m\varepsilon}$, where p and ε are the momentum and energy of the particle. Thermal wavelength gets its name because $2\pi kT$ is roughly equal to the mean thermal energy $\frac{3}{2}kT$ for a particle in an environment of temperature T. We have used the classical limit condition (10.60) in Chapters 1 and 9.

The condition in inequality (10.60) for the classical limit to be satisfied requires that the system to be in a state of low pressure and high temperature. The reason is that low pressure (low density) will make $\overline{d}$ large, and high temperature will make λ_{th} small, but there is no absolute value for how low the pressure must be or how high the temperature must be for the classical limit to hold. The reliable criterion for the classical limit is the condition of inequality (10.60). We may use the available data at STP (T=273 K, P=1 atm) to compute the numerical values of the relevant quantities, $\overline{d} = (kT/P)^{1/3} \approx 3.3 \times 10^{-9}$ m, and $\lambda_{\text{th}} \leq 7.0 \times 10^{-11}$ m, for m the mass of an H_2 molecule. Therefore we may say that, except for very high pressure or very low temperature (relative to STP),[10] in general a system of gaseous particles may be considered

[10] **Very high pressure makes it easy for the vapor to condense to a liquid, and the condition (10.60) is no longer applicable. If a system is in a state of very low pressure, it may still be stable in a gaseous phase even down to very low temperatures and the condition (10.60) does not hold. The system is then a quantum gas of extremely low density which we have discussed in Sec. 6.5.**

as a classical gas, because the condition of inequality (10.60) holds for most cases.

10.10 Localized Independent Particles

We have derived the three distributions, MB, FD, and BE, in which we have assumed that the particles are indistinguishable. However if we are considering a system composed of *localized particles*, then the particles are distinguishable (traceable), even though they may be bosons or fermions. If the particles can be approximated as weakly interacting and the total energy is the sum of the energy of each particle, which has the same energy levels, then *Maxwell-Boltzmann statistics* may be used to describe this system even at very low temperatures. However we have to note that the correct number of possible microstates for a given distribution is now Eq. (10.27), rather than Eq. (10.28), i.e., the factor $N!$ should be retained, because a permutation of particles will give a different microstate. The distribution $\overline{n}_j$ given in Eq. (10.40) is still valid,[11] i.e., it is given by the **Boltzmann distribution**. There are two things which are worth mentioning. (1) Although the system obeys classical statistics, the energy levels ε_j in Eq. (10.40) have to be found quantum mechanically. (2) The expression for $e^{-\alpha}$ given in Eq. (10.35) cannot be applied in this case, because it is derived for a classical ideal gas system.

10.11 Relation with Macroscopic Heat and Work

When the most probable distribution for a system is $\{\overline{n}_i\}$, the internal energy U is

$$U = \sum_j \overline{n}_j \, \varepsilon_j.$$

If the system is interacting with its surroundings, $\overline{n}_j$ and ε_j may change, and the corresponding change of the internal energy dU is

$$dU = \sum_j \varepsilon_j \, d\overline{n}_j + \sum_j \overline{n}_j \, d\varepsilon_j, \tag{10.61}$$

where $d\overline{n}_j$ is due to the change of the distribution of particles among the energy levels, and $d\varepsilon_j$ is the change of the energy level due to the

[11] Note that $\delta \ln W'/\delta n_j = \delta \ln W_{\mathrm{MB}}/\delta n_j$, because $\delta \ln N!/\delta n_j = 0$.

configuration change. For example, the configuration for a PVT system is the volume V, while in a magnetic system the configuration is the magnetic field B. In the following we use X to denote the configuration of the system, then

$$d\varepsilon_j = \frac{d\varepsilon_j}{dX}\, dX. \tag{10.62}$$

The second term on the right hand side of Eq. (10.61) can be written as

$$\sum_j \overline{n}_j\, d\varepsilon_j = \left(\sum_j \overline{n}_j \frac{d\varepsilon_j}{dX} \right) dX = -Y dX, \tag{10.63}$$

where we have defined the variable Y which is canonically conjugate to X,

$$Y \equiv -\sum_j \overline{n}_j \frac{d\varepsilon_j}{dX}. \tag{10.64}$$

If we take $X = V$ then $Y = P$ and $YdX = PdV$. Hence Eq. (10.61) becomes

$$dU = \sum_j \varepsilon_j\, d\overline{n}_j - PdV.$$

By comparing the above equation with the first law of thermodynamics $dU = đQ - đW$, because $PdV = đW$, we obtain the relation between the macroscopic heat $đQ$ and work $đW$ with the microscopic quantities:

$$đQ = \sum_j \varepsilon_j\, d\overline{n}_j, \quad đW = -\sum_j \overline{n}_j\, d\varepsilon_j. \tag{10.65}$$

The above equations tell us that when the system absorbs or rejects heat, the only change of the system is the distribution of particles among the energy levels $\{\overline{n}_i\}$, which is associated with a change of entropy S of the system (for a reversible process $đQ = TdS$). However when work is done on or by the system, the values of the energy levels ε_j change, with the number of particles in each level remaining unchanged.

10.12 The Microcanonical Ensemble

We have derived the distribution functions for free particles which obey one of the three different statistics, MB, FD, or BE statistics. From the distribution function, we can obtain most of the thermodynamic properties of an ideal gas whose particles obey a specific statistics. However the distribution functions were derived for free particles only, and their applicability is rather limited. Therefore we need to develop a method or a theory which can treat a general system, that can be as simple as an ideal gas or as complicated as a liquid, a solid, or a magnetic system, etc. This is what we will do in the rest of this chapter.

We consider an isolated system whose number of particles N, internal energy U, and volume V are all constants. The set of the state variables (N, U, V) defines a macrostate, but there are very large number of microstates which are consistent with the macrostate. These microstates are called the microscopically accessible states. Statistical mechanics assumes that every microscopically accessible state has equal probability to appear in the corresponding macrostate. This is known as the **postulate of equal probability**, which we have discussed briefly in Sec. 10.4. Based on this postulate we may define a microcanonical ensemble. An *ensemble* is an abstract concept, which is a collection of a very large number of (hypothetical) systems and how the systems are chosen depends on the ensemble considered. There are three different ensembles which are frequently used in statistical mechanics, the *microcanonical ensemble*, the *canonical ensemble*, and the *grand canonical ensemble*. The microcanonical ensemble is the basis for the other two ensembles, therefore we discuss the microcanonical ensemble first.

A **microcanonical ensemble** contains a very large number of (hypothetical) identical thermodynamic systems, whose macrostates are all the same as (N, U, V), but each system has a different microstate. **Each accessible microstate of the macrostate (N, U, V) has one and only one representative system in the microcanonical ensemble.** Here we assume that all the microstates in the ensemble are time independent, i.e., the microstate of each representative system in the ensemble will not change with time. This includes both the quantum systems and the classical systems. We may write the probability p_r that a *microstate* r appears in the microcanonical ensemble as

$$p_r = \begin{cases} 1 & \text{if } U \leq E_r \leq U + dU, \\ 0 & \text{otherwise,} \end{cases} \tag{10.66}$$

where E_r is the energy for the microstate r, which is allowed to lie in an infinitesimal (but macroscopic scale) energy range from U to $U + dU$. The probability p_r is chosen such that the microstates in the ensemble satisfy the postulate of equal probability. When we want to calculate a macroscopic physical quantity, we calculate the value of the quantity in each microstate, and sum over the values of all the microstates and then divide it by the total number of systems (microstates) in the ensemble. This is known as the **ensemble average**, because it is the average value for all the systems in the ensemble. This is the result of the theoretical calculation. The calculated results for a successful theory should be in good agreement with the experimental measurements. However in the laboratory there is only one sample for the experimental measurement. The macrostate of the experimental sample is the same as that of the systems in the ensemble, but the microstate of the sample is unknown to the observer. Then why can the theoretical value (the ensemble average) be in good agreement with the experimental value? The reason is that the microstate of the sample is not time independent; the state changes rapidly during the interval of experimental measurement.[12] The postulate of equal probability assumes that every accessible microstate has an equal probability (including time duration) in a real physical system. In any experimental measurement it takes some time interval to obtain the experimental value, therefore we may say that the experiment measurement is the result of a *time average*.

10.13 The Canonical Ensemble

For a microcanonical ensemble, the systems are required to have a fixed internal energy U, which is usually not the experimental situation. Hence it is desirable to extend the concept of the microcanonical ensemble to an ensemble which is consistent with the experimental situation. Therefore the **canonical ensemble** is defined to meet this requirement. In this ensemble, the macrostate of the systems is *equivalent* to specifying the state variables (N, T, V), which means that the internal energy U is replaced by the temperature T. In the laboratory, temperature can easily be controlled, so the theoretical values for the canonical ensemble are much easier to compare with the experimental values. We will see that when the number of particles in the system N is very large, the condition for keeping T constant is equivalent to keeping U constant. Now we will

[12] We have said that the microstate of a classical system will change with time, and that of a quantum system which is not a pure stationary state, will also change with time.

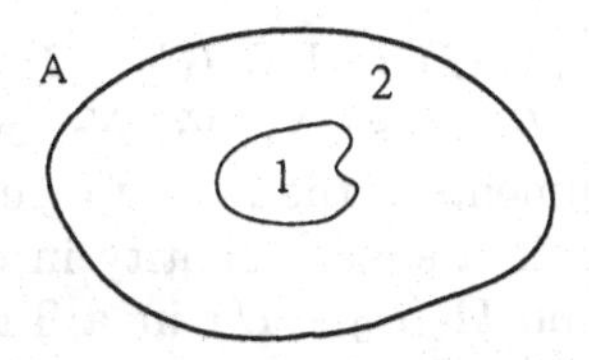

Figure 10.3: A diagram to illustrate the canonical ensemble. An isolated system A is composed of subsystems 1 and 2, in which 1 is the system under consideration. Subsystem 2 is much larger than subsystem 1 and is considered as a heat reservoir.

use the microcanonical ensemble as the basis for developing the theory for the canonical ensemble.

We consider an isolated system A which has the number of particles N_0, internal energy U_0, and volume V_0. Since A is isolated, all the values N_0, U_0, and V_0 are constants. We divide A into two subsystems 1 and 2, in which 1 is the system we will study. Subsystem 2 is much larger than subsystem 1 and plays the role of a heat reservoir, as shown in Fig. 10.3. The number of particles, internal energy, and volume for subsystems 1 and 2 are, respectively, (N_1, U_1, V_1) and (N_2, U_2, V_2). We assume that (N_1, V_1) and (N_2, V_2) are fixed values, but subsystems 1 and 2 can exchange energy. The internal energy U_0 of A is a constant, which is the sum of the internal energies of 1 and 2, i.e.,

$$U_1 + U_2 = U_0 = \text{constant}.$$

Since 1 and 2 can exchange energy, U_1 and U_2 are not constants, but their sum is a constant. In thermal equilibrium, 1 and 2 have the same temperature ($T_1 = T_2$), and their internal energies in this equilibrium state are, respectively, $\overline{U}_1$ and $\overline{U}_2$. Since subsystem 2 is a heat reservoir for subsystem 1, we have

$$\overline{U}_1 \ll \overline{U}_2, \quad \overline{U}_1 + \overline{U}_2 = U_0. \tag{10.67}$$

The internal energy of subsystem 1 is not a constant, we want to know the probability $p(U)$ that the internal energy of 1 is U (for brevity, the subscript 1 of U_1 is dropped). When the internal energy of 1 is U, the internal energy of 2 will be $U_2 = U_0 - U$. For a system with a constant internal energy, the thermodynamic probability is proportional to the multiplicity of the corresponding macrostate, therefore

$$p(U) = c' \, \Omega_1(U) \, \Omega_2(U_0 - U), \tag{10.68}$$

where c' is a constant, $\Omega_1(U)$ and $\Omega_2(U_0 - U)$ are, respectively, the multiplicities for 1 and 2 (because N_1, V_1, N_2 and V_2 are all constants, we omit them in the arguments of the Ω's). In view of inequality (10.67), we may presume that U is a small quantity in comparison with[13] U_0. Therefore we may expand $\Omega_2(U_0 - U)$ in a Taylor series around U_0. Instead of expanding $\Omega_2(U_0 - U)$ directly, we use the Boltzmann relation to obtain the entropy $S_2 = k \ln \Omega_2$ and expand S_2 around U_0. We keep only the first order term in the expansion,

$$\begin{aligned} S_2(U_0 - U) &\approx S_2(U_0) - \left(\frac{\partial S_2}{\partial U_2}\right)_{V_2}\Bigg|_{U_2=U_0} U \\ &= k \ln \Omega_2(U_0) - \frac{U}{T}, \end{aligned} \tag{10.69}$$

where we have used $(\partial S/\partial U)_V = 1/T$, and $T = T_2$ is the temperature of the heat reservoir. Since $\Omega(U) = e^{S(U)/k}$, from the above equation we have

$$\Omega_2(U_0 - U) = e^{S_2(U_0-U)/k} \approx \Omega_2(U_0)\, e^{-U/kT}.$$

By substituting the above equation into Eq. (10.68), we obtain the probability $p(U)$ that subsystem 1 has internal energy U as

$$p(U) = c\,\Omega(U)\, e^{-U/kT}, \tag{10.70}$$

where $\Omega(U) = \Omega_1(U)$ (the subscript 1 is dropped), and $c = c'\,\Omega_2(U_0)$ is a constant independent of U. We may use this equation to establish a **canonical ensemble.**

All the systems in this ensemble have the same macroscopic variables N, the number of particles, and V, the volume. Both N and V are fixed and the same for all systems, but the internal energy U is distributed from 0 (the possible minimum energy) to ∞. In the canonical ensemble we still require that the state of each system is a fixed microstate, which does not change with time, but the number of systems which represent a given microstate may be different for different microstates. This means that the probability of finding a microstate to appear in the systems in the ensemble may be different for different microstates. If r denotes a microstate with energy E_r, then the probability of finding the state r appearing in the systems in the canonical ensemble is

$$p_r = c\, e^{-E_r/kT}, \tag{10.71}$$

[13]The probability that U is much larger than $\overline{U}$ is very small, therefore we can always treat U as much smaller than U_0.

where c is a constant. Summing over all microstates r whose energy E_r satisfies $U \leq E_r \leq U + dU$, we get the probability $p(U)$ in Eq. (10.70). By comparing Eqs. (10.71) and (10.66), we find that there are two main differences between the canonical ensemble and the microcanonical ensemble. (1) Microstates with all possible energies will appear in the canonical ensemble, but in the microcanonical ensemble only the microstates with energies within a small range appear in the ensemble. (2) In the microcanonical ensemble, the probability of finding a microstate is either 0 or 1, but in the canonical ensemble the probability is proportional to $e^{-E_r/kT}$, where E_r is the energy of the microstate and T the temperature of the heat reservoir. We see that $e^{-E_r/kT}$ is the Boltzmann factor we have mentioned in Chapter 9.

It is worth mentioning that all systems in the canonical ensemble are isolated systems, whose macroscopic variables N, U and V are all fixed quantities. However different systems may have different internal energy U. The quantity T in the Boltzmann factor $e^{-U/kT}$ is a parameter controlled externally, which is independent of the thermodynamic properties of the system under consideration. In the next section we will show that the probability given in Eq. (10.70) will make almost all of the systems in the ensemble have the same thermal equilibrium temperature T^*, which is equal to the external parameter T. Therefore **the canonical ensemble is equivalent to the collection of systems whose macroscopic variables (N, T, V) are fixed.**

10.14 Energy Fluctuation in the Canonical Ensemble

In the canonical ensemble, the internal energy U of the systems varies from 0 to ∞. We try to understand the physical properties of a system in terms of the systems in the ensemble. The question is then can we say anything about the internal energy of the system? The answer is, *when the number of particles in the system N is macroscopically large, then the Boltzmann factor $e^{-U/kT}$ will make almost all the systems in the ensemble have the same internal energy U^*, and therefore the same equilibrium temperature T^*, which will be equal to the parameter T in the Boltzmann factor.* In the following we will show why we can obtain this result.

In the canonical ensemble, the probability $p(U)$ that a system has an internal energy U is given by Eq. (10.70), where $\Omega(U)$ increases rapidly as U increases, but the factor $e^{-U/kT}$ decreases rapidly as U increases. These two distinct behaviors of the two multiplying factors will make $p(U)$ have a *very sharp* peak. We assume that the sharp peak of $p(U)$ occurs at $U = U^*$, which implies that the function $\ln p(U)$ will also have a maximum at the same value of U. In order to avoid confusion with the macroscopic temperature T, in the following discussion we denote the temperature of the heat reservoir as T_0 (instead of T), which is a constant independent of the thermodynamic properties of the system under consideration. The Boltzmann factor is then rewritten as e^{-U/kT_0}. We expand $\ln p(U)$ around $U = U^*$ keeping N and V constant,

$$\ln p(U) = \ln p(U^*) + \frac{1}{2}\left(\frac{\partial^2 \ln p}{\partial U^2}\right)_V\bigg|_{U=U^*}(U-U^*)^2 + \cdots, \qquad (10.72)$$

where we have used the condition that the maximum occurs at $U = U^*$, so $(\partial \ln p/\partial U)|_{U=U^*} = 0$. From

$$k\,\ln p(U) = k\,\ln\left[c\,\Omega(U)\,e^{-U/kT_0}\right] = k\,\ln c + S(U) - \frac{U}{T_0},$$

and $(\partial S/\partial U)_{N,V} = 1/T$, the condition for a maximum is

$$\begin{aligned} k\left(\frac{\partial \ln p}{\partial U}\right)_V\bigg|_{U=U^*} &= \left(\frac{\partial S}{\partial U}\right)_V\bigg|_{U=U^*} - \frac{1}{T_0} \\ &= \frac{1}{T^*} - \frac{1}{T_0} = 0. \end{aligned} \qquad (10.73)$$

Therefore we get the result that $T^* = T_0$, where T^* is the equilibrium temperature of the system when the internal energy of the system is U^*, and U^* is the energy which makes $p(U)$ a maximum. Therefore, when $p(U)$ has the maximum value, the temperature of the system is equal to that of the heat reservoir. Now we consider the second order derivatives. The factor e^{-U/kT_0} has no contribution, because T_0 is a constant independent of U, therefore

$$\begin{aligned} k\left(\frac{\partial^2 \ln p}{\partial U^2}\right)_V\bigg|_{U=U^*} &= \left(\frac{\partial^2 S}{\partial U^2}\right)_V\bigg|_{U=U^*} = \left(\frac{\partial}{\partial U}\frac{1}{T}\right)_V\bigg|_{U=U^*} \\ &= -\frac{1}{T^{*2}C_V}, \end{aligned} \qquad (10.74)$$

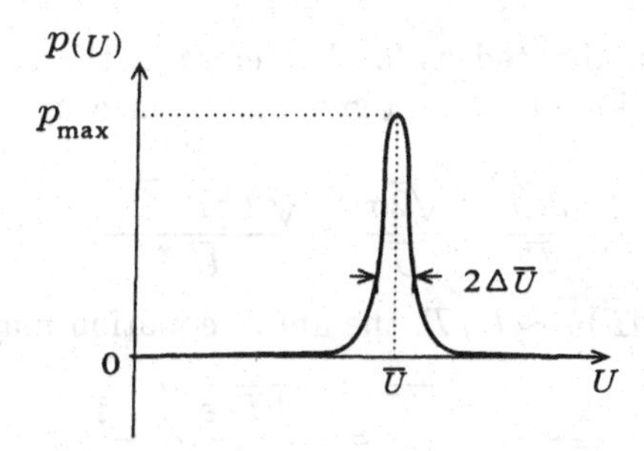

Figure 10.4: The probability distribution $p(U)$ vs. U plot in the canonical ensemble, where $p(U)$ is a Gaussian.

where $C_V = (\partial U/\partial T)_V$ is the constant volume heat capacity of the system. By substituting the above equation into Eq. (10.72), we obtain

$$p(U) = p(U^*)\, e^{-(U-U^*)^2/2\sigma^2}, \quad \sigma^2 = kT^2 C_V, \tag{10.75}$$

where $T = T^* = T_0$ is the temperature of the heat reservoir. This probability is a **Gaussian**[14] **distribution**; the value of the mean energy $\overline{U}$ is

$$\overline{U} \equiv \frac{\displaystyle\int_{-\infty}^{\infty} U\, p(U) dU}{\displaystyle\int_{-\infty}^{\infty} p(U) dU} = U^*. \tag{10.76}$$

Therefore in the canonical ensemble, the mean energy $\overline{U}$ is equal to the most probable energy U^* of the system. The constant σ in Eq. (10.75) is the *standard deviation*[15] of the Gaussian distribution. One way to understand the significance of the standard deviation is to plot the distribution $p(U)$ as a function of U, as shown in Fig. 10.4. We see that the region where $p(U)$ is not equal to zero occurs only near the peak at $U = \overline{U}$, and the width of the peak is very small. This is because $p(U)$ has a very sharp peak. We take an energy U_1 such that $p(U_1) = e^{-1}\, p(\overline{U}) \approx 0.37\, p(\overline{U})$, therefore

$$|\overline{U} - U_1| \equiv \Delta\overline{U} = \sqrt{2}\sigma. \tag{10.77}$$

This may be roughly interpreted as saying that $p(U)$ has a peak centered at $\overline{U}$, with a half-width $\Delta\overline{U}$. When U is outside the width,

[14]J. Carl F. Gauss, German mathematician (1777–1855).

[15]See Problem 10.21 or 10.22 for the proof. The standard deviation $\Delta^* U$ for the probability distribution $p(U)$ is defined as $\Delta^* U \equiv (\overline{U^2} - \overline{U}^2)^{1/2}$.

$p(U)$ may be approximated as 0. Therefore in the canonical ensemble the energy fluctuation of the systems may be measured by the following quantity:

$$\frac{\Delta \overline{U}}{\overline{U}} = \frac{\sqrt{2}\sigma}{\overline{U}} = \frac{\sqrt{2\,kT^2\,C_V}}{\overline{U}}. \tag{10.78}$$

Since $C_V = (\partial \overline{U}/\partial T)_V \sim \overline{U}/T$, the above equation may be written as

$$\frac{\Delta \overline{U}}{\overline{U}} \sim \sqrt{\frac{kT}{\overline{U}}} = \frac{\sqrt{kT/\overline{\varepsilon}}}{\sqrt{N}} \sim \frac{1}{\sqrt{N}}. \tag{10.79}$$

The last form is obtained because the internal energy of the system is proportional to the number of particles N in the system, $\overline{U} = N\overline{\varepsilon}$. Since $kT/\overline{\varepsilon}$ is a fixed number for a given T, when N is macroscopically large (roughly 10^{20}), the energy fluctuation in the canonical ensemble may be considered as 0. Therefore in practice, the canonical ensemble may be considered as being equivalent to a microcanonical ensemble whose systems have fixed internal energy. *In the following we will use the abbreviation U to denote the internal energy of a system instead of the mean energy $\overline{U}$, because in practice they have the same meaning.*

10.15 The Partition Function

In statistical mechanics there is an important and useful function known as the **partition function**. It is a function of the number of particles N, temperature T, and volume V. We use $Z_N(T,V)$ to denote this function, whose definition is

$$Z_N(T,V) = \sum_r e^{-\beta E_r} \quad \left(\beta = \frac{1}{kT}\right), \tag{10.80}$$

where r *represents one of the microstates of the N-particle system*; E_r is the energy of the state r, which is a function of the volume V. In the above equation the summation is over all microstates r with weight $p_r \propto e^{-E_r/kT}$, which is proportional to the probability that the microstate r is present in the ensemble. The function Z_N is also called the **sum over states**.

In Eq. (10.80) we used the language and notations of quantum mechanics to define Z_N. In classical statistical mechanics Z_N has the same definition, but we have to replace the sum symbol by an integral, because the microstates of classical systems are continuous. Also the energy level E_r has to be replaced by the Hamiltonian H of the system. In classical mechanics, a microstate is defined by the $3N$ generalized coordinates $q = (q_1, q_2, \cdots, q_{3N})$ and the $3N$ generalized momenta $p = (p_1, p_2, \cdots, p_{3N})$. The space spanned by these sets of variables is called the $6N$-dimensional *phase space*. Each set of $6N$ numbers (q, p) denotes a microstate. The value of the Hamiltonian $H(q, p)$ of the microstate is the energy of the state, therefore $H(q, p)$ corresponds to the quantum mechanical energy level[16] E_r. In Chapter 3, we said that each microstate occupies an infinitesimal phase space volume such that we can evaluate the number of microstates. If we take h^{3N} (h is the Planck constant) as the volume occupied by each microstate, and replace the sum in Eq. (10.80) by an integral, and divide the integral by $N!$ to obtain the correct counting of the number of states (for a gas and liquid systems),[17] we can define the classical partition function as

$$Z_N(T, V) = \frac{1}{N!\, h^{3N}} \int e^{-\beta H(q,p)}\, d^{3N}q\, d^{3N}p. \tag{10.81}$$

Since Eq. (10.80) is more compact than Eq. (10.81), and both of them have the same physical content, in the following we will use Eq. (10.80) to represent the partition function Z_N, no matter whether the system is a classical system or a quantum system.

In the following we will see that if we know the partition function Z_N of a system, then we can obtain most of the thermodynamic properties of the system. At first we note that E_r is a function of the volume V and is independent of the temperature T, i.e., $E_r = E_r(V)$. Also the variable β ($\beta = 1/kT$) is more convenient than the variable T, therefore we may use β as the independent variable instead of T. The first quantity we would like to evaluate is the internal energy U (the mean energy of the systems in the ensemble), which can be obtained from Z_N,

$$U = \frac{\sum\limits_r E_r\, e^{-\beta E_r}}{\sum\limits_r e^{-\beta E_r}} = -\left(\frac{\partial \ln Z_N}{\partial \beta}\right)_V.$$

[16]In quantum mechanics H is an operator which has no numerical value. The eigenvalue E_r of H is the energy of the state r.

[17]For solids, there is no need to divide Z_N by $N!$, because the particles are localized in solids.

If we consider Z_N as a function of β and V, then

$$
\begin{aligned}
d\ln Z_N &= \left(\frac{\partial \ln Z_N}{\partial \beta}\right)_V d\beta + \left(\frac{\partial \ln Z_N}{\partial V}\right)_\beta dV \\
&= -U\,d\beta + \frac{-\beta \sum_r \frac{dE_r}{dV} e^{-\beta E_r}}{Z_N} dV \\
&= -U\,d\beta + \beta \left(-\overline{\frac{dE_r}{dV}}\right) dV \\
&= -U\,d\beta + \beta\, PdV \\
&= \beta\,(PdV + dU) - d(\beta U),
\end{aligned}
\tag{10.82}
$$

where in the second term on the right hand side of the fourth equality, we used Eq. (10.64) for the microscopic definition of the pressure P (take $Y = P$). Rearrange the terms of the above equation to obtain

$$d\,[\ln Z_N(T,V) + \beta U] = \beta\,(PdV + dU) = \frac{1}{k}\,dS. \tag{10.83}$$

By integration we have[18]

$$S = k\,[\ln Z_N(T,V) + \beta U]\,, \tag{10.84}$$

therefore

$$kT \ln Z_N(T,V) = -(U - TS) = -F,$$

where F is Helmholtz free energy. Therefore

$$Z_N(T,V) = e^{-\beta F}. \tag{10.85}$$

From the definition of the partition function and the above relation, we know that if the the partition Z_N of a system is known, then we can calculate most of the thermodynamic properties of the system. For example, the internal energy U, the Helmholtz free energy F, the entropy S, the pressure P, and the chemical potential μ, can all be obtained from Z_N. From the internal energy U we can calculate the constant volume

[18]There is an integration constant on the right hand side of Eq. (10.84), which can be taken to be 0 because of the third law. The reason is that when $T \to 0$ ($\beta \to \infty$), $Z_N \to e^{-\beta E_0}$, $U \to E_0$, which is the ground state energy, therefore $S \to 0$ if the constant is zero.

heat capacity C_V. In the following we write down the formulas for U, F, S, P and μ:

$$U = -\left(\frac{\partial \ln Z_N}{\partial \beta}\right)_{N,V} = kT^2\left(\frac{\partial \ln Z_N}{\partial T}\right)_{N,V}, \tag{10.86}$$

$$F = -kT \ln Z_N(T, V), \tag{10.87}$$

$$S = -\left(\frac{\partial F}{\partial T}\right)_{N,V} = kT\left(\frac{\partial \ln Z_N}{\partial T}\right)_{N,V} + k \ln Z_N, \tag{10.88}$$

$$P = -\left(\frac{\partial F}{\partial V}\right)_{N,T} = kT\left(\frac{\partial \ln Z_N}{\partial V}\right)_{N,T}, \tag{10.89}$$

$$\mu = \left(\frac{\partial F}{\partial N}\right)_{T,V} = -kT\left(\frac{\partial \ln Z_N}{\partial N}\right)_{T,V}. \tag{10.90}$$

The above equations are the thermodynamic properties which can be obtained from the partition function Z_N. In the next chapter we will apply the above formulas to several simple systems whose partition functions can be evaluated, at least approximately.

10.16 The Grand Canonical Ensemble

We will see in the next chapter that the approach of using the partition function, cannot solve the problem of quantum ideal gases. This means that the FD distribution, Eq. (10.50), and the BE distribution, Eq. (10.55), cannot be derived from the partition function. This can also easily be understood from the structure of the partition function, which contains a Boltzmann factor $e^{-E_r/kT}$, but the chemical potential μ does not play any role. In the MB distribution, the chemical potential μ is contained in the proportionality constant, and therefore has the same form as the partition function. In the FD and BE distributions, however, the chemical potential μ cannot be separated from the energy term ε, and therefore the partition function approach cannot be used for the quantum gases. It is therefore necessary for us to extend the canonical ensemble to allow the number of particles in the system to vary, which is controlled by the chemical potential. The extension of the canonical ensemble is called the **grand canonical ensemble**. In extending the microcanonical ensemble to the canonical ensemble, we allowed the system under consideration to exchange energy with its surroundings (a heat reservoir), but the number of particles in the system was kept constant.

Now we extend this further to let the system under consideration exchange particles with its surroundings, but the volume V of the system is still assumed to be constant. Therefore we are considering an open system, with its surroundings acting both as a heat reservoir and a particle reservoir. For simplicity we call it a *reservoir*. We assume that the temperature and the pressure of the reservoir are kept constant at T and P, therefore the chemical potential $\mu = \mu(T, P)$ of the reservoir is a constant. In the grand canonical ensemble we find that the probability $p(N, U)$ that the system has the particle number N and the internal energy U has the form

$$p(N, U) = c\,\Omega_N(U)\, e^{-(U-\mu N)/kT}. \tag{10.91}$$

The derivation of the above equation is similar to the derivation of Eq. (10.70) for the canonical ensemble; we will not give the details here. From the above probability we can define the *grand canonical ensemble* as the collection of a very large number of systems, and the probability for a microstate (N, r) with particle number N, and energy $E_{N,r}$ to appear in the systems is

$$p_{N,r} = c\, e^{-(E_{N,r}-\mu N)/kT}. \tag{10.92}$$

The corresponding equation for the canonical ensemble is Eq. (10.71), in which N is a constant. Although the energy $E_{N,r}$ of a system in the ensemble varies from $-\infty$ to ∞, and the particle number N varies from 0 to ∞, yet, as in the case of the canonical ensemble, we can show that almost all the systems in the ensemble have the same energy $\overline{U}$, and the same particle number $\overline{N}$, with negligible fluctuations for a macroscopic system.[19] This implies that almost all the systems in the ensemble have the same temperature T^*, and the same chemical potential μ^*.[20] Moreover by using the same procedure[21] we used to show $T^* = T$ in the canonical ensemble, we can also show that $T^* = T$ and $\mu^* = \mu$ in the grand canonical ensemble, where T and μ are, respectively, the

[19]This means that the conditions of the reservoir will make $\overline{N}$ of the system a macroscopically large number.

[20]The particle density is controlled by the chemical potential. Same chemical potential implies same particle density.

[21]The procedure is to consider the condition for the maximum of the total entropy of the system and the reservoir. In the grand canonical ensemble, the system can exchange energy and particles with the reservoir.

temperature and the chemical potential of the reservoir. We will not give the details here.

For the grand canonical ensemble we define the **grand partition function**, which plays a role like the partition function in the canonical ensemble. Most of the thermodynamic properties of the system can be obtained from the grand partition function $\mathcal{Z}$, which is defined as

$$\begin{aligned} \mathcal{Z}(T,V,\mu) &= \sum_N \sum_r e^{-\beta(E_{N,r}-\mu N)} \quad \left(\beta = \frac{1}{kT}\right) \\ &= \sum_N e^{\beta\mu N} Z_N(T,V), \end{aligned} \tag{10.93}$$

where Z_N is the N-particle partition function defined in Eq. (10.80). We note that $\mathcal{Z}$, like Z_N, is a sum over states, each term in the sum is proportional to the probability that the microstate (N,r) will appear in the ensemble. From $\mathcal{Z}$ we can calculate the internal energy U and the mean number of particles of the system $\overline{N}$:

$$\begin{aligned} U = \overline{U} &= \frac{1}{\mathcal{Z}} \sum_{N,r} E_{N,r} e^{-\beta(E_{N,r}-\mu N)} \\ &= -\left(\frac{\partial}{\partial\beta} \ln \mathcal{Z}\right)_{V,\mu}, \end{aligned} \tag{10.94}$$

$$\begin{aligned} \overline{N} &= \frac{1}{\mathcal{Z}} \sum_{N,r} N e^{-\beta(E_{N,r}-\mu N)} \\ &= kT \left(\frac{\partial}{\partial\mu} \ln \mathcal{Z}\right)_{T,V}. \end{aligned} \tag{10.95}$$

Finally we define a **grand potential** Φ, such that

$$\Phi = kT \ln \mathcal{Z}, \quad \text{or} \quad \mathcal{Z} = e^{\Phi/kT}. \tag{10.96}$$

It can be shown that the grand potential $\Phi = PV$. One way to do this is to write the grand partition function as (cf. Eq. (10.93))

$$\mathcal{Z} = \sum_N e^{\beta\mu N} Z_N(T,V) = e^{\beta\mu\overline{N}} Z_{\overline{N}}(T,V)\Delta^* N. \tag{10.97}$$

The last form is obtained by a proper choice of $\Delta^* N$. We note that $\overline{N}$ is the mean number of particles, which is also the most probable number of particles for the system, which makes $\exp(\beta\mu N)Z_N(T, V)$ a maximum. Therefore the sum in $\mathcal{Z}$ will be dominated by the term with $N = \overline{N}$, and we should have

$$\frac{\Delta^* N}{\overline{N}} \to 0 \text{ for macroscopic } \overline{N}. \tag{10.98}$$

When the above condition is satisfied, the relation

$$\Phi = PV \tag{10.99}$$

can easily be proved; this will be left as an exercise for the reader (see Problem 10.26).

10.17 Problems

10.1. Show that the density of states for a two-dimensional system is independent of the energy, i.e., $g^{(2)}(\varepsilon) = c_2$, and that of a one-dimensional system is proportional to $\varepsilon^{-1/2}$, i.e., $g^{(1)}(\varepsilon) = c_1\,\varepsilon^{-1/2}$. Find the constants c_2 and c_1.

10.2. There are 60 distinguishable particles distributed among three nondegenerate energy levels labeled 1, 2, 3, such that $n_1 = 30$, $n_2 = 20$, and $n_3 = 10$. The energies of the levels are $\varepsilon_1 = \varepsilon$, $\varepsilon_2 = 2\varepsilon$, $\varepsilon_3 = 3\varepsilon$.

(a) If the change in the occupation number of level 2 is $\delta n_2 = -2$, find δn_1 and δn_3 such that $\delta U = 0$.

(b) Find the thermodynamic probability for the distribution before and after the change.

10.3. Consider a system of four identical but *localized* particles. Each particle can be in any of the nondegenerate states with energy $j\varepsilon$, where $j = 0, 1, 2, \cdots$. Consider the macrostate that the system has a total energy $U = 6\varepsilon$.

(a) Tabulate all the possible distributions for the four particles among the energy levels $j\varepsilon$. How many are they?

(b) Evaluate the thermodynamic probability $W(\{n_j\})$ for each distribution and calculate the multiplicity $\Omega = \sum_{\{n_j\}} W(\{n_j\})$.

(c) Calculate the mean occupation number for each level $\overline{n}_j = \sum_{\{n_j\}} n_j W(\{n_j\})/\Omega$ of the four particles in the energy states.

10.4. Repeat the previous problem if

(a) the particles are gaseous bosons.

(b) the particles are gaseous fermoins.

10.5. Two distinguishable particles are to be distributed among three nondegenerate energy levels 0, ε, and 2ε, such that the total energy $U = 2\varepsilon$.

(a) What is the entropy of the system?

(b) If a distinguishable particle with zero energy is added to the system, show that the entropy of the system is increased by a factor of 1.63.

10.6. Consider a system which has 6 identical particles occupying the jth energy level, which is eight-fold degenerate. How many possible microstates are there if

(a) the particles are gaseous bosons.

(b) the particles are gaseous fermoins.

10.7. Show that Eq. (10.41) for W_{FD} and Eq. (10.53) for W_{BE} both reduce to Eq. (10.28) for W_{MB} in the limit $g_j \gg n_j$.

10.8. Show that it is possible to write the thermodynamic probability in the general form

$$W = \prod_{j=1} \frac{g_j(g_j - a)(g_j - 2a)\cdots[g_j - (n_j - 1)a]}{n_j!},$$

where

$$a = \begin{cases} +1 & \text{for FD statistics,} \\ -1 & \text{for BE statistics,} \\ 0 & \text{for MB statistics.} \end{cases}$$

10.9. Let p_r be the probability that a system is in a state r with energy E_r. Show that if the entropy S is defined by

$$S = -k \sum_r p_r \ln p_r \ \text{ with } \sum_r p_r = 1,$$

then the values of the p_r which make S a maximum, under the condition that the total energy of the system U is fixed, give the canonical distribution, i.e., p_r has the same form as in the canonical ensemble, $p_r \propto e^{-\beta E_r}$.

10.10. Show that for the canonical distribution

$$p_r = \frac{1}{Z_N} e^{-\beta E_r},$$

the entropy can be written in the form

$$S = -k \sum_r p_r \ln p_r.$$

[*Hint*: You may use Eq. (10.84) as the starting point.]

10.11. Substitute the Maxwell-Boltzmann distribution function into Eq. (10.65) for $đQ$, and use the relation $đQ = TdS$, to show that the entropy S has the form

$$S = k \left(N - \sum_j \overline{n}_j \ln \frac{\overline{n}_j}{g_j} \right).$$

Note that this expression is exactly the same as we have obtained in the second equality of Eq. (10.37), which is derived from the Boltzmann relation .

10.12. (a) Show that

$$\mu = -T \left(\frac{\partial S}{\partial N} \right)_{U,V}.$$

(b) From (a) show that

$$\Delta \ln \Omega \approx -\frac{\mu}{kT} \Delta N,$$

where Ω is the multiplicity of the macrostate. Consider a system with a chemical potential $\mu = -0.25$ eV. By what factor is Ω increased when a single particle is added to it at $T = 300$ K?

10.13. Consider an ideal Fermi gas at $T = 300$ K. Calculate the average number of particles in a nondegenerate single-particle state with energy ε such that (a) $\varepsilon - \mu = -0.1$ eV; (b) $\varepsilon - \mu = -0.01$; (c) $\varepsilon - \mu = 0$; and (d) $\varepsilon - \mu = 0.01$ eV; (e) $\varepsilon - \mu = 0.1$ eV.

10.14. Consider two single-particle states, 1 and 2, in an ideal Ferm gas, where $\varepsilon_1 = \mu - \varepsilon_0$ and $\varepsilon_2 = \mu + \varepsilon_0$. Show that the probability of ε_2 is *occupied* is equal to the probability that ε_1 being *unoccupied*. This implies that the Fermi-Dirac distribution is "symmetrical" about the chemical potential μ.

10.15. Consider an ideal Bose gas at $T = 300$ K. Calculate the average number of particles in a nondegenerate single-particle state and the probability of the state containing 0, 1, or 2 particles, if the energy of the state ε satisfies (a) $\varepsilon - \mu = 0.001$ eV; (b) $\varepsilon - \mu = 0.01$ eV; and (c) $\varepsilon - \mu = 0.1$ eV. [*Hint*: The probability that there are n particles in the state ε is equal to $e^{-nx} / \sum_n e^{-nx}$, where $x = (\varepsilon - \mu)/kT$.]

10.16. Consider the classical limit of an ideal gas. What is the smallest value for $-\mu/kT$ such that the Fermi-Dirac, Bose-Einstein, and Maxwell-Boltzmann distributions agree to within 0.1%? Use ^{4}He gas as an example to discuss whether the condition on $-\mu/kT$ can be satisfied at $T = 300$ K, and at $T = 200$ K.

10.17. Consider a system consisting of 2 identical gaseous particles, which can occupy any of the following 3 energy levels, $\varepsilon_n = n\,\varepsilon$, $n = 0, 1, 2$. The $n = 0$ level is doubly degenerate, and the other two are nondegenerate. The system is in thermal equilibrium at temperature T. For each of the following cases determine the partition function and the energy of the system.

(a) The particles obey FD statistics.
(b) The particles obey BE statistics.
(c) The particles obey MB statistics.

10.18. Consider a system of N non-interacting particles obeying MB statistics which is in thermodynamic equilibrium at temperature T.

(a) Show that the chemical potential μ is given by

$$\mu = -kT(\ln Z_1 - \ln N),$$

where $Z_1 = \sum_j g_j e^{-\varepsilon_j/kT}$ is the one-particle partition function.

(b) Show that the mean occupation number for the jth energy level is given by

$$\overline{n}_j = -NkT \left(\frac{\partial \ln Z_1}{\partial \varepsilon_j} \right)_T .$$

10.19. Consider a system of N non-interacting particles obeying MB statistics which is in thermodynamic equilibrium at temperature T.

(a) From Eq. (10.90) and the formula μ given in Part (a) of the preceding problem, show that the relation between the N-particle partition function Z_N and the one-particle partition function Z_1 is given by

$$Z_N = \frac{Z_1^N}{N!}.$$

(b) Show that the factor $1/N!$ in Part (a) does not affect the result for the internal energy U and the equation of state obtained by calculating the pressure P.

(c) Show that the factor $1/N!$ in Part (a) will affect the expression for the entropy S.

(d) Show that Z_1 is proportional to the volume V, and without the factor $1/N!$ the formula for S will have the Gibbs paradox, which has been discussed in Chapter 3.

10.20 Configuration work $đW$ and heat $đQ$ can be expressed in terms of the partition function Z_N of a system. To be specific we consider a PVT system .

(a) Show that

$$đW = \frac{1}{\beta}\left(\frac{\partial \ln Z_N}{\partial V}\right)_T dV,$$

where $\beta = 1/kT$.

(b) Show that

$$đQ = \frac{1}{\beta}\left(\frac{\partial \ln Z_N}{\partial V}\right)_T dV - \left[\frac{\partial}{\partial \beta}\left(\frac{\partial \ln Z_N}{\partial V}\right)_T\right]_V dV - \left(\frac{\partial^2 \ln Z_N}{\partial \beta^2}\right)_V d\beta.$$

10.21. Consider a Gaussian distribution

$$p\,(x) = c\,e^{-(x-x_0)^2/2\sigma^2}, \quad -\infty \leq x \leq \infty.$$

(a) Find the normalization constant c, such that

$$\int_{-\infty}^{\infty} p\,(x)\,dx = 1.$$

(b) Find $\overline{x}$.

(c) Show that $(\Delta^* x)^2 \equiv \overline{x^2} - \overline{x}^2 = \sigma^2$, where $\Delta^* x$ is known as the *standard deviation.*

10.22. In the canonical ensemble show that (with $\beta = 1/kT$)

$$(\Delta^* U)^2 \equiv \overline{U^2} - \overline{U}^2 = \frac{\partial^2 \ln Z_N}{\partial \beta^2} = kT^2 C_V,$$

where C_V and Z are, respectively, the constant volume heat capacity and the partition function of the system. Note that $(\Delta^* U)^2$ is the square of the standard deviation of U.

10.23. Find the equation of state and the constant volume heat capacity for the following two systems.

(a) The partition function of the system is

$$Z_N = V^N (2\pi mkT)^{5N/2}.$$

(b) The partition function of the system is

$$Z_N = (V - Nb)^N (2\pi mkT)^{3N/2} e^{aN^2/VkT}.$$

10.24. (a) Show that the Gibbs free energy G is given by

$$G = kTV^2 \left[\frac{\partial}{\partial V} \left(\frac{\ln Z_N}{V} \right) \right]_T,$$

where Z_N is the partition function.

(b) Show that the enthalpy H is given by

$$H = \frac{V}{\beta} \left(\frac{\partial \ln Z_N}{\partial V} \right)_T - \left(\frac{\partial \ln Z_N}{\partial \beta} \right)_V,$$

where $\beta = 1/kT$.

10.25. Consider thermodynamic systems in the grand canonical ensemble with the grand partition function $\mathcal{Z}$ defined in Eq. (10.93). We have already derived the formula for the average number of particles in the system $\overline{N}$ in Eq. (10.95).

(a) Show that the mean square number of particles is

$$\overline{N^2} = \frac{(kT)^2}{\mathcal{Z}} \left(\frac{\partial^2 \mathcal{Z}}{\partial \mu^2} \right)_{T,V}.$$

(b) Show that the standard deviation of N is

$$\sigma_N = \sqrt{kT \left(\frac{\partial \overline{N}}{\partial \mu} \right)_{T,V}}.$$

Find σ_N for a classical ideal gas in terms of $\overline{N}$.

10.26. Consider the grand potential Φ defined in Eq. (10.96). Show that Eq. (10.99) can be obtained if the condition in Eq. (10.98) holds.

Chapter 11

Application to Simple Systems

11.1 Introduction

In the previous chapter we have seen that if the partition function of a system can be calculated, then most of the equilibrium thermodynamic properties of the system can be obtained from this partition function. Unfortunately, in general, the calculation of the partition function of a thermodynamic system is not easy, and approximation methods or simplified models have to be used. In this chapter we will introduce several relatively simple systems whose partition functions can be approximately reduced to the products of the *one-particle partition function.* This will simplify the calculations of the partition function greatly, so that the thermodynamic properties of these systems can easily be obtained. Although these are simple systems, nevertheless they do include many practical and important ones. Therefore the study of these systems is useful and important in both thermodynamics and statistical mechanics.

11.2 The One-Particle Partition Function

Except for some relatively simple systems, the partition functions for most of the thermodynamic systems are not easy to calculate. The main reason is that in calculating Z_N in Eq. (10.80), first, we have to know all the energy states E_r of the N-particle system, and, second, we have to sum over all the exponential functions of E_r. Both calculations are not easy for large N. In this section we will consider a class of systems which are relatively easy to treat. These are the systems in which the interaction between the particles is weak enough that the total Hamiltonian H of the N-particle system can be approximated as the sum of N identical *one-particle Hamiltonians* h_1. Since each particle has the same one-particle Hamiltonian h_1, each particle has the same eigenstates ε_j for all possible quantum numbers[1] j of h_1. This means that E_r may be written as the sum of the one-particle energies ε_j,

$$H = \sum_{n=1}^{N} h_1(n), \quad E_r = \sum_{n=1}^{N} \varepsilon_{j_n^{(r)}}, \tag{11.1}$$

where $(j_1^{(r)}, j_2^{(r)}, \cdots, j_N^{(r)})$ denotes the set of quantum numbers for the rth microstate of the system, in which the quantum number for the nth particle is $j_n^{(r)}$. In fact there are also systems which are *not weakly interacting*, yet their H can be written in the form of Eq. (11.1) by proper transformations of the dynamic variables. An important and well-known example is the lattice vibrations in a crystal. In a crystal, neighboring atoms interact strongly while each atom can be displaced only a small distance around its equilibrium position. After a proper expansion of the interaction potentials and proper transformations of the coordinates and momenta, the Hamiltonian of a crystal can be expressed approximately as the sum of N one-particle Hamiltonians h_1. We will discuss this problem in Sec. 11.8.

When E_r is the sum of N one-particle energies ε_j, the N-particle partition function Z_N can be decomposed into products of the N one-particle partition functions Z_1:

$$\begin{aligned} Z_N &= \sum_r e^{-\beta E_r} = \sum_r e^{-\beta \sum_n \varepsilon_{j_n^{(r)}}} \\ &= \sum_{j_1, j_2, \cdots, j_N} e^{-\beta(\varepsilon_{j_1} + \varepsilon_{j_2} + \cdots + \varepsilon_{j_N})} \\ &= \left(\sum_{j_1} e^{-\beta\varepsilon_{j_1}}\right)\left(\sum_{j_2} e^{-\beta\varepsilon_{j_2}}\right) \cdots \left(\sum_{j_N} e^{-\beta\varepsilon_{j_N}}\right) \\ &\equiv Z_1^N, \quad \beta = \frac{1}{kT}, \end{aligned} \tag{11.2}$$

[1]The symbol j may denote a set of quantum numbers, not necessarily just one number.

where Z_1 is the **one-particle partition function**, which is defined by,

$$Z_1 = \sum_j e^{-\beta \varepsilon_j}, \quad \ln Z_1 = \frac{1}{N} \ln Z_N. \tag{11.3}$$

On the right hand side of the third equality of Eq. (11.2), the quantum numbers j_1, j_2, j_3, $\cdots$, j_N under the $\sum$ sign are all independent of each other,[2] therefore the fourth equality holds. From the definition of Z_1, we see that Z_1 is the partition function for a system with one particle, so we call it the *one-particle partition function*. Apparently the problem of calculating Z_1 is much easier than the calculation of Z_N. Therefore for systems whose Z_N can be expressed in terms of Z_1, Z_N can be obtained easily. **However, the expression $Z_N = Z_1^N$ is not correct for gaseous systems.** For a classical ideal gas, we have to consider the problem of the Gibbs paradox, therefore for a classical ideal gas the correct relation between Z_1 and Z_N is (see Eq. (10.81) for the definition of Z_N for a classical system with non-localized particles)

$$Z_N = \frac{Z_1^N}{N!}. \tag{11.4}$$

Note that for quantum gases, the particles are indistinguishable and cannot be traced, therefore Eqs. (11.2)–(11.3) *do not hold for quantum gases.* For non-localized particles which obey Fermi statistics, any microstate with two or more particles having the same j_n cannot exist, because of the Pauli exclusion principle, i.e., there is a restriction $j_{n'} \neq j_n$, if $n' \neq n$. Therefore the expression for E_r in Eq. (11.1) does not hold for Fermi gases. For Bose gases Eq. (11.1) holds, but Eq. (11.2) does not hold, because the particles cannot be traced and each particle cannot be considered as an independent particle. This is because the system must be symmetric under an exchange of any two particles. This implies that there exist relations between the j_n's for different n. Therefore Eq. (11.2) does not hold for either fermions or bosons. This can also be understood from the FD and the BE distribution functions, which cannot be written in the form $ce^{-\beta\varepsilon_j}$ (see Eq. (10.56)). Therefore the partition function approach cannot be applied to quantum gases. This difficulty can be resolved by using the grand partition function, which will be given in Sec. 12.2. **However Eqs. (11.2)–(11.3) can be applied to quantum particles which are localized**, such as solids. In these cases, the particles can be traced and can be considered as distinguishable.

[2]This statement does not hold for a quantum gas system.

For many systems, the one-particle energy ε may be the sum of several independent terms. For example, the energy for a diatomic molecule is the sum of the following four terms: the kinetic energy of the displacement of the center of mass, the rotational kinetic energy, the energy of the relative motion of the two atoms, and the excitation energy of the orbital electrons. In this case the one-particle partition function is the products of several sub-particle partition functions. In the following we consider the simple case where ε is the sum of two independent terms as an example to illustrate what we have said above. It is straightforward to extend this to the case where ε is the sum of several independent terms. Suppose ε is the sum of ε_a and ε_b, which are independent of each other, then ε_j can be written in the form

$$\varepsilon_j = \varepsilon_{a,j_a} + \varepsilon_{b,j_b}. \tag{11.5}$$

Therefore the quantum number j is actually a set of two quantum numbers j_a and j_b. We consider the case where j_a and j_b are independent of each other, therefore the one-particle partition function Z_1 may be written as

$$\begin{aligned} Z_1 &= \sum_j e^{-\beta\varepsilon_j} = \sum_{j_a,j_b} e^{\beta(\varepsilon_{a,j_a}+\varepsilon_{b,j_b})} \\ &= \left(\sum_{j_a} e^{-\beta\varepsilon_{a,j_a}}\right)\left(\sum_{j_b} e^{-\beta\varepsilon_{b,j_b}}\right) \\ &= Z_1^{(a)} Z_1^{(b)}, \end{aligned} \tag{11.6}$$

$$Z_1^{(\alpha)} = \sum_{j_\alpha} e^{-\beta\varepsilon_{\alpha,j_\alpha}}, \quad \alpha = a \text{ or } b. \tag{11.7}$$

From Eq. (11.6) we have $\ln Z_1 = \ln Z_1^{(a)} + \ln Z_1^{(b)}$, and the mean energy $\overline{\varepsilon}$ and the constant volume heat capacity c_v are both the sum of the contributions from a and b,

$$\overline{\varepsilon} = \overline{\varepsilon}_a + \overline{\varepsilon}_b, \quad c_v = c_v^{(a)} + c_v^{(b)}. \tag{11.8}$$

Many other physical quantities, such as the entropy and free energy will also be the sum of the contributions from a and b. Equation (11.6) can easily be extended to the case where ε is the sum of several independent terms.

11.3 The Equipartition Theorem

We have mentioned in Chapter 9 that if the Hamiltonian for a *classical system* contains a term like aq^2 or bp^2, where q is a coordinate variable and p a momentum variable, then the mean energy for that term is equal to $\frac{1}{2}kT$, where T is the temperature of the system. This is known as the **equipartition theorem**. It is important to note that *this theorem holds only for classical systems, it does not hold for quantum systems, including those systems whose particles are localized.* We are going to prove this theorem in the following.

We consider the mean kinetic energy first. Suppose the Hamiltonian H of an N-particle system can be written as

$$H = H' + bp_j^2,$$

where p_j represents a certain component of the momenta, and H' is the rest of H *which is independent of* p_j. For example p_j may be the x-component of the momentum of the nth particle $p_x^{(n)}$. From the definition of mean values in the canonical ensemble, we have

$$\overline{bp_j^2} = \frac{\int bp_j^2\, e^{-\beta H} d^{3N}q\, d^{3N}p}{\int e^{-\beta H} d^{3N}q\, d^{3N}p} = \frac{\int_{-\infty}^{\infty} bp_j^2\, e^{-\beta b p_j^2}\, dp_j}{\int_{-\infty}^{\infty} e^{-\beta b p_j^2}\, dp_j}.$$

We obtain the above result, because both in the numerator and the denominator, the integration involving H' and that involving p_j can be separated. Therefore the integrations involving H' in both the numerator and the denominator can be cancelled. From the above equation and the formula for integration in Eq. (9.32), we obtain

$$\begin{aligned} \overline{bp_j^2} &= -\frac{1}{\partial\beta} \ln\left(\int_{-\infty}^{\infty} e^{-\beta b p_j^2}\, dp_j\right) = -\frac{1}{\partial\beta} \ln\left(\frac{\pi}{b\beta}\right)^{1/2} \\ &= \frac{1}{2\beta} = \frac{1}{2}kT, \end{aligned} \tag{11.9}$$

which is the result we want to prove.

Now we consider the mean value for a potential energy. Suppose the N-particle Hamiltonian H can be written as

$$H = H' + aq_j^2,$$

where q_j represents a certain component of the coordinates, and H' is the rest of H, *which is independent of* q_j. For example q_j may be the x-component of the position vector of the nth particle $x^{(n)}$. Following the same procedure in deriving Eq. (11.9), the mean value of aq_j^2 can be written as

$$\overline{aq_j^2} = \frac{\displaystyle\int_{q_j^{\min}}^{q_j^{\max}} aq_j^2\, e^{-\beta aq_j^2}\, dq_j}{\displaystyle\int_{q_j^{\min}}^{q_j^{\max}} e^{-\beta aq_j^2}\, dq_j}. \tag{11.10}$$

We have to note that in the above equation, the lower and upper limits of integration are no longer $-\infty$ and ∞, respectively. This is because the particles are confined in a container, and therefore q_j is confined to the range, $q_j^{\min}$ to $q_j^{\max}$. Then the formula given in Eq. (9.32) cannot be applied to the above integration, and a result, as in Eq. (11.9), cannot be obtained in this way. We need to use another method to evaluate the integrations.

The reason that the particles are confined in the container is due to the fact that the wall of the container provides a very large potential barrier for the particles; the particles will be reflected back when they strike the wall. Therefore we may imagine the Hamiltonian $h(q_j)$ for the potential barrier which provides the effect of the wall to be:

$$h(q_j) = \begin{cases} \infty, & q_j \geq q_j^{\max} \text{ or } q_j \leq q_j^{\min}, \\ aq_j^2, & q_j^{\min} < q_j < q_j^{\max}. \end{cases} \tag{11.11}$$

By using the method of integration by parts, and $u = -q_j/(2\beta)$, $dv = d(e^{-\beta aq_j^2})$, the numerator of Eq. (11.10) can be integrated once to get

$$\int_{q_j^{\min}}^{q_j^{\max}} aq_j^2\, e^{-\beta h(q_j)}\, dq_j = -\frac{q_j}{2\beta}\, e^{-\beta h(q_j)}\Big|_{q_j^{\min}}^{q_j^{\max}} + \frac{1}{2\beta}\int_{q_j^{\min}}^{q_j^{\max}} e^{-\beta aq_j^2}\, dq_j.$$

The integrated term on the right hand side of the above equation is 0, because $h(q_j)$ has the value ∞ at the boundaries. The integral in the second term is exactly the same as the integral in the denominator of the right hand side of Eq. (11.10). Hence the integrals in the numerator and in the denominator can be cancelled, and only the factor $1/2\beta$ in front of the integral is left. The mean potential energy is thus also equal to $\frac{1}{2}kT$,

$$\overline{aq_j^2} = \frac{1}{2\beta} = \frac{1}{2}kT. \tag{11.12}$$

11.4 One-Dimensional Harmonic Oscillators

The Hamiltonian of a one-dimensional harmonic oscillator h_{ho} has the form

$$h_{\text{ho}} = \frac{p_x^2}{2m} + \frac{1}{2}Kx^2, \tag{11.13}$$

where m is the mass of the particle which moves in the x-direction and K is the force constant. There are many physical systems in nature whose motion can be approximated as a harmonic oscillation. For example, the vibrational motion in a diatomic molecule, and the lattice vibration in a crystal, etc. The harmonic oscillator is one of the important dynamic models. If there are N weakly interacting harmonic oscillators which constitute a thermodynamic system at temperature T, the mean energy and the heat capacity of an oscillator of the system can be calculated. There are two approaches for the calculations.

1. Classical oscillators:

By using the methods of classical mechanics to solve Eq. (11.13), we will get the solution that the particle is doing a harmonic oscillation with frequency ν,

$$\nu = \frac{1}{2\pi}\sqrt{\frac{K}{m}}. \tag{11.14}$$

The main quantities we want to calculate are the mean energy and the heat capacity per particle. Since the Hamiltonian h_{ho} in Eq. (11.13) has one term of the form aq^2 and the other term of the form bp^2, therefore we can use the equipartition theorem directly, without doing any calculations. The mean energy $\bar{\varepsilon}$ and the constant volume heat capacity[3] c_v are, respectively,

$$\bar{\varepsilon} = \overline{\frac{p_x^2}{2m}} + \overline{\frac{1}{2}Kx^2} = \frac{1}{2}kT + \frac{1}{2}kT = kT, \tag{11.15}$$

$$c_v = \left(\frac{\partial \bar{\varepsilon}}{\partial T}\right)_v = \frac{d\bar{\varepsilon}}{dT} = k. \tag{11.16}$$

Therefore c_v is a constant independent of the temperature T. This is the classical result.

[3]The mean energy for a quantum oscillator $\bar{\varepsilon}$ (see Eq. (11.19)) depends on the frequency ν, which is a function of the volume. Therefore the heat capacity obtained is the constant volume heat capacity c_v.

2. Quantum oscillators:

We consider only the case where the oscillators *can be traced*, such as the diatomic molecules in a classical gas, or are *localized*, such as the particles in a crystal, therefore there is no need to consider the problem of quantum BE or FD statistics. *The approach of quantum oscillators means that we have to solve for the energy levels of the Hamiltonian* h_{ho} *quantum mechanically.* It is well known in quantum mechanics that the energy of a harmonic oscillator is quantized with values given by

$$\varepsilon_j = \left(j + \frac{1}{2}\right) h\nu, \quad j = 0, 1, 2, \cdots, \tag{11.17}$$

where h is the Planck constant, ν is the frequency in Eq. (11.14), and j the quantum number which can be 0 or a positive integer. Therefore the one-particle partition function is

$$Z_1 = \sum_{j=0}^{\infty} e^{-\beta\varepsilon_j} = \sum_{j=0}^{\infty} e^{-(j+\frac{1}{2})\beta h\nu} = \frac{e^{-\beta h\nu/2}}{1 - e^{-\beta h\nu}}. \tag{11.18}$$

The mean energy $\bar{\varepsilon}$ and the constant volume heat capacity c_v per oscillator are, respectively,

$$\bar{\varepsilon} = -\frac{\partial}{\partial\beta} \ln Z_1 = h\nu \left(\frac{1}{e^{\beta h\nu} - 1} + \frac{1}{2}\right), \tag{11.19}$$

$$c_v = \left(\frac{\partial\bar{\varepsilon}}{\partial T}\right)_v = \frac{(h\nu)^2}{kT^2} \frac{e^{h\nu/kT}}{(e^{h\nu/kT} - 1)^2}. \tag{11.20}$$

Obviously, the thermodynamic properties of the quantum oscillators are quite different from those of the classical oscillators. For example the constant volume heat capacity c_v is no longer a constant, it depends strongly on the temperature T. We expect that in the classical limit, the behavior of a quantum oscillator will approach that of a classical oscillator. From Eq. (11.20) we find that when $h\nu/kT \ll 1$ the heat capacity c_v of the quantum oscillators can be approximated as the constant k, which is the classical result. This is because $e^{h\nu/kT} \approx 1 + h\nu/kT$ ($h\nu/kT \ll 1$), Eq. (11.20) becomes

$$c_v \approx \frac{(h\nu)^2}{kT^2} \frac{1}{(h\nu/kT)^2} = k, \quad \frac{h\nu}{kT} \ll 1. \tag{11.21}$$

Therefore for high temperatures such that $T \gg h\nu/k$, the result of the classical limit is obtained.

At very low temperatures, such that $h\nu/kT \gg 1$, then $e^{h\nu/kT} \gg 1$ and c_v will approach 0,

$$c_v \approx \frac{(h\nu)^2}{kT^2}\, e^{-h\nu/kT} \approx 0, \quad \frac{h\nu}{kT} \gg 1. \tag{11.22}$$

That the heat capacity c_v approaches 0 as $T \to 0$ can be understood as follows. At $T = 0$ all the oscillators are in the lowest energy level $j = 0$. When T is slightly greater than 0, with $kT \ll h\nu$ and $e^{-h\nu/kT} \approx 0$, this means that the probability for an oscillator to be excited from the level $j = 0$ to the level $j = 1$ is almost zero. Therefore the mean energy is almost the same as the ground state energy ($T = 0$), although T is not exactly 0, this implies that $c_v \approx 0$ (T increases without absorbing heat). This result is rather different from the classical result $c_v = k$.

From the above result, we see that *when a thermodynamic system can be approximated as a collection of weakly interacting harmonic oscillators, then we have to use quantum mechanical approach to solve for the energy levels of the harmonic oscillators in order to get the correct thermodynamic properties for all temperatures.* However if we know that the system is in a classical limit state, then we can use the equipartition theorem to obtain the result directly; there is no need to do any calculations.

11.5 Classical Monatomic Ideal Gas

The energy of a monatomic gas molecule is the sum of the translational kinetic energy and the energy of the orbital electrons. However the energy difference between the first excited state and the ground state of the orbital electrons is of the order of a few electron volts (eV), which corresponds to a few 10^4 K of thermal energy. Therefore we may neglect the contributions of the orbital electrons, because they will always remain in the ground state if the temperature is a few hundred kelvin or below.[4]

[4] If we take the ground state and the first excited state energies of the orbital electrons to be 0 and $\varepsilon_1^{(\mathrm{el})}$, respectively, then the partition function due to an orbital electron is $Z_1^{(\mathrm{el})} = 1 + e^{-\varepsilon_1^{(\mathrm{el})}/kT} + \cdots$. If we compute $Z_1^{(\mathrm{el})}$ at $T = 300$ K, then we will get $Z_1^{(\mathrm{el})} = 1$ (note the equal sign), because $\varepsilon_1^{(\mathrm{el})}/kT > 100$ and $e^{-\varepsilon_1^{(\mathrm{el})}/kT} = 0$, with an error of less than 10^{-30}.

Therefore the one-particle Hamiltonian h_1 for a monatomic ideal gas molecule contains only the translational kinetic energy, which has three terms, because of the three dimensional motion. Thus we have

$$h_1 = \frac{\boldsymbol{p}^2}{2m} = \frac{p_x^2}{2m} + \frac{p_y^2}{2m} + \frac{p_z^2}{2m}. \tag{11.23}$$

From the partition function for the classical gas, Eq. (10.81), taking $N = 1$ and $H = h_1$,

$$\begin{aligned} Z_1 &= \frac{1}{h^3}\int e^{-\beta h_1}\, d^3q\, d^3p = \frac{V}{h^3}\left(\int_{-\infty}^{\infty} e^{-\beta p_x^2/2m}\, dp_x\right)^3 \\ &= \frac{V}{h^3}\left(\frac{2\pi m}{\beta}\right)^{3/2} = V\left(\frac{2\pi mkT}{h^2}\right)^{3/2}. \end{aligned} \tag{11.24}$$

Since h_1 is independent of the spatial variables $\boldsymbol{q}$, the integration over $\boldsymbol{q}$, $\int d^3q$, will get a volume factor V.

From Z_1 we can obtain the thermodynamic properties for a monatomic classical gas. From Eqs. (11.4) and (11.24)

$$\ln Z_N = N\ln\frac{V}{N} + \frac{3}{2}N\ln T + \frac{3}{2}N\ln\left(\frac{2\pi mk}{h^2}\right) + N.$$

By using Eqs. (10.86)–(10.90) we obtain

$$PV = NkT,\ \ U = N\bar{\varepsilon} = \frac{3}{2}NkT,\ \ c_v = \frac{3}{2}k, \tag{11.25}$$

$$\mu = -kT\left(\frac{\partial \ln Z_N}{\partial N}\right)_{T,V} = kT\ln\left[\frac{N}{V}\left(\frac{h^2}{2\pi mkT}\right)^{3/2}\right], \tag{11.26}$$

$$S = Nk\left[\ln\frac{V}{N} + \frac{3}{2}\ln T + \frac{3}{2}\ln\left(\frac{2\pi mk}{h^2}\right) + \frac{5}{2}\right]. \tag{11.27}$$

Here c_v and μ are, respectively, the constant volume heat capacity and the chemical potential *per particle*. We note that: (1) the mean energy and the heat capacity can also be obtained directly from the equipartition theorem with the same result; (2) the chemical potential obtained is exactly the same as that obtained from Eqs. (10.35) and (10.39); (3) the entropy obtained is exactly the same as Eq. (3.45), which has no Gibbs paradox, because we have correctly included the factor $1/N!$ in Eq. (11.4).

11.6 The Specific Heat of a Diatomic Gas

In the classical limit we may consider each molecule as moving independently. We consider a diatomic molecule which consists of two atoms with mass m_1 and m_2. The one-particle Hamiltonian contains four parts: (1) the translational motion of the center of mass; (2) the rotational motion of the molecule; (3) the relative motion of the two atoms (i.e., vibration); (4) the energy of the orbital electrons. Therefore the one-particle Hamiltonian h_1 has the form

$$h_1 = h_{\text{tr}} + h_{\text{rot}} + h_{\text{vib}} + h_{\text{el}}.$$

As in the monatomic gas, the term h_{el} can be considered as a constant and may be neglected, because the first excited state energy is at least 10^4 K above the ground state, so no electrons can be excited at ordinary temperatures. Therefore, in practice, h_1 has only three parts. We discuss each of them separately.

1. The translational motion term h_{tr}:

We may use either the classical or the quantum approach to solve for the energy levels for h_{tr}, both give the same result $\boldsymbol{p}^2/2(m_1+m_2)$. Moreover, $|\boldsymbol{p}|$ can be considered as a continuous quantity even in the quantum approach if the particle is moving in a macroscopic volume. Therefore we may use the equipartition theorem to calculate the average energy $\bar{\varepsilon}$ and the constant volume heat capacity c_v per particle,

$$\bar{\varepsilon}_{\text{tr}} = \frac{3}{2}kT, \quad c_v^{\text{tr}} = \frac{3}{2}k. \tag{11.28}$$

2. The rotational motion of the molecule h_{rot}:

Rotational motion involves angular momentum, which is quantized in quantum mechanics. Therefore we have to use the quantum approach to find the energy levels for h_{rot} in order to get the correct results. From quantum mechanics we know that the eigenvalues of the Hamiltonian h_{rot} are

$$\varepsilon_j = j(j+1)\frac{\hbar^2}{2I}, \quad j = 0, 1, 2, \cdots, \tag{11.29}$$

where I is the moment of inertia about an axis which passes through the center of mass and perpendicular to the line connecting the two centers of the atoms; $\hbar = h/2\pi$, and j is the quantum number (positive integers including 0). It is helpful to define a **characteristic temperature for rotation,**

$$\theta_{\rm rot} \equiv \frac{\hbar^2}{2Ik}, \tag{11.30}$$

then Eq. (11.29) can be rewritten as

$$\varepsilon_j = j(j+1)\, k\theta_{\rm rot}. \tag{11.31}$$

Since the degrees of degeneracy for each energy level is $g_j = 2j+1$, the partition function for the rotational motion is

$$Z_1^{\rm rot} = \sum_{j=0}^{\infty} g_j\, e^{-\varepsilon_j/kT} = \sum_{j=0}^{\infty} (2j+1)\, e^{-j(j+1)\theta_{\rm rot}/T}. \tag{11.32}$$

This sum is not easy to calculate, so we consider only the following two limits: (a) the high temperature limit $\theta_{\rm rot} \ll T$, and (b) the low temperature limit $\theta_{\rm rot} \gg T$.

(a) At high T such that $\theta_{\rm rot} \ll T$ holds, the difference between the successive terms in the sum is very small, and thus can be considered as quasi-continuous. Therefore the *sum* may be approximately replaced by an *integral*. Let $x = j(j+1)$, $dx = (2j+1)dj$. We then get for the high temperature limit ($\theta_{\rm rot} \ll T$),

$$Z_1^{\rm rot} \approx \int_0^{\infty} e^{-x\theta_{\rm rot}/T} dx = \frac{T}{\theta_{\rm rot}}, \tag{11.33}$$

$$\bar{\varepsilon}_{\rm rot} = kT^2 \left(\frac{\partial \ln Z_1^{\rm rot}}{\partial T} \right)_V = kT,$$

$$c_v^{\rm rot} = k, \quad T \gg \theta_{\rm rot}. \tag{11.34}$$

This is the same as the classical result, which can be obtained from the equipartition theorem for the rotational Hamiltonian $h_{\rm rot}$, which contains two rotational kinetic energy terms of the form bp_j^2. There are only two, not three, rotational kinetic energy terms, because the moment of inertia along the axis which passes through the centers of the two atoms is very small and can be considered as zero. Hence $\bar{\varepsilon}_{\rm rot} = 2 \times \frac{1}{2}kT = kT$ by the equipartition theorem. For all the diatomic gases except H_2, the rotational characteristic temperature $\theta_{\rm rot}$ is of the order of 10 K (see Table 11.1), therefore in practice $T \gg \theta_{\rm rot}$ always holds for diatomic gases, because all these gases have liquefied (in fact solidified) at such low temperatures.

It should be noted that the partition function given in Eq. (11.33) is *too large by a factor of 2 for a diatomic molecule composed of two identical atoms,*[5] such as N_2, O_2, and H_2. This is because turning the molecular axis into the opposite direction does not give a new state. However, this factor does not affect the results in calculating the mean energy and heat capacity. Therefore this factor is not important except in the calculation of the entropy of the system (cf. Problem 11.7).

(b) At low T such that $\theta_{\rm rot} \gg T$, the terms $e^{-j(j+1)\theta_{\rm rot}/T}$ will get smaller and approach 0 rapidly as j increases. Therefore in calculating Z_1 we may just keep only the $j = 0$ and $j = 1$ terms and neglect all higher j terms, thus (for $\theta_{\rm rot} \gg T$)

$$\begin{aligned} Z_1^{\rm rot} &= 1 + 3e^{-2\theta_{\rm rot}/T} + \cdots , \\ \bar{\varepsilon}_{\rm rot} &= 6k\,\theta_{\rm rot} e^{-2\theta_{\rm rot}/T} , \\ c_v^{\rm rot} &= 3k \left(\frac{2\theta_{\rm rot}}{T}\right)^2 e^{-2\theta_{\rm rot}/T}, \quad T \ll \theta_{\rm rot} . \end{aligned} \tag{11.35}$$

Hence when $T \to 0$, $c_v^{\rm rot} \to 0$ rapidly according to $e^{-2\theta_{\rm rot}/T}/T^2$.

Table 11.1 $\theta_{\rm rot}$ and $\theta_{\rm vib}$ for some diatomic molecules

Molecule	$\theta_{\rm rot}$(K)	$\theta_{\rm vib}$(K)	Molecule	$\theta_{\rm rot}$(K)	$\theta_{\rm vib}$(K)
H_2	85.4	6140	CO	2.8	3080
O_2	2.1	2239	HCl	15.2	4300
N_2	2.9	3352	Cl_2	0.36	810

3. The relative motion of the two atoms (vibration) $h_{\rm vib}$:

The Hamiltonian for the relative motion of the two atoms $h_{\rm vib}$ can be approximately written as Eq. (11.13), where the mass m in this case is the *reduced mass* of the two atoms $m_1m_2/(m_1+m_2)$. Consequently the relative motion of the two atoms is a vibration which can be approximated as a one-dimensional harmonic oscillation. In quantum mechanics

[5]A factor of 2 correction is valid in the classical limit only. The situation may be much more complicated in the low temperature limit, where quantum effect becomes important.

the energy levels of the simple harmonic oscillation are quantized. Therefore we have to treat the problem quantum mechanically in order to get the correct result. In Eqs. (11.19)–(11.20) we have the respective formulas for the mean energy $\bar{\varepsilon}_{\text{vib}}$ and the constant volume heat capacity c_v^{vib} of a harmonic oscillator. To simplify the discussions we define a **characteristic temperature for vibration** θ_{vib},

$$\theta_{\text{vib}} \equiv \frac{h\nu}{k}. \tag{11.36}$$

From Table 11.1 we see that the characteristic temperatures θ_{vib} for most of the diatomic molecules are larger than 2000 K, and therefore the condition $\theta_{\text{vib}}/T \gg 1$ can be satisfied at room temperatures. We may then use the result of Eq. (11.22) to obtain

$$c_v^{\text{vib}} \approx k \left(\frac{\theta_{\text{vib}}}{T}\right)^2 e^{-\theta_{\text{vib}}/T} \approx 0. \tag{11.37}$$

We may say that the vibrational degrees of freedom have no contribution to the heat capacity. This is because the first excited state energy of vibration ($\sim k\theta_{\text{vib}}$) is much larger than the mean energy of each molecule, and therefore the probability that a molecule can be excited to the first excited state is almost zero. Almost all molecules stay in the ground state of the vibrational motion, as if the vibrational degrees of freedom do not exist. Thus the classical equipartition theorem predicts the incorrect result; it would predict the total c_v to be $7k/2$. However, the correct value of c_v should be $5k/2$, rather than $7k/2$, since there is no contribution from the vibrational motion, thus

$$c_v = c_v^{\text{tr}} + c_v^{\text{rot}} + c_v^{\text{vib}} \approx \frac{3}{2}k + k + 0 = \frac{5}{2}k. \tag{11.38}$$

This result holds at room temperature for almost all diatomic gases, including those molecules with low θ_{vib}, such as Cl_2. For Cl_2, the characteristic temperature $\theta_{\text{vib}} = 810$ K, and $c_v^{\text{vib}} < 0.01k$ if we take $T = 300$ K. Therefore the error in the prediction of Eq. (11.38) is less than 1%.

11.7 The Statistics of a Photon Gas

According to Planck's theory the energy of an electromagnetic wave is not continuous but quantized. If the frequency of the electromagnetic wave is ν, then the energy of the wave has the following form,

$$\varepsilon = nh\nu, \quad n = 0, 1, 2, 3, \cdots, \tag{11.39}$$

where h is the Planck constant, and n is a positive integer including 0. The electromagnetic wave therefore can be considered as a collection of **photons**. The energy for each photon is $h\nu$, and n is the number of photons. When the electromagnetic wave and the wall of the container are in thermal equilibrium (blackbody radiation), we may use the relations of macroscopic thermodynamics to study the thermodynamic properties of the photons. This is what we have done in Chapter 8. The macroscopic theory, however, cannot derive the *spectral energy density* of blackbody radiation, as given in Eq. (8.78). In this section we will use the partition function method to derive the Planck distribution and the spectral energy density of blackbody radiation.

Photons are indistinguishable, non-interacting, and cannot be traced; they can be treated as an *ideal quantum gas.* It has only one energy level $h\nu$, and there is no restriction on the number of particles in the quantum state. Therefore the photons are bosons. However there is a big difference between the photons and other bosons: *the number of photons is not conserved.* Consequently the restrictions imposed by Eqs. (10.14)–(10.15) do not apply to the photon gas. Photon statistics are therefore simpler than Bose statistics.

Since photons with different frequencies do not interfere with each other, we can consider a single frequency ν and use the partition function approach to study its thermodynamic properties. We may consider Eq. (11.39) as the energy levels for a photon gas with frequency ν, i.e., *we may consider n as a quantum number.* Therefore the single-frequency partition function for the photons with frequency ν in equilibrium with the wall at temperature T has the form

$$Z(\nu) = \sum_{n=0}^{\infty} e^{-nh\nu/kT} = \frac{1}{1 - e^{-h\nu/kT}}. \tag{11.40}$$

From $Z(\nu)$ we may calculate the average energy $\overline{\varepsilon}$ for photons with frequency ν at temperature T. We write the mean energy in the form $\overline{\varepsilon} = \overline{n}h\nu$ to obtain

$$\overline{\varepsilon} = kT^2 \frac{\partial \ln Z(\nu)}{\partial T} = \frac{h\nu}{e^{h\nu/kT} - 1} = \overline{n}h\nu, \tag{11.41}$$

$$\overline{n} = \frac{1}{e^{h\nu/kT} - 1}. \tag{11.42}$$

By comparing Eqs. (11.41) with (11.39), *we may interpret $\overline{n}$ as the average number of photons with frequency ν at temperature T*. The dependence of the average number of photons $\overline{n}$ on the temperature T is called the **Planck distribution**, which is given in Eq. (11.42).

Compare the Planck distribution with the BE distribution given in Eq. (10.55), we see that photons belong to a special class of bosons. It is a Bose gas with zero chemical potential $\mu = 0$. The reason for this special property is due to the fact that the number of photons is not conserved. We may also understand it from the fact that the Lagrange multiplier α for the photon gas must be 0, because the restriction of Eq. (10.14) does not exist for the photons. Since $\mu = -\alpha kT$, therefore one should have $\mu = 0$. In Chapter 8 we have also derived the result $\mu = 0$ for the photon gas from macroscopic thermodynamics considerations. Thus both the macroscopic and the microscopic theories get the same result.

Now we are going to derive Eq. (8.78), the spectral energy density for blackbody radiation with frequency between ν and $\nu + d\nu$. First we calculate the number of modes $g(\nu)d\nu$ for an electromagnetic wave with frequencies between ν and $\nu + d\nu$, where $g(\nu)$ is known as the *density of modes*. We may use either standing waves, such as in Eq. (10.3), or plane waves, such as in Eq. (10.6), to describe the electromagnetic waves. In either case, the volume occupied by each mode (p_x, p_y, p_z) in momentum space is h^3/V, where V is the volume of the container. This is the result of Eq. (10.11). We note that for each mode (p_x, p_y, p_z) there are two polarizations for the electric field, which are perpendicular to the direction of propagation. By using the relation $p = h\nu/c$, and $\Delta p_x \Delta p_y \Delta p_z \to 4\pi p^2\, dp$, we get

$$g(\nu)d\nu = 2\,\frac{V}{h^3}\,4\pi p^2\, dp = V\frac{8\pi}{c^3}\,\nu^2\, d\nu. \tag{11.43}$$

The *spectral energy density* $u(\nu)$ for the photon gas is therefore

$$u(\nu) = \frac{1}{V}\,g(\nu)\,\overline{n}\,h\nu = \frac{8\pi h}{c^3}\,\frac{\nu^3}{e^{h\nu/kT} - 1}. \tag{11.44}$$

This is the result of Eq. (8.78). From this we can obtain thermodynamic properties of a photon gas as in Sec. 8.8. We will not repeat here.

11.8 The Specific Heat of a Solid: Phonons

This is one of the important examples where the classical theory fails, the prediction is not in agreement with the experimental results. Therefore it is worthwhile for us to give it special attention. Classical theory predicts that the constant volume heat capacity C_V of a solid is a constant independent of the temperature T. The experimental results show that at high temperatures C_V is a constant, in good agreement with the theory, but at lower temperatures C_V will decrease and eventually approach 0 when $T \to 0$. We note that the experimental result at $T \to 0$ is in agreement with the third law. Now we will see why the classical theory fails. However we should note that *in this section we study only the contributions from the crystal vibrations, not including the contributions of the conduction electrons in a metal.*

We consider a crystal consisting of N identical atoms (ions) at N lattice points. In the crystal, each atom (ion) will interact strongly with its nearest neighbors, and the motions of the atoms are very complicated. However each atom can make only small amplitude vibrations around its equilibrium position, and thus approximations can be made about the vibrations. By using the properties of the periodic structure of the crystal and the small amplitude vibrations, a proper transformation of the coordinates can be found, which will transform the total Hamiltonian of the system into a sum of $3N$ independent one-dimensional simple harmonic oscillators. These $3N$ independent motions are called the *normal modes*. In this picture the crystal is a collection of $3N$ weakly interacting one-dimensional harmonic oscillators. Classical theory uses the equipartition theorem to calculate the average energy of each oscillator, which is kT when the temperature is T. Therefore, classically, the total energy U^{cl} and the constant volume heat capacity C_V^{cl} are, respectively,

$$U^{\text{cl}} = 3NkT, \quad C_V^{\text{cl}} = 3Nk. \tag{11.45}$$

Hence C_V is a constant independent of the temperature. This result is known as the **Dulong-Petit[6] law**. However this law is valid only at high enough temperatures; it fails at lower temperatures. We need a quantum mechanical approach to solve this difficulty.

In quantum mechanics the allowed energy levels of a simple harmonic oscillator are $(n + \frac{1}{2})h\nu$, where ν is the frequency of the oscillator and

[6]Pierre L. Dulong, French chemist (1785–1838); Alexis T. Petit, French physicist (1791–1820).

$n = 0, 1, 2, 3, \cdots$. We may say that the oscillator is quantized, and each quantum has an energy $h\nu$. The quantized crystal vibrations are called **phonons**, which represent quantized sound waves. If we do not include the ground state energy $\frac{1}{2}h\nu$, then at temperature T, the mean energy $\bar{\varepsilon}$ for each phonon mode has the same form as that of a photon mode Eq. (11.41),

$$\bar{\varepsilon} = kT^2 \frac{\partial \ln Z(\nu)}{\partial T} = \frac{h\nu}{e^{h\nu/kT} - 1} = \bar{n}h\nu, \tag{11.46}$$

$$\bar{n} = \frac{1}{e^{h\nu/kT} - 1}. \tag{11.47}$$

Here $\bar{n}$ may be interpreted as the average number of phonons for frequency ν at temperature T, just like the photon mean number. Therefore phonons, just like the photons, obey the Planck distribution, in which the chemical potential $\mu = 0$, and the total number is not conserved. However there are two important differences between phonons and photons. The photon frequency ν varies from 0 to ∞, but the phonon frequency ν has a finite maximum frequency. Moreover the velocity of light c is a constant independent of the frequency and the direction of polarization, but the sound velocity v_s is not a constant; it may depend on the frequency, and also on the direction of propagation and polarization.

Einstein was the first to use the theory of the quantized oscillator Eqs. (11.19)–(11.20) to describe the heat capacity of a solid. He assumed that all the phonons have the same frequency $\nu_{\rm E}$, and defined a characteristic temperature, now known as the **Einstein temperature**, $\theta_{\rm E} \equiv h\nu_{\rm E}/k$, therefore

$$U = 3Nk\theta_{\rm E}\left(\frac{1}{e^{\theta_{\rm E}/T} - 1} + \frac{1}{2}\right),$$

$$C_V = 3Nk\left(\frac{\theta_{\rm E}}{T}\right)^2 \frac{e^{\theta_{\rm E}/T}}{(e^{\theta_{\rm E}/T} - 1)^2}. \tag{11.48}$$

At high temperatures such that $\theta_{\rm E}/T \ll 1$, the heat capacity C_V will approach the classical result $3Nk$, as expected. At the low temperature limit, such that $\theta_{\rm E}/T \gg 1$, the above equation becomes

$$C_V \approx 3Nk\left(\frac{\theta_{\rm E}}{T}\right)^2 e^{-\theta_{\rm E}/T}. \tag{11.49}$$

This result tells us that $C_V \to 0$ as $T \to 0$, which is consistent with the third law. However the agreement with the experimental result is poor.

Experimentally $C_V \to 0$ according to the T^3 law, but in the Einstein model C_V approaches 0 according to $e^{-\theta_{\rm E}/T}/T^2$, which is much faster than the T^3 law.

The main discrepancy of the Einstein model is that it assumes that all the phonons have the same frequency. Actually the $3N$ *normal modes* of the crystal are *not identical* to one another. Different modes have different frequencies. Therefore after quantization, phonons have different frequencies for the different modes.[7] We use a simple model to study the heat capacity of a solid. Let the sound velocity v_s be a constant independent of the frequency and the direction of propagation. This is known as the **Debye[8] model**. Sound waves, like electromagnetic waves, can be described by standing waves or by plane waves, which will give the same result for the *density of modes.* Therefore the the density of modes for the photons in Eq. (11.43) can also be used in the case of phonons. However there are three places where it should be changed: (1) The velocity of light c should be replaced by the velocity of sound v_s. (2) The factor 2 should be replaced by 3, because for each mode there are only 2 polarizations for photons (transverse wave), but there are 3 modes for the phonons (transverse and longitudinal). (3) The photon frequency varies from 0 to ∞, but there is a maximum phonon frequency $\nu_{\rm m}$. This means that the density of modes is 0 for $\nu > \nu_{\rm m}$ for phonons. The frequency $\nu_{\rm m}$ is called the *cut-off frequency.* Hence the density of modes for phonons is

$$g(\nu) = \begin{cases} \dfrac{12\pi V}{\overline{v}_s^3}\nu^2, & 0 \le \nu \le \nu_{\rm m}, \\ 0, & \nu > \nu_{\rm m}. \end{cases} \tag{11.50}$$

Here we have taken into account the fact that the sound velocity may be different for the longitudinal and transverse modes, and defined an

[7]It may be worth emphasizing that the oscillators in a solid are the normal modes, which can be considered to be independent of each other. Each normal mode has its own frequency. This means that we cannot consider each atom as an independent harmonic oscillator. If we could, then each oscillator must have the same frequency, because all the atoms are identical. Since the oscillators have different frequencies, we have to modify Eq. (11.2) in order that it can be applied to this problem. We should write $Z_N = Z_1^{(1)} Z_1^{(2)} \cdots Z_1^{(3N)}$, where the frequency for $Z_1^{(n)}$ may be different for different n.

[8]Peter J. W. Debye, Dutch chemist (1884–1966), Nobel prize laureate in chemistry in 1936.

average sound velocity $\overline{v}_s$ by

$$\frac{1}{\overline{v}_s^3} = \frac{1}{3}\left(\frac{1}{v_l^3} + \frac{2}{v_t^3}\right), \tag{11.51}$$

where v_l and v_t are, respectively, the sound velocity of the longitudinal and transverse modes. Note that there are one longitudinal and two transverse modes for a given direction of propagation.

The reason that there is a maximum frequency for the phonons is due to the fact that the total number of phonon modes is finite; it is equal to $3N$. The maximum frequency ν_m satisfies

$$3N = \int_0^\infty g(\nu)d\nu = \frac{12\pi V}{\overline{v}_s^3}\int_0^{\nu_m} \nu^2\, d\nu = \frac{4\pi V}{\overline{v}_s^3}\nu_m^3, \tag{11.52}$$

$$\nu_m = \left(\frac{3}{4\pi}\right)^{1/3}\left(\frac{N}{V}\right)^{1/3}\overline{v}_s. \tag{11.53}$$

Thus the maximum phonon frequency ν_m is proportional to $(N/V)^{1/3}\,\overline{v}_s$, where N/V is the number density of the crystal. Since the ground state energy of the oscillators is a constant, we may set it equal to 0, then the total internal energy U of the system is

$$U = \int g(\nu)\overline{n}h\nu\, d\nu = \frac{9N}{\nu_m^3}\int_0^{\nu_m} \frac{h\nu^3\, d\nu}{e^{h\nu/kT} - 1}. \tag{11.54}$$

We define a characteristic temperature called the **Debye temperature** $\theta_{\rm D}$, such that $k\theta_{\rm D}$ represents the maximum phonon energy,

$$\theta_{\rm D} \equiv \frac{h\nu_m}{k}. \tag{11.55}$$

By changing the variable $x = h\nu/kT$, $x_m = h\nu_m/kT = \theta_{\rm D}/T$, Eq. (11.54) becomes

$$U = 9NkT\left(\frac{T}{\theta_{\rm D}}\right)^3\int_0^{x_m}\frac{x^3\, dx}{e^x - 1}. \tag{11.56}$$

This is an integral which cannot be evaluated easily. Therefore we evaluate it in the following two limits only: (1) at the high T limit $x_m = \theta_{\rm D}/T \ll 1$, and (2) at the low T limit $x_m = \theta_{\rm D}/T \gg 1$.

(1) At high enough T, such that $x \le x_m \ll 1$, $e^x - 1 \approx x$, therefore

$$U = 9NkT\left(\frac{T}{\theta_{\rm D}}\right)^3\frac{x_m^3}{3} = 3NkT, \;\; C_V = 3Nk.$$

This is the classical result, which can be obtained directly from the equipartition theorem .

(2) At low enough T, such that $x_m \gg 1$, we may let the upper limit of the integration x_m be extended to ∞, which will introduce negligible errors. The integration in Eq. (11.54) is then independent of T and can be evaluated exactly (see Appendix E), we obtain

$$\int_0^\infty \frac{x^3\,dx}{e^x - 1} = \frac{1}{15}\pi^4,$$

$$U = \frac{3\pi^4}{5} NkT\left(\frac{T}{\theta_D}\right)^3, \quad T \ll \theta_D, \tag{11.57}$$

$$C_V = \frac{12\pi^4}{5} Nk\left(\frac{T}{\theta_D}\right)^3, \quad T \ll \theta_D. \tag{11.58}$$

Therefore at very low temperatures, C_V is proportional to T^3, in agreement with the experimental result. This is known as the **Debye T^3 law.**

We see that in discussing the heat capacity of a solid, *high* or *low* temperature is compared with the *Debye temperature* θ_D. Since θ_D varies in a wide range for different materials, a high temperature for one substance may be just a low temperature for another substance. Therefore high temperature or low temperature is relative, not an absolute. Since the theoretical value of θ_D for a given substance is not easy to calculate accurately,[9] in practice, the Debye temperature is determined experimentally by measuring C_V, and then Eq. (11.58) is used to fit the data to obtain the value of θ_D. We tabulate the values of θ_D for some solids in Table 11.2.

Table 11.2 Debye temperature θ_D for some solids

solid	θ_D(K)	solid	θ_D(K)	solid	θ_D(K)
Lithium	344	Diamond	2230	Zinc	327
Sodium	158	Iron	467	Silver	225
Potassium	91	Cobalt	445	Gold	165
Silicon	640	Copper	343	Boron	1250

[9]The Debye model simplifies the calculations by using the average sound velocity $\overline{v}_s$ to replace the complicated frequency and directional dependence of the sound velocity; thus the result of the Debye model is just a model calculation.

11.9 Paramagnetism

A paramagnetic system is a system which has no net total magnetic moment when there is no external magnetic field, but when there is an external field, the system will have an induced total magnetic moment parallel to the external field. If all or part of the atoms (molecules) in a crystal possess *permanent magnets*, then the crystal will exhibit *paramagnetism* at a high enough temperature.[10] Paramagnetism occurs when the interactions between the neighboring magnets may be neglected, i.e., the pair interaction energy between magnets is much smaller than the thermal energy kT. At lower temperatures when the interaction between magnets is no longer negligible, paramagnetism will be transformed either to ferromagnetism or to antiferromagnetism, which will be briefly discussed in Sec. 11.11. In this section we are going to study some magnetic properties of a paramagnetic material.

We consider a crystal which has N magnetic particles (atoms or molecules) of equal magnetic moment. We assume that the temperature is high enough so that the interaction between the magnets may be considered as zero. Hence, when there is no external magnetic field, the directions of the magnets are random, and the *net* total magnetic moment is zero. Suppose the magnetic moment of each magnetic particle is $\boldsymbol{\mu}$, and there is an external magnetic field $\boldsymbol{B}$, then each magnet has a *magnetic potential energy* ε,

$$\varepsilon = -\boldsymbol{\mu} \cdot \boldsymbol{B} = -\mu_z B. \tag{11.59}$$

In the last form we have chosen the direction of $\boldsymbol{B}$ to be in the $+z$ direction. Since there is no interaction between the magnets, therefore in the presence of the external field $\boldsymbol{B}$, the internal energy of the system can be written as

$$E = E' + E_{\mathrm{M}}, \quad E_M = -\sum_{i=1}^{N} \mu_{iz} B \tag{11.60}$$

where E' is the non-magnetic part of the energy, and μ_{iz} is the magnitude of μ_z for the ith particle. Therefore E_{M} is the sum of all the magnetic potentials of the $\boldsymbol{\mu}$'s in $\boldsymbol{B}$. The partition function Z_N of the system can therefore be written as the product of two factors,

$$Z_N = Z'_N Z_N^{\mathrm{M}}, \quad Z_N^{\mathrm{M}} = (Z_1^{\mathrm{M}})^N, \tag{11.61}$$

[10]The spins of the conduction electrons in a metal are another source of paramagnetism.

where $Z_N^{\rm M}$ and Z'_N are the magnetic and non-magnetic parts of the partition functions, and $Z_1^{\rm M}$ is the **magnetic one-particle partition function.** If $Z_1^{\rm M}$ can be calculated, then the magnetic properties of the system can be obtained.

In order to calculate $Z_1^{\rm M}$, we have to know the energy levels of ε in Eq. (11.59). The permanent magnetic moment $\boldsymbol{\mu}$ of a particle is related to the total angular momentum $\boldsymbol{J}$ of the particle,

$$\boldsymbol{\mu} = -g\mu_{\rm B}\boldsymbol{J}, \quad |\boldsymbol{\mu}| = g\mu_{\rm B}\sqrt{J(J+1)},$$

where g is an angular momentum dependent constant, called the *Landé*[11] *g-factor*. The constant $\mu_{\rm B}$ is called the *Bohr*[12] *magneton*, which is the magnetic moment unit, defined by[13]

$$\mu_{\rm B} = \frac{e\hbar}{2m_{\rm e}} = 9.27 \times 10^{-24}\,{\rm J\,T^{-1}} = 5.788 \times 10^{-5}\,{\rm eV\,T^{-1}}, \qquad (11.62)$$

where $\hbar = h/2\pi$, $m_{\rm e}$ and e are the mass and charge of an electron ($e > 0$ is taken), and T (tesla) is the unit of B. In quantum mechanics J is quantized, which can be either in integers $(1, 2, \cdots)$ or half-intergers $(1/2, 3/2, \cdots)$. The z-component of $\boldsymbol{J}$ is denoted as j ($J_z = j$), which is also quantized $j = -J, -J+1, \cdots, J-1, J$, thus j can take $2J+1$ different values. Equation (11.59) may be written as

$$\varepsilon_j = j\,g\mu_{\rm B}B, \quad j = -J, -J+1, \cdots, J-1, J. \qquad (11.63)$$

Therefore the one-particle partition function $Z_1^{\rm M}$ at temperature T is

$$Z_1^{\rm M} = \sum_{j=-J}^{J} e^{-\varepsilon_j/kT} = \frac{\sinh(J+\frac{1}{2})x}{\sinh\frac{1}{2}x}, \quad x = \frac{g\mu_{\rm B}B}{kT}. \qquad (11.64)$$

When $B = 0$, the mean value of the magnetic moment is 0, i.e., $\overline{\mu_x} = \overline{\mu_y} = \overline{\mu_z} = 0$; we expect that when $B \neq 0$, $\overline{\mu_z} \neq 0$ but $\overline{\mu_x} = \overline{\mu_y} = 0$. From $Z_1^{\rm M}$ we may calculate the average magnetic moment per particle $\overline{\mu_z}$ ($\varepsilon_j = -\mu_z B$),

$$\overline{\mu_z} = \frac{\sum_j \mu_z\, e^{-\varepsilon_j/kT}}{\sum_j e^{-\varepsilon_j/kT}} = kT\left(\frac{\partial \ln Z_1^{\rm M}}{\partial B}\right)_T = g\mu_{\rm B}J\mathcal{B}_J(x), \qquad (11.65)$$

[11] Alfred Landé, German physicist (1888–1976).

[12] Niels H. D. Bohr, Danish physicist (1885–1962), Nobel prize laureate in physics in 1922.

[13] In CGS (Gaussian) units, $\mu_{\rm B} = e\hbar/2m_{\rm e}c = 0.927 \times 10^{-20}$ erg Oe^{-1}, where c is the speed of light.

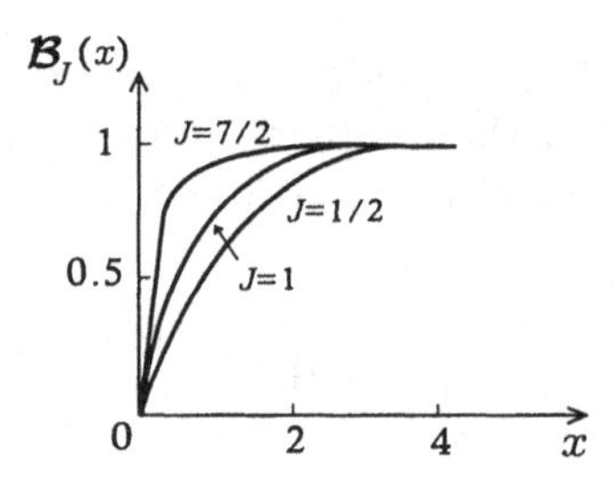

Figure 11.1: The plot of the Brioullin function $\mathcal{B}_J(x)$ vs. x for several different values of J.

$$\mathcal{B}_J(x) \equiv \frac{1}{J}\left[\left(J+\frac{1}{2}\right)\coth\left(J+\frac{1}{2}\right)x - \frac{1}{2}\coth\frac{1}{2}x\right]. \tag{11.66}$$

The function $\mathcal{B}_J(x)$ is called the *Brillouin*[14] *function.* The x-dependence of the function $\mathcal{B}_J(x)$ depends on the value of J, however the dependence is a *quantitative one, not a qualitative one*, as can be seen from Fig. 11.1. Therefore we may discuss the simplest case with[15] $J = 1/2$, and the cases for other values of J will be qualitatively the same.

For $J = 1/2$, j can take only two values, $-1/2$ or $1/2$. Therefore the one-particle partition function Z_1^{M} contains only two terms,

$$Z_1^{\mathrm{M}} = e^{-x/2} + e^{+x/2} = 2\cosh\frac{x}{2} = 2\cosh\left(\frac{\mu_{\mathrm{B}}B}{kT}\right). \tag{11.67}$$

From Z_1^{M} many magnetic properties of a paramagnetic material can be obtained. In the rest of this section we will discuss three topics: (1) the magnetization, (2) the constant magnetic field heat capacity, and (3) the magnetic entropy.

1. The magnetization m

The magnetization m is defined as the magnetic moment per unit volume. It is related to the total magnetic moment M by the relation $m = (N_0/N)M$, where N is the total number of magnetic particles, and N_0 is the number of magnetic particles *per unit volume.* From Eq. (11.65)

---

[14]Leon N. Brillouin, French physicist (1889–1969).

[15]The J value for the orbital angular momentum is an integer, hence $J = 1/2$ comes from the electron spin with the degeneracy $g = 2s + 1 = 2$.

we obtain

$$M = N\overline{\mu_z} = NkT\left(\frac{\partial \ln Z_1^{\mathrm{M}}}{\partial B}\right)_T = N\mu_{\mathrm{B}} \tanh\left(\frac{\mu_{\mathrm{B}}B}{kT}\right). \tag{11.68}$$

This is the equation of state for a paramagnetic system $M = M(T, B)$. Now we discuss the relation between M and T in the following two limits:

(1) At low T and/or high B, such that $\mu_{\mathrm{B}}B \gg kT$ holds. In this limit, letting $\mu_{\mathrm{B}}B/kT = x \gg 1$, $\tanh x \approx 1 - 2\exp(-2x)$, therefore

$$M = M_0\left(1 - 2e^{-2\mu_{\mathrm{B}}B/kT}\right), \quad \frac{\mu_{\mathrm{B}}B}{kT} \gg 1, \tag{11.69}$$

where $M_0 \equiv N\mu_{\mathrm{B}}$ is the *saturation magnetic moment* when all the magnetic moments are aligned in the same direction.

(2) At high T and/or low B, such that $\mu_{\mathrm{B}}B \ll kT$ holds. In this limit $\tanh(\mu_{\mathrm{B}}B/kT) \approx \mu_{\mathrm{B}}B/kT$, consequently

$$M = \left(\frac{N\mu_{\mathrm{B}}^2}{k}\right)\frac{B}{T} \equiv C_c\frac{B}{T}, \quad \frac{\mu_{\mathrm{B}}B}{kT} \ll 1. \tag{11.70}$$

The last form is **Curie's law** we have mentioned in Chapter 8, with the Curie constant $C_c = N\mu_{\mathrm{B}}^2/k$ for the case $J = 1/2$. In a magnetic system an important measurable quantity is the *magnetic susceptibility* χ, which is defined as

$$\chi = \mu_0\left(\frac{\partial M}{\partial B}\right)_T, \tag{11.71}$$

where μ_0 is the permeability of the vacuum.[16] For a paramagnetic system,

$$\chi = \mu_0 C_c \frac{1}{T}. \tag{11.72}$$

Equation (11.72) is also known as Curie's law, because it is equivalent to Eq. (11.70).

We plot the graph of M vs. T/B in Fig. 11.2(a) for all values of T/B, including both cases (1) and (2) discussed above. The graph shows that M decreases from the maximum value M_0 (at $T = 0$) to 0 as T/B increases to a large value.

[16]If Gaussian units are used, then μ_0 is taken to be 1.

2. The constant magnetic filed heat capacity C_B

The heat capacity C_B is defined as $C_B = (\partial U_{\rm M}/\partial T)_B$, where $U_{\rm M}$ is the magnetic energy of the system. From Eq. (11.68) and $\varepsilon = -\mu_z B$, we have

$$U_{\rm M} = N\bar{\varepsilon} = -N\overline{\mu_z}B = -N\mu_{\rm B}B\tanh\left(\frac{\mu_{\rm B}B}{kT}\right), \tag{11.73}$$

$$\begin{aligned} C_B &= \left(\frac{\partial U_M}{\partial T}\right)_B = Nk\left(\frac{\mu_{\rm B}B}{kT}\right)^2 \mathrm{sech}^2\left(\frac{\mu_{\rm B}B}{kT}\right) \\ &= Nk\left(\frac{2\mu_{\rm B}B}{kT}\right)^2 \frac{e^{2\mu_{\rm B}B/kT}}{(e^{2\mu_{\rm B}B/kT}+1)^2}, \end{aligned} \tag{11.74}$$

We discuss only the two limits we have studied above:

(1) At low T and/or high B, such that $\mu_{\rm B}B/kT \gg 1$ holds,

$$C_B \approx Nk\left(\frac{2\mu_{\rm B}B}{kT}\right)^2 e^{-2\mu_{\rm B}B/kT}, \quad \frac{\mu_{\rm B}B}{kT} \gg 1. \tag{11.75}$$

Hence $C_B \to 0$ as $T/B \to 0$. This is in agreement with the third law.

(2) At high T and/or low B, such that $\mu_{\rm B}B/kT \ll 1$ holds,

$$C_B \approx Nk\left(\frac{\mu_{\rm B}B}{kT}\right)^2, \quad \frac{\mu_{\rm B}B}{kT} \ll 1. \tag{11.76}$$

Therefore at high T and/or low B ($\mu_{\rm B}B/kT \ll 1$), the heat capacity C_B will also approach 0, but with a much slower pace. Thus C_B will have a maximum as T/B varies from 0 to a large value, as shown in Fig. 11.2(b).

3. The magnetic entropy $S_{\rm M}$

The magnetic entropy $S_{\rm M}$ due to N magnetic moments can be obtained from Eq. (10.84),

$$\begin{aligned} S_{\rm M} &= \frac{U_{\rm M}}{T} + Nk\ln Z_1^{\rm M} \\ &= Nk\ln\left[2\cosh\left(\frac{\mu_{\rm B}B}{kT}\right)\right] - N\frac{\mu_{\rm B}B}{T}\tanh\left(\frac{\mu_{\rm B}B}{kT}\right). \end{aligned} \tag{11.77}$$

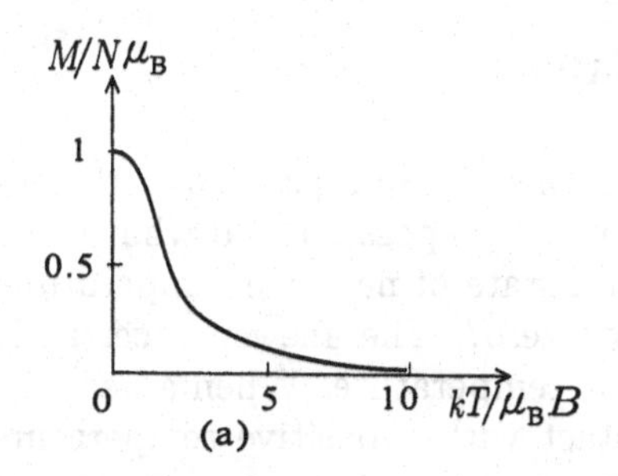

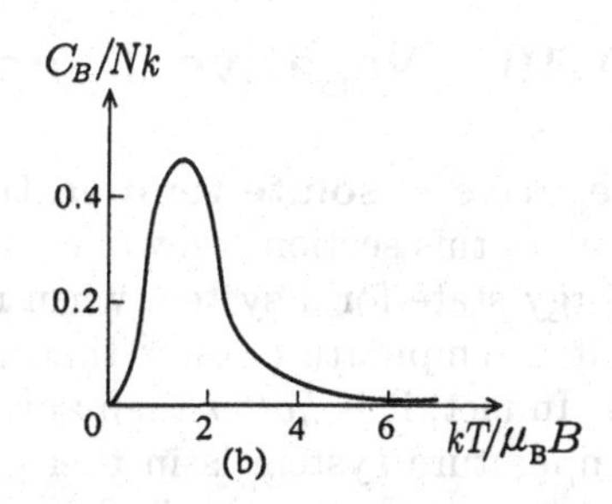

Figure 11.2: The plots for the magnetic moment M and the constant field heat capacity C_B for a paramagnetic system with $J = 1/2$. (a) $M/N\mu_{\mathrm{B}}$ vs. $kT/\mu_{\mathrm{B}}B$. (b) C_B/Nk vs. $kT/\mu_{\mathrm{B}}B$.

For large x ($x \gg 1$), $2\cosh x \approx e^x$ and $\tanh x \approx 1$, hence at very low T, such that $\mu_{\mathrm{B}}B/kT$ is very large, we have

$$S_{\mathrm{M}} \approx Nk\ln(e^{\mu_{\mathrm{B}}B/kT}) - N\frac{\mu_{\mathrm{B}}B}{T} \approx 0,$$

which implies that S_{M} satisfies the third law. For small x ($x \ll 1$), $\cosh x \approx 1$ and $\tanh x \approx x$, consequently at very high T, such that $\mu_{\mathrm{B}}B/kT \ll 1$, $S_{\mathrm{M}} = Nk\ln 2 = k\ln 2^N$, which shows that 2^N is the total number of microstates. Now 2^N is the total number of microstates for N random magnets with $J = 1/2$. This means that the number of microstates is the same as the case with $B = 0$. In other words, in the high T and low B limits the thermal energy is much larger than the magnetic energy, and the effect of the magnetic field is negligible.

Next we consider how the magnetic entropy changes with the magnetic field at constant temperature. From Eq. (11.77)

$$\left(\frac{\partial S_{\mathrm{M}}}{\partial B}\right)_T = -N\frac{\mu_{\mathrm{B}}^2 B}{kT^2}\cosh^{-2}\left(\frac{\mu_{\mathrm{B}}B}{kT}\right) < 0. \qquad (11.78)$$

Since $(\partial S/\partial B)_T < 0$, a paramagnetic salt can be used for cooling by the following processes. (1) A magnetic field is applied at constant T, which will decrease the entropy by Eq. (11.78). (2) A reversible adiabatic demagnetization (S constant, and B decreasing) which will decrease T, as shown in Fig. 6.4. The above result provides the principle for cooling by adiabatic demagnetization, which we have discussed in Chapter 6.

11.10 Negative Temperatures

Negative absolute temperature may happen in a paramagnetic system. In this section, we will explain why this is possible and what is the energy state for a system when it is in a state of negative temperature. Is it a temperature lower than absolute zero? The answer is certainly no. In fact, it is *hotter* than any positive temperature. When a negative temperature system is in thermal contact with a positive temperature system, energy (heat) will flow from the negative temperature system to the positive temperature system.

In Chapter 8 we have defined a *magnetic enthalpy* $H_\mathrm{M} = U + U_\mathrm{M} \equiv E$, where U is the part of the internal energy which is independent of the magnetic moment, and U_M is the magnetic energy. Thus the physical meaning of H_M is the *total internal energy* E. Since U is the non-magnetic part of the energy, we may keep only the magnetic part of the energy and rewrite Eq. (8.8) as

$$dU_\mathrm{M} = TdS_\mathrm{M} - MdB, \tag{11.79}$$

where S_M is the magnetic entropy. In thermal equilibrium, the temperature T in the above equation is equal to the temperature of the vibrational motion of the system. However *if the magnetic part of the system can be decoupled from the vibrational part for a certain period of time, then the temperatures for these two parts may be different.* Therefore from Eq. (11.79) we may define a *magnetic temperature* T_M by the following equation,[17]

$$T_\mathrm{M} = \left(\frac{\partial U_\mathrm{M}}{\partial S_\mathrm{M}}\right)_B, \quad \text{or} \quad \frac{1}{T_\mathrm{M}} = \left(\frac{\partial S_\mathrm{M}}{\partial U_\mathrm{M}}\right)_B. \tag{11.80}$$

We see that if we can find a situation where the energy U_M increases but the entropy S_M decreases, then $T_\mathrm{M} < 0$ occurs. This can happen for a system with a finite number of energy states, such as the paramagnetic system we are considering.[18] The reason is simple. For this class of systems, if its energy U_M is greater than a critical value, then as U_M is increased further, its entropy S_M will decrease, because most of its particles will occupy the highest energy level, and the multiplicity

[17] We use T_M, instead of T, to denote the temperature emphasizing that it is the temperature for *an isolated magnetic system.*

[18] For a system with an infinite number of energy states, such as the translational or vibrational energy states, S always increases as U increases.

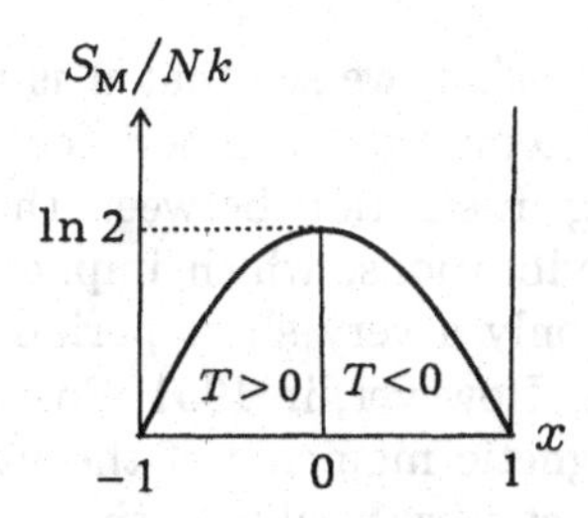

Figure 11.3: The plot of the magnetic entropy S_M vs. U_M, the total magnetic energy. In the figure $x = U_M/N\mu_B B$.

decreases. As an example, we calculate S_M explicitly for the $J = 1/2$ paramagnetic salt.

We consider a paramagnetic salt of N magnets with $J = 1/2$ in an applied magnetic field $\boldsymbol{B}$. Suppose the number of particles with magnetic moment $\boldsymbol{\mu}$ parallel to the magnetic field $\boldsymbol{B}$ is N_+, and for $\boldsymbol{\mu}$ antiparallel to $\boldsymbol{B}$ it is N_-, then the magnetic entropy may be obtained from the Boltzmann relation

$$S_M = k \ln \frac{N!}{N_+!\,N_-!}, \quad N_+ + N_- = N. \tag{11.81}$$

The magnetic energy for $\boldsymbol{\mu}$ in $\boldsymbol{B}$ is $-\boldsymbol{\mu} \cdot \boldsymbol{B}$, which is $-\mu_B B$ (for $\boldsymbol{\mu} \parallel \boldsymbol{B}$), or $+\mu_B B$ (for $\boldsymbol{\mu} \parallel (-\boldsymbol{B})$). Therefore the total magnetic energy is

$$U_M = -\mu_B B(N_+ - N_-). \tag{11.82}$$

From the above two equations the entropy can be calculated to be

$$\frac{S_M}{Nk} = \ln 2 - \frac{1}{2}[(1-x)\ln(1-x) + (1+x)\ln(1+x)], \tag{11.83}$$

where $x = U_M/N\mu_B B$. We see that S_M is maximum at $x = 0$, as shown in Fig. 11.3. We note that at $x = 0$ ($U_M = 0$ and $N_+ = N_-$), $T_M = \infty$; and for $U_M > 0$ ($N_+ < N_-$) the temperature is negative $T_M < 0$. However, for $U_M < 0$ ($N_+ > N_-$) the temperature is positive $T_M > 0$. Clearly the state of $T < 0$ is *hotter* (higher energy) than the state of $T > 0$. Therefore when a system at $T < 0$ (spin system) is in thermal contact with a system at $T > 0$ (usually the crystal itself), heat will flow from the system at $T < 0$ to the system at $T > 0$. Finally these two systems will reach thermal equilibrium with an equal temperature at $T > 0$.

From the above discussion, we see that it is not easy to observe the phenomenon of negative temperature experimentally. The reason is that there is always a strong interaction between the atomic magnetic moments and the crystal vibrations, which implies that the phenomenon of $T_{\rm M} < 0$ will last for only a very short period of time and cannot be detected experimentally. However, in 1951, Pound, Purcell and Ramsey discovered that the magnetic moments of the *nucleus* of lithium atoms in the LiF crystal interact very weakly with the crystal, therefore they observed the phenomenon of $T < 0$ in a LiF crystal for about 2–5 minutes. They applied a strong magnetic field to the LiF crystal at $T > 0$ ($N_+ > N_-$), and then reversed the magnetic field, which reverses the numbers N_+ and N_-, thus $N_+ < N_-$ when the magnetic field is reversed. This is the condition for *negative magnetic temperature.* The number of particles in the higher energy state (μ anti-parallel to $\mathbf{B}$) N_- is larger than the number of particles in the lower energy state (μ parallel to $\mathbf{B}$) N_+. This is known as a **population inversion**; radiation of an electromagnetic wave with an angular frequency of $2\mu_{\rm B}B/\hbar$ was observed in the experiment.[19] In the canonical ensemble at temperature T, the number of particles N_+ and N_- are given by (just like the Boltzmann distribution)

$$N_+ = \frac{N}{Z_1^M}\, e^{\mu_{\rm B}B/kT}, \quad N_- = \frac{N}{Z_1^M}\, e^{-\mu_{\rm B}B/kT}, \tag{11.84}$$

where $\mu_{\rm B}B$ is always considered as a positive quantity. We see that population inversion $N_+ < N_-$ requires that $T < 0$. The phenomenon of population inversion also occurs in the action of a laser, which is sustained by external electromagnetic fields.

11.11 Mean Field Theory for Ferromagnetism

When the interactions between neighboring magnetic moments are no longer negligible, a paramagnetic material will transform either to a ferromagnetic or to an antiferromagnetic material. In nature there are several materials, usually the transitions metals such as Fe and Ni, which exhibit ferromagnetism at temperatures as high as room temperatures or even much higher. Consequently the phenomenon of ferromagnetism

[19]Since the energy levels of the spin system are discrete, it can change its temperature only by radiation, i.e., by changing its population distribution.

is well known. In this section we will use a simple model and a simple approximation to study the properties of ferromagnetism.

We consider a system consisting of N permanent magnetic moments μ, whose z-component can have only two values, $-\mu_B$ or μ_B, i.e., a $J = 1/2$ system. An external magnetic field B is applied to the system. We consider a simple model whose Hamiltonian H has the form

$$H = -\epsilon\,\mu_B^2 \sum_{<i,j>} s_i\, s_j - \mu_B B \sum_i s_i, \tag{11.85}$$

where $< i, j >$ denotes that the summation is over the *nearest neighbor* sites i and j only; ϵ represents the interaction energy between the nearest neighbor magnets; and each s_i can have only two values $+1$ or -1. This model is known as the **Ising**[20] **model**. For $\epsilon > 0$, the neighboring magnets prefer to align in the same direction; thus the system may have a net magnetic moment at temperatures below a *critical temperature* T_c, even when the external field $B = 0$. This is known as *spontaneous magnetization*, the phenomenon of **ferromagnetism**. For $\epsilon < 0$, the neighboring magnets prefer to align in the opposite directions, and this is the phenomenon of **antiferromagnetism**. In this section we will use the Ising model to study the properties of ferromagnetism (with $\epsilon > 0$).

This is an example of a strong interaction between particles, such that the N-particle partition function cannot be expressed as a product of the one-particle partition functions. Nevertheless, the one- and two-dimensional Ising models can be solved exactly. For the one-dimensional case, the exact solution is not hard to obtain (see Problem 11.22). The two-dimensional Ising model was first solved in closed form (for $N \to \infty$ and $B = 0$) by Onsager[21] in the 1940s. The mathematics involved in the solution is very complicated; it will not be discussed here. The important feature is that there is no spontaneous magnetization for the one-dimensional case (see Problem 11.23); however there is spontaneous magnetization for the two-dimensional case[22] with a non-zero critical temperature T_c, which is proportional to ϵ. For the three-dimensional case, nobody has ever found an exact solution. It is necessary to solve the problem by approximate methods. A simple and yet reliable approximate

[20]Ernst Ising, German physicist (1900–1998).

[21]Lars Onsager, Norwegian chemist (1903–1976), Nobel prize laureate in chemistry in 1968.

[22]However, for a more refined *Heisenberg model*, $H = -\epsilon\,\mu_B^2 \sum_{<i,j>} \mathbf{s}_i \cdot \mathbf{s}_j$ $(B = 0)$, there is no spontaneous magnetization for both the one- and two-dimensional cases. See Mermin and Wagner, *Phys. Rev. Lett.* **17**, 1133 (1966).

method does exist for the *three-dimensional* case. This is known as the **mean field** or **molecular field approximation** (for brevity, the MFA). For any three-dimensional model Hamiltonian, the MFA usually predicts correct results, at least qualitatively, though not quantitatively. We will proceed to study the thermodynamic properties of Eq. (11.85) by using the MFA.

The main spirit of the MFA is to reduce the N-site problem to a one-site problem. Choose an arbitrary site j and forget all other sites, except that, in addition to the external field B, there is an *internal magnetic field* B_{int} produced by all the other sites, which also acts on the magnet at site j. In the MFA, the internal magnetic field has the form

$$B_{\text{int}} = q\epsilon\mu_{\text{B}}\overline{s} = q\epsilon\overline{m}, \tag{11.86}$$

where q is the number of nearest neighbor sites,[23] and $\overline{m} \equiv \mu_{\text{B}}\overline{s} = M/N$ is the average magnetic moment *per site*. This means that the value of s_i for each site, except site j, is replaced by the mean value $\overline{s}$. Therefore the one-site Hamiltonian in the MFA is

$$h_{\text{MF}} = -(B_{\text{int}} + B)\mu_{\text{B}}s = -(q\epsilon\overline{m} + B)\mu_{\text{B}}s, \tag{11.87}$$

where s (the subscript j is omitted) can have two possible values $+1$ or -1. We can see immediately that this problem is exactly the same as a paramagnetic system with $J = 1/2$, except that the magnetic field B has to be replaced by $B_{\text{int}} + B = q\epsilon\overline{m} + B$. In particular, Eq. (11.68) now becomes

$$\overline{m} \equiv \frac{M}{N} = \mu_{\text{B}} \tanh\left[\frac{\mu_{\text{B}}(q\epsilon\overline{m} + B)}{kT}\right]. \tag{11.88}$$

This is the main result for the MFA in ferromagnetism. We will use it to study the magnetic properties of a ferromagnetic material. We are interested in the phenomenon of spontaneous magnetization, and therefore set $B = 0$ in Eq. (11.88) to obtain

$$\overline{m} = \mu_{\text{B}} \tanh\left(\frac{q\epsilon\mu_{\text{B}}\overline{m}}{kT}\right). \tag{11.89}$$

By solving this equation we obtain $\overline{m}$ as a function of T. Apparently $\overline{m} = 0$ is always a solution, however it is also possible to have a solution with $\overline{m} \neq 0$. It is instructive to solve this equation graphically. To simplify the notation, we use $\overline{s} = \overline{m}/\mu_{\text{B}}$, the above equation becomes

[23]In three dimensions, $q = 6$ for a simple cubic crystal, $q = 8$ for a body-centered cubic crystal, and $q = 12$ for a face-centered cubic crystal.

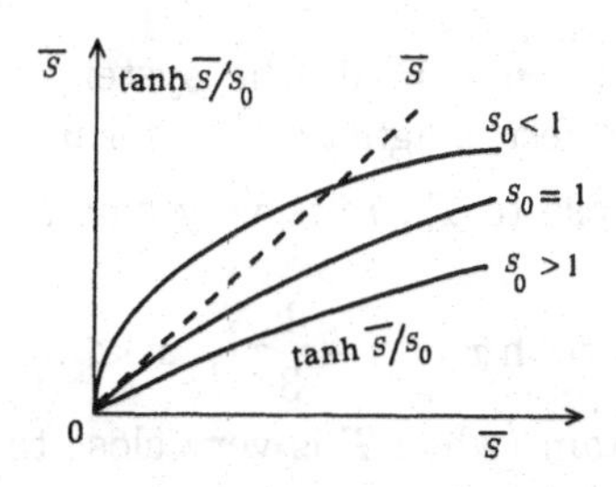

Figure 11.4: The plots of $\overline{s}$ vs. $\overline{s}$ and $\tanh(\overline{s}/s_0)$ vs. $\overline{s}$. Three different values of s_0, $s_0 > 1$, $s_0 = 1$ and $s_0 < 1$, are plotted for $\tanh(\overline{s}/s_0)$.

$$\overline{s} = \tanh\left(\frac{\overline{s}}{s_0}\right), \quad s_0 = \frac{kT}{q\epsilon\mu_B^2}. \tag{11.90}$$

We plot $\overline{s}$ and $\tanh(\overline{s}/s_0)$ vs. $\overline{s}$ on the same graph, as shown in Fig. 11.4. The intersection points of these two curves are the solutions for Eq. (11.90). We see that when $s_0 > 1$, there is only one solution with $\overline{s} = 0$, which is the paramagnetic state, $\overline{m} = 0$ when $B = 0$. However[24] when $s_0 < 1$ there is a solution with $\overline{s} \neq 0$ in addition to $\overline{s} = 0$. We have to decide which state is the stable one, $\overline{s} = 0$ or $\overline{s} \neq 0$? It can easily be shown that the magnetic free energy[25] F for the state with $\overline{s} \neq 0$ is smaller than the that of the state with $\overline{s} = 0$. Therefore the state with $\overline{s} \neq 0$ is the stable state. This implies that the system exhibits the phenomenon of **spontaneous magnetization**, because the system has a net magnetic moment $\overline{m} \neq 0$ even when $B = 0$. Hence there exists a **critical temperature** T_c ($s_0 = 1$ when $T = T_c$),

$$T_c = \frac{q\epsilon\mu_B^2}{k} \text{ (mean field)}, \tag{11.91}$$

such that $T < T_c$ the system is in a ferromagnetic state, while for $T > T_c$, the system is in a paramagnetic state. Thus there is a *phase transition* at $T = T_c$. This is a second-order phase transition, because at T_c both the ferromagnetic and the paramagnetic states have the same magnetization $\overline{m} = 0$. The critical temperature is also called the **Curie temperature**. Note that the MFA predicts that T_c is proportional to the magnetic interaction energy ϵ, which is consistent with more refined

[24] This is the condition that the slope of $\tanh(\overline{s}/s_0)$ is greater than that of $\overline{s}$.

[25] The free energy is related to the partition function by $F = -kT \ln Z$.

methods. Therefore when $\epsilon = 0$ (the system is then paramagnetic) $T_c = 0$, and there is no ferromagnetic transformation as expected.

When T is very close to T_c, $\overline{m}$ is very small for $T < T_c$, so we can use the expansion

$$\tanh x \approx x - \frac{1}{3}x^3, \quad x \ll 1,$$

in Eq. (11.89) and obtain (when T is very close to T_c)

$$\overline{m} \approx \sqrt{3}\,\mu_B \left(1 - \frac{T}{T_c}\right)^{1/2}, \quad T < T_c, \tag{11.92}$$

and $\overline{m} = 0$ for $T > T_c$. Note that the curve of $\overline{m}$ vs. T has an infinite slope at $T = T_c$, $(d\overline{m}/dT \to \infty)$ which implies that, when $B = 0$, $\overline{m}$ increases rapidly from 0 to a finite value as T is decreased from $T > T_c$ to $T < T_c$. For $T > T_c$, $\overline{m} = 0$ when $B = 0$ (paramagnetism). However when $B \neq 0$, $\overline{m}$ will be nonzero. For small $\mu_B B/kT$ we may use the approximation $\tanh x \approx x$ in Eq. (11.88) and get

$$\overline{m} = \frac{\mu_B^2 B}{k(T - T_c)}, \quad T > T_c. \tag{11.93}$$

The magnetic susceptibility χ is easily calculated to be

$$\chi = \mu_0 \left(\frac{\partial M}{\partial B}\right)_T = \mu_0 \frac{N\mu_B^2}{k(T - T_c)}, \quad T > T_c. \tag{11.94}$$

This is known as the **Curie-Weiss**[26] **law**. When $T_c = 0$, the Curie-Weiss law becomes Curie's law, Eq. (11.72). If we retain B in Eq. (11.88) and use Eq. (11.92) and the small x expansion for $\tanh x$, we may calculate the magnetic susceptibility χ for $T < T_c$:

$$\chi = \mu_0 \frac{N\mu_B^2}{2k(T_c - T)}, \quad T < T_c. \tag{11.95}$$

We see that χ approaches $+\infty$ when T approaches T_c both from below and from above. This is one of the main features of the **critical phenomena** of a second order phase transition. Near the *critical point* the system experiences large fluctuations; a small amount of an external field will produce a large effect on the system, such as χ diverges at the critical point. The theory of critical phenomena is one of the more important subjects in statistical mechanics. However it is beyond the scope of this book, so we will not discuss it further.

---

[26]Pierre Ernest Weiss, French physicist (1865–1940).

11.12 Problems

11.1. Consider an ideal gas consisting of N monatomic molecules obeying MB statistics which is confined in a two-dimensional area A at temperature T. Calculate the one-particle partition function Z_1 and then find the following thermodynamic functions of the system: the internal energy U, entropy S, Helmholtz free energy F, pressure P, enthalpy H, and the heat capacities C_V and C_P.

11.2. Consider an ideal gas consisting of N monatomic molecules obeying MB statistics which is confined in a volume V at temperature T. Suppose that the energy of each particle ε is proportional to the magnitude of the momentum p, $\varepsilon = cp$. Calculate the one-particle partition function of the system Z_1 and then find the following thermodynamic functions of the system: the internal energy U, entropy S, Helmholtz free energy F, pressure P, enthalpy H, and the heat capacities C_V and C_P.

11.3. Consider a system consisting of N weakly interacting localized one-dimensional *classical* harmonic oscillators. Each oscillator has the Hamiltonian

$$h = \frac{p^2}{2m} + \frac{1}{2}m\omega^2 x^2,$$

where $\omega = 2\pi\nu$ is the angular frequency. Calculate the *classical* one-particle partition function Z_1. Starting from Z_1, calculate the average internal energy u and the specific heat c_v per oscillator. Show that the results are the same as those obtained from the equipartition theorem.

11.4. Consider a particle moving in a one-dimensional potential well of the form

$$u(x) = u_0 + a(x - x_0)^2 + b(x - x_0)^3 + c(x - x_0)^4,$$

where u_0, x_0, a, b and c are constants ($a > 0$, $c > 0$). The particle is in thermal equilibrium with a heat reservoir at temperature T. The motion of the particle in this potential well is known as an *anharmonic oscillator*. We consider the case that T is high enough such that the motion of the particle can be treated classically, but T is not too high. More specifically we assume that the inequality $a(x - x_0)^2 \gg |b(x - x_0)^3| \gg c(x - x_0)^4 > 0$ always holds for the particle displacement x. Note that when the above inequality holds, then one may set $c = 0$ in the calculation of the lowest order corrections due to $b \neq 0$.

(a) Find the average position $\overline{x}$ when $b = c = 0$. This is the *harmonic* case.

(b) Repeat the same problem as (a) when $b \neq 0$. Note that in this case $\overline{x}$ is a function of T, which is the origin of the thermal expansion of a solid. [*Hint*: The expansion $e^{-\beta b(x-x_0)^3} \approx 1 - \beta b(x-x_0)^3 + \cdots$ may be used in the calculation.]

(c) Calculate the partition function of the particle. Note that it contains a kinetic energy part and a potential energy part.

(d) Calculate the mean energy and the heat capacity of the particle by using the partition function.

11.5. Consider a system of N localized, weakly interacting, one-dimensional quantum harmonic oscillators, whose one-particle partition function is given by Eq. (11.18). Show that the entropy of the system is given by

$$S = \frac{U}{T} + Nk \ln Z_1.$$

By expressing U and Z_1 in terms of ν and T, show that S becomes

$$S = Nk \left[\frac{h\nu/kT}{e^{h\nu/kT} - 1} - \ln(1 - e^{-h\nu/kT}) \right].$$

11.6. Using Eq. (10.89)

$$P = kT \left(\frac{\partial \ln Z_N}{\partial V} \right)_{N,T},$$

show that the equation of state of a diatomic gas is the same as that of a monatomic gas. [*Hint*: Both θ_{rot} and θ_{vib} are independent of the volume V.]

11.7. Consider a diatomic gas of N molecules in a volume V near room temperature T. Show that the entropy is

$$S = S(T,V) = Nk \left[\ln \left(\frac{V}{N} T^{5/2} \right) + s_0 \right],$$

where s_0 is a constant. Find the respective expressions of s_0 for CO and O_2.

11.8. Consider a gas of N weakly interacting molecules. Each molecule consists of two atoms with a Hamiltonian of the form:

$$h = \frac{p_1^2}{2m_1} + \frac{p_2^2}{2m_2} + \frac{1}{2}K|r_1 - r_2|^2, \quad K > 0.$$

The system is in thermal equilibrium at temperature T. Find c_v and the mean square diameter $\overline{|r_1 - r_2|^2}$ for a molecule in the *classical limit.* Explain why this is not a realistic model for a real diatomic gas in nature.

11.9. Consider a photon gas in thermal equilibrium at temperature T. Photons with different frequencies are independent of each other, hence the total partition function Z_{tot} can be written as

$$Z_{\text{tot}} = Z(\nu_1)Z(\nu_2)Z(\nu_3)\cdots, \quad \text{including all } \nu's,$$

where we have treated ν as a discrete variable, and $Z(\nu)$ is given by Eq. (11.40).

(a) Show that if we treat ν as a continuous variable, then the total partition function can be written as

$$\ln Z_{\text{tot}} = -\frac{8\pi V}{c^3}\int_0^\infty \nu^2 \ln(1 - e^{-h\nu/kT})d\nu,$$

where Eq. (11.43) has been used.

(b) Show that an integration by parts leads to the result

$$\ln Z_{\text{tot}} = \frac{8\pi V}{3}\left(\frac{kT}{hc}\right)^3 \int_0^\infty \frac{x^3 dx}{e^x - 1}, \quad x = \frac{h\nu}{kT}.$$

Note that the integral contains no parameters, and can be integrated to give $\pi^4/15$ (see Appendix E).

11.10. The partition function of a photon gas Z is known to have the form $\ln Z = aT^3V$, where a is a constant. By using Eqs. (10.86)-(10.90), show that: (a) $U = 3aT^4V$; (b) $\mu = 0$; (c) $S = 4akT^3V$; and (d) $P = \frac{1}{3}\left(\frac{U}{V}\right)$.

11.11. The characteristic Debye temperature for diamond is 1860 K and the characteristic Einstein temperature is 1450 K. The experimental value of c_v for diamond, at temperature of 207 K, is 2.68×10^3 J kilomole^{-1} K^{-1}. Calculate c_v at 207 K from the Einstein and Debye approximations and compare the results with the experiment.

11.12. Show that the constant volume heat capacity C_V of a Debye solid can be written as

$$C_V = 12Nk\,D(\theta_D/T) - 9Nk\left(\frac{\theta_D}{T}\right)\frac{1}{e^{\theta_D/T}-1},$$

where $D(\theta_D/T)$ is one of several Debye functions, defined by

$$D(\theta_D/T) = 3\left(\frac{T}{\theta_D}\right)^3\int_0^{\theta_D/T}\frac{x^3\,dx}{e^x-1}.$$

11.13. Consider a solid in the Debye approximation.

(a) Show that the partition function can be written as

$$\ln Z = -9\left(\frac{T}{\theta_D}\right)^3\int_0^{\theta_D/T} x^2\ln(1-e^{-x})dx.$$

(b) Show that the Helmholtz free energy can be written as

$$F = 3NkT\ln\left(1-e^{-\theta_D/T}\right) - NkT\,D(\theta_D/T),$$

where $D(\theta_D/T)$ is one of the Debye function defined in Problem 11.12.

(c) Show that the internal energy can be written as

$$U = \frac{9}{8}Nk\theta_D + 3NkT\,D(\theta_D/T).$$

11.14. A paramagnetic salt contains 10^{25} magnetic ions per cubic meter, each with a magnetic moment of 1 Bohr magneton. Calculate the difference between the number of ions aligned parallel to the applied magnetic field of 1 T and those aligned antiparallel at the temperature (a) 300 K, and (b) 10 K, if the volume of the sample is 1 cm^3. Calculate the magnetic moment of the sample, in units of J T^{-1}, at these two temperatures.

11.15. Consider a system consisting of N weakly interacting localized particles in thermal equilibrium at temperature T. Each particle has two nondegenerate energy levels ε_0 and ε_1 ($\varepsilon_1 > \varepsilon_0 > 0$).

(a) Calculate the one-particle partition function Z_1.

(b) Show that the average energy per particle is given by

$$u \equiv \frac{U}{N} = \frac{\varepsilon_0 + \varepsilon_1 \, e^{-\Delta\varepsilon/kT}}{1 + e^{-\Delta\varepsilon/kT}}, \qquad \Delta\varepsilon = \varepsilon_1 - \varepsilon_0.$$

(c) Show that when $T \to 0$,

$$u \approx \varepsilon_0 + \Delta\varepsilon \, e^{-\Delta\varepsilon/kT},$$

and when $T \to \infty$,

$$u \approx \frac{1}{2}(\varepsilon_0 + \varepsilon_1) - \frac{1}{4}\frac{\Delta\varepsilon^2}{kT}.$$

(d) Show that the constant volume heat capacity per particle is

$$c_v = k\left(\frac{\Delta\varepsilon}{kT}\right)^2 \frac{e^{-\Delta\varepsilon/kT}}{(1 + e^{-\Delta\varepsilon/kT})^2}.$$

(e) Compute c_v in the limits $T \to 0$ and $T \to \infty$, and make a careful sketch of c_v versus $kT/\Delta\varepsilon$.

11.16. The energy levels of a localized particle are 0, ε, and 2ε. The middle level is doubly degenerate and the other two levels are nondegenerate. The particle is in thermal equilibrium with a heat reservoir at temperature T.

(a) Write and simplify the partition function of the particle.

(b) Find the mean energy, the heat capacity, and the entropy of the particle.

11.17. In a magnetic system, the magnetic free energy is defined as $F_{\mathrm{M}} = U - TS - MB$, $dF_{\mathrm{M}} = -SdT - M\,dB$. We can consider F_{M} as the Gibbs free energy G in a PVT system, because $dG = -SdT + VdP$, and both B and P are intensive parameters.

(a) Consider the simplest case, that the magnetic partition function is the product of the one-particle partition functions, $Z_N^{\mathrm{M}} = (Z_1^{\mathrm{M}})^N$. Show that

$$M = kT\left(\frac{\partial \ln Z_N^{\mathrm{M}}}{\partial B}\right)_T.$$

Note that from this the equation of state can be obtained.

(b) From the result of part (a) show that

$$Z_N^{\mathrm{M}} = e^{-F_{\mathrm{M}}/kT}.$$

(c) Show that the magnetic entropy is

$$S = kT\left(\frac{\partial \ln Z_N^{\mathrm{M}}}{\partial T}\right)_B + k\ln Z_N^{\mathrm{M}} = \frac{U}{T} + k\ln Z_N^{\mathrm{M}}.$$

11.18. By using Eqs. (11.65)–(11.66), find the total magnetic moment M as a function of T and B for a system with a general value of J in the following two limits: (a) $Jg\mu_{\mathrm{B}}B/kT \ll 1$ (low-field and high T limit), (b) $Jg\mu_{\mathrm{B}}B/kT \gg 1$ (high-field and low T limit). Show that for case (a) M obeys Curie's law. Find the Curie constant in this case.

11.19. Consider a solid or a liquid, whose molecules have permanent dipole moments. In an applied electric field $\boldsymbol{E}$, each dipole moment has an electrostatic energy $-\boldsymbol{p}\cdot\boldsymbol{E}$, where $\boldsymbol{p}$ is the dipole moment for each molecule. The dipole-dipole interaction is weak and can be neglected. The system is in thermal equilibrium at temperature T.

(a) Show that the one-particle partition function for the dipole moment is

$$Z_1 = 4\pi\frac{\sinh\beta pE}{\beta pE}, \quad \beta = \frac{1}{kT}.$$

(b) Show that the total dipole moment of N such dipoles is

$$\mathcal{P} = Np\,\mathcal{L}(\beta pE),$$

where $\mathcal{L}(x)$ is the Langevin function, defined as

$$\mathcal{L}(x) = \coth x - \frac{1}{x}.$$

(c) Show that in the high T and/or low E limit ($pE/kT \ll 1$), $\mathcal{P} = \alpha E$, where $\alpha = \alpha(T)$ is independent of E and is known as the polarizability. Find the expression of α.

11.20. Consider a paramagnetic salt of N identical magnetic ions in thermal equilibrium at temperature T.

(a) Show that for the case $J = 1/2$,

$$\overline{\mu_z^2} = \mu_{\rm B}^2.$$

(b) Show that

$$(\Delta^* M)^2 = N(\overline{\mu_z^2} - \overline{\mu_z}^2) = \frac{kT}{\mu_0}\chi,$$

where $\Delta^* M$ is the standard deviation of the total magnetic moment $M = N\overline{\mu_z}$, χ is the magnetic susceptibility, and μ_0 is the permeability of the vacuum. Note that this relation holds for the case with a general value of J.

11.21. Derive the expression for the entropy $S_{\rm M}(x)$ given in Eq. (11.83). From this expression show that $T < 0$ for $U > 0$ (i.e., $x > 0$), and $T > 0$ for $U < 0$ (i.e., $x < 0$).

11.22. Consider a one-dimensional Ising model with N lattice sites in thermal equilibrium at temperature T. The partition function Z is defined as (setting $\mu_{\rm B} = 1$)

$$Z = \sum_{\substack{s_1, s_2, \cdots s_N \\ = \pm 1}} \exp\left[\sum_{i=1}^{N}(\beta\epsilon\, s_i s_{i+1} + \beta B s_i)\right],$$

where B is the external magnetic field and $\beta = 1/kT$. There are only N lattice sites, we may set $s_{N+1} = 0$. Show that when $B = 0$, Z is given by

$$Z = 2^N(\cosh\beta\epsilon)^{N-1} \approx [2\cosh\beta\epsilon]^N.$$

[*Hint:* Consider the sum over s_1, $\sum_{s_1} e^{\beta\epsilon s_1 s_2} = e^{\beta\epsilon} + e^{-\beta\epsilon} = 2\cosh\beta\epsilon$, which holds for both $s_2 = 1$ and $s_2 = -1$. Then extend the consideration to the sum over s_2, etc.]

11.23. Consider the same problem as Problem 11.22 with $B \neq 0$ using the periodic boundary condition, $s_{i+N} = s_i$. The problem can be solved by the transfer matrix method.

(a) Show that Z can be expressed as

$$Z = \mathrm{Tr}\; t^N,$$

where Tr denotes trace, and $\boldsymbol{t}$ is the transfer matrix

$$\boldsymbol{t} = \begin{pmatrix} e^{\beta(\epsilon - B)} & e^{-\beta\epsilon} \\ e^{-\beta\epsilon} & e^{\beta(\epsilon + B)} \end{pmatrix}.$$

[*Hint*: In terms of quantum mechanics, Z may be written as $Z = \mathrm{Tr}\, e^{-\beta H} \equiv \sum_r < r|e^{-\beta H}|r> = \sum_r e^{-\beta E_r}$. The last form holds if all the microstate states $|r>$ are eigenstates of H, $H|r> = E_r|r>$. In this problem, H may be expressed as the sum of terms which are symmetrical with respect to the indices i and $i+1$ by replacing Bs_i by $B(s_i + s_{i+1})/2$.]

(b) By noting that the trace of a matrix is independent of the representation, show that Z can be expressed as

$$Z = \lambda_1^N + \lambda_2^N,$$

where λ_1 and λ_2 are the eigenvalues of the matrix $\boldsymbol{t}$.

(c) Determine these two eigenvalues and show that in the limit $N \to \infty$, for $\lambda_1 > \lambda_2$,

$$\frac{\ln Z}{N} = \ln \lambda_1 = \beta\epsilon + \ln\left\{\cosh(\beta B) + \left[\sinh^2(\beta B) + e^{-4\beta\epsilon}\right]^{1/2}\right\}.$$

(d) Evaluate the average magnetization and show that the magnetization vanishes as $B \to 0$. This implies that there is no *spontaneous magnetization.* [*Hint*: $\overline{s} \propto (\partial \ln Z/\partial B)_T$.]

11.24. Show that, in the mean field approximation of Eq. (11.87) with $B = 0$, the magnetic free energy for the ferromagnetic state ($\overline{s} \neq 0$) is smaller than that of the paramagnetic state ($\overline{s} = 0$) at $T < T_c$.

11.25. Starting from Eq. (11.88), derive Eqs. (11.92)–(11.95) for a ferromagnetic material.

Chapter 12

Ideal Quantum Gases

12.1 Introduction

In this chapter we will study the low temperature thermodynamic properties of ideal quantum gases, both Fermi and Bose. First, we will use the grand partition function to derive the Fermi-Dirac and Bose-Einstein distribution functions which cannot be derived by using the method of the one-particle partition function discussed in the previous chapter. Then we will study the properties of an ideal Fermi gas at absolute zero, introducing the *Fermi energy* and the concept of the *Fermi surface*, both of which are important in a Fermi system at low temperatures. The properties of an ideal Fermi gas at non-zero but low temperatures are studied by using a series expansion method. The results for the ideal Fermi gas are applied to the conduction electrons in a metal, which is one of the most important Fermi systems. For an ideal Bose gas, we discuss the existence and the properties of the Bose-Einstein condensation phenomenon, which occurs when the temperature of the system is below a critical temperature. We then study the thermodynamic properties of an ideal Bose gas below the critical temperature, including the phase transition of the Bose-Einstein condensation.

12.2 The Grand Partition Function Approach

In Sec. 11.2 we pointed out that the one-particle partition function approach cannot be applied to both the ideal Fermi and Bose gases. The solution to this problem is the application of the grand partition function. This approach will solve the difficulty due to the indistinguishability and the symmetry requirement of the quantum particles. In this approach the chemical potential μ will be properly introduced in the calculations. The grand partition function $\mathcal{Z}$ for a system with a constant volume V, in contact with a heat reservoir having the temperature T and the chemical potential μ was defined in Eq. (10.93):

$$\mathcal{Z}(T,V,\mu) = \sum_{N=0}^{\infty} \sum_{r} e^{-\beta(E_{N,r}-\mu N)}, \quad \beta = \frac{1}{kT}, \tag{12.1}$$

where $E_{N,r}$ is the energy of the rth state for the system of N particles. Now we consider one of the microstates (N,r) under the summation sign in Eq. (12.1). The state has N particles with energy $E_{N,r}$. We assume that the particles are monatomic and that the energy of each particle contains only the translational kinetic energy $\varepsilon_{\boldsymbol{p}}$, where $\boldsymbol{p}$ is the momentum of the particle. At the moment we do not consider the spin degeneracy, so the energy state $\varepsilon_{\boldsymbol{p}}$ can be considered as nondegenerate. If we denote the number of particles in the state $\boldsymbol{p}$ by $n_{\boldsymbol{p}}$, then the microstate (N,r) is represented by a set of numbers $\{n_{\boldsymbol{p}}^{(r)}\}$ for all $\boldsymbol{p}$, which satisfies the following two constraints:

$$\sum_{\boldsymbol{p}} n_{\boldsymbol{p}}^{(r)} = N, \tag{12.2}$$

$$\sum_{\boldsymbol{p}} n_{\boldsymbol{p}}^{(r)} \varepsilon_{\boldsymbol{p}} = E_{N,r}. \tag{12.3}$$

A set of numbers $\{n_{\boldsymbol{p}}^{(r)}\}$ which satisfies the above two constraints, is called *a distribution.* Note that different r's should have different distributions. Substituting Eq. (12.3) into Eq. (12.1), and using Eq. (12.2) for the condition on $\{n_{\boldsymbol{p}}^{(r)}\}$, the grand partition function can be written as

$$\mathcal{Z}(T,V,\mu) = \sum_{N=0}^{\infty} \left(\sum_{\substack{r \\ \sum n_{\boldsymbol{p}}^{(r)}=N}} e^{\beta\mu N} e^{-\beta \sum_{\boldsymbol{p}} n_{\boldsymbol{p}}^{(r)} \varepsilon_{\boldsymbol{p}}} \right).$$

We note that the sum over r means summing over all possible microstates, which is equivalent to summing over all possible distributions

$\{n_p\}$ satisfying the constraint Eq. (12.2). Hence we may drop the superscript (r) and rewrite the above equation as

$$
\begin{aligned}
\mathcal{Z}(T, V, \mu) &= \sum_{N=0}^{\infty} \left(\sum_{\substack{\{n_p\} \\ \sum n_p = N}} e^{\beta \mu N} \, e^{-\beta \sum_p n_p \varepsilon_p} \right) \\
&= \sum_{N=0}^{\infty} \left[\sum_{\substack{\{n_p\} \\ \sum n_p = N}} \prod_p (e^{\beta\mu} \, e^{-\beta \varepsilon_p})^{n_p} \right] \\
&= \sum_{n_{p_1}=0}^{n_{\max}} \sum_{n_{p_2}=0}^{n_{\max}} \cdots (e^{\beta\mu} \, e^{-\beta\varepsilon_{p_1}})^{n_{p_1}} \, (e^{\beta\mu} \, e^{-\beta\varepsilon_{p_2}})^{n_{p_2}} \cdots \\
&= \prod_p \left[\sum_{n_p=0}^{n_{\max}} (e^{\beta\mu} \, e^{-\beta\varepsilon_p})^{n_p} \right] ,
\end{aligned}
\tag{12.4}
$$

where $n_{\max}$ in the third equality is the maximum number of particles that a quantum state p can accommodate. The second equality is straightforward. One way to prove the third equality is to check that every term on the left hand side of the equality will also be found on the right hand side, and vice versa. We note that there is no correlation between all the n_{p_i}'s for different p_i's. Therefore the fourth equality follows immediately.

For Fermi particles, $n_{\max} = 1$ due to the restriction of the Pauli principle, hence the grand partition function $\mathcal{Z}_{\text{Fermi}}$ is

$$
\mathcal{Z}_{\text{Fermi}} = \prod_p \left[1 + e^{-\beta(\varepsilon_p - \mu)} \right] . \tag{12.5}
$$

For Bose particles, there is no restriction on the number of particles in each state, $n_{\max} = \infty$, so the grand partition function $\mathcal{Z}_{\text{Bose}}$ is

$$
\mathcal{Z}_{\text{Bose}} = \prod_p \frac{1}{1 - e^{-\beta(\varepsilon_p - \mu)}} . \tag{12.6}
$$

In the grand canonical ensemble the average number (or the most probable number) of particles in the system $\overline{N}$ is given by Eq. (10.95):

$$\overline{N} = \frac{1}{\mathcal{Z}} kT \left(\frac{\partial \mathcal{Z}}{\partial \mu}\right)_{T,V}$$

$$= \begin{cases} \displaystyle\sum_{p} \frac{1}{e^{(\varepsilon_p - \mu)/kT} + 1} & \text{for fermions,} \\ \displaystyle\sum_{p} \frac{1}{e^{(\varepsilon_p - \mu)/kT} - 1} & \text{for bosons.} \end{cases}$$

We may write $\overline{N} = \sum_p \overline{n}(p)$, where $\overline{n}(p)$ is the average number of particles in the quantum state p, which is known as the *distribution*. Thus from the above equation we have the **Fermi distribution**

$$\overline{n}_{\text{FD}}(p) = \frac{1}{e^{(\varepsilon_p - \mu)/kT} + 1}, \quad \text{fermions,} \tag{12.7}$$

which is exactly the same as what we had obtained in Eq. (10.50), except that the state is now denoted by p instead of j. For Bose particles, the **Bose distribution** is

$$\overline{n}_{\text{BE}}(p) = \frac{1}{e^{(\varepsilon_p - \mu)/kT} - 1}, \quad \text{bosons,} \tag{12.8}$$

which is exactly the same as what we had obtained in Eq. (10.55).

12.3 Ideal Fermi Gases at $T = 0$

We consider an ideal Fermi gas of N particles at temperature T in a volume V. For Fermi statistics we usually write the Fermi distribution (12.7) in the form

$$f(\varepsilon) \equiv \overline{n}_{\text{FD}}(p) = \frac{1}{e^{(\varepsilon - \mu)/kT} + 1}, \tag{12.9}$$

where $f(\varepsilon)$ is called the **Fermi function**. Note that we have replaced $\overline{n}_{\text{FD}}(p)$ by $f(\varepsilon)$ (the state label p is replaced by the energy label ε). Apparently the Fermi function $f(\varepsilon)$ has the meaning of the average number of particles in the state ε (assumed to be nondegenerate) at temperature T. Due to the restriction of the Pauli principle, the condition $0 \leq f(\varepsilon) \leq 1$ is always satisfied. Since the total number of particles in the system is N (dropping the overline), the Fermi function $f(\varepsilon)$ must satisfy the following relation:

$$N = \sum_{p} \overline{n}_{\text{FD}}(p) = \int_0^\infty f(\varepsilon)\, g(\varepsilon)\, d\varepsilon = \int_0^\infty \frac{g(\varepsilon)\, d\varepsilon}{e^{(\varepsilon-\mu)/kT}+1},$$

$$= g_s\, 2\pi V \left(\frac{2m}{h^2}\right)^{3/2} \int_0^\infty \frac{\varepsilon^{1/2}\, d\varepsilon}{e^{(\varepsilon-\mu)/kT}+1}, \tag{12.10}$$

where we have used the density of states $g(\varepsilon)$ given in Eq. (10.13), which applies to a three dimensional system. Apparently, for a fixed N/V, the chemical potential μ must be a function of T in order that Eq. (12.10) holds for all T. Therefore $\mu = \mu(T, N/V)$. One has to solve this relation before the thermodynamic properties can be obtained. However at $T = 0$ the mathematics involved is rather simple, we therefore study the case of $T = 0$ first.

At $T = 0$, the chemical potential is called the **Fermi energy** ε_{F}, i.e., $\mu(T = 0) = \varepsilon_{\text{F}}$. The physical significance of the Fermi energy is easily understood from the Fermi distribution. When $T \to 0$ the Fermi function $f(\varepsilon)$ is almost a constant, except for those ε which are close to ε_{F} where $f(\varepsilon)$ changes abruptly. This is because for ε close to ε_{F}, $e^{(\varepsilon-\varepsilon_{\text{F}})/kT}$ may be either 0 or ∞, depending on whether ε is less or greater than ε_{F}. Therefore at $T = 0$ we have

$$f(\varepsilon) = \begin{cases} 1 & \text{if } \varepsilon \le \varepsilon_{\text{F}}, \\ 0 & \text{if } \varepsilon > \varepsilon_{\text{F}}. \end{cases} \tag{12.11}$$

At $T = 0$ the Fermi function $f(\varepsilon)$ is therefore a step function, as shown in Fig. 12.1(a). This means that all states with $\varepsilon \le \varepsilon_{\text{F}}$ are occupied, and all states with $\varepsilon > \varepsilon_{\text{F}}$ are empty. Thus ε_{F} is the highest energy state which is occupied. Consequently the most important feature for a Fermi system at $T = 0$ is that there is a sharp division between the occupied and unoccupied states at the Fermi energy $\varepsilon = \varepsilon_{\text{F}}$. When T increases, the sharp edge of the step will be rounded off a little bit. This means that some of the particles in the states with $\varepsilon < \varepsilon_{\text{F}}$ will be excited to the states with $\varepsilon > \varepsilon_{\text{F}}$, as shown in Fig. 12.1(b).

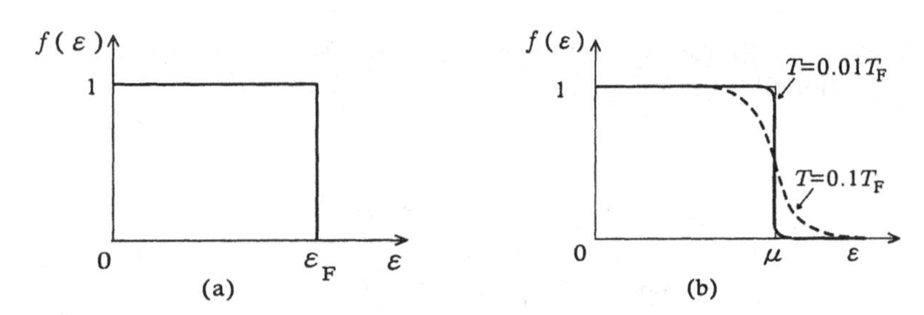

Figure 12.1: The Fermi function $f(\varepsilon)$ vs. ε for three different T: (a) $T = 0$, (b) $T = 0.01\,T_{\mathrm{F}}$, and $T = 0.1\,T_{\mathrm{F}}$, where $\mu \approx \varepsilon_{\mathrm{F}}$ for $T/T_{\mathrm{F}} < 0.1$.

If we substitute the Fermi function given in Eq. (12.11) into Eq. (12.10) we obtain

$$\begin{aligned} N &= g_s 2\pi V \left(\frac{2m}{h^2}\right)^{3/2} \int_0^{\varepsilon_{\mathrm{F}}} \varepsilon^{1/2} d\varepsilon \\ &= g_s \frac{4\pi V}{3} \left(\frac{2m}{h^2}\right)^{3/2} \varepsilon_{\mathrm{F}}^{3/2}. \end{aligned} \tag{12.12}$$

By solving the above equation for ε_{F} we get

$$\varepsilon_{\mathrm{F}} = \frac{h^2}{2m} \left(\frac{3N}{g_s 4\pi V}\right)^{2/3}. \tag{12.13}$$

We see that the Fermi energy is proportional to two thirds power of the number density, i.e., $\varepsilon_{\mathrm{F}} \propto (N/V)^{2/3}$. The ground state energy of the system is

$$\begin{aligned} U_0 &= \sum_{\boldsymbol{p}} \overline{n}_{\mathrm{FD}}(\boldsymbol{p}) \varepsilon_{\boldsymbol{p}} = \int_0^{\varepsilon_{\mathrm{F}}} \varepsilon\, g(\varepsilon)\, d\varepsilon = g_s 2\pi V \left(\frac{2m}{h^2}\right)^{3/2} \int_0^{\varepsilon_{\mathrm{F}}} \varepsilon^{3/2} d\varepsilon \\ &= g_s \frac{4\pi V}{5} \left(\frac{2m}{h^2}\right)^{3/2} \varepsilon_{\mathrm{F}}^{5/2} = \frac{3}{5} N \varepsilon_{\mathrm{F}}. \end{aligned} \tag{12.14}$$

Thus at $T = 0$ the average energy per particle $\overline{\varepsilon} = U_0/N = \frac{3}{5}\varepsilon_{\mathrm{F}}$, which is greater than $\frac{1}{2}\varepsilon_{\mathrm{F}}$. This implies that the higher energy states have a larger density of states than the lower energy states, which can be seen from the formula of the density of states, $g(\varepsilon) \propto \varepsilon^{1/2}$.

In a Fermi system there is a very important quantity (or concept) called the Fermi surface. At $T = 0$ the Fermi particles will fill up all the states between $\varepsilon = 0$ and ε_{F}, and all states with $\varepsilon > \varepsilon_{\mathrm{F}}$ are unoccupied.

The surface, *in momentum space*, which separates the occupied and unoccupied states is called the **Fermi surface.** Since $\varepsilon = \mathbf{p}^2/2m$, we can define a *Fermi momentum* p_{F}:

$$p_{\mathrm{F}}^2 = p_x^2 + p_y^2 + p_z^2 = 2m\varepsilon_{\mathrm{F}}. \tag{12.15}$$

The Fermi surface for an ideal non-interacting Fermi gas is therefore a *spherical surface* in momentum space, with radius p_{F}. For a weakly interacting Fermi gas, such as the conduction electrons in a metal, a sharp Fermi surface still exists at $T = 0$, however its shape may be non-spherical.[1] The physical significance of the Fermi surface is that when a *weak* external field or a *small* thermal energy is applied to the system (at $T = 0$), only particles with their states very close to the Fermi surface can be excited, and the states of the excited particles will also be very close to the Fermi surface. Therefore thermodynamic and transport properties of a Fermi system will depend on the structure of the Fermi surface. The shape and structure of the Fermi surface of the conduction electrons is therefore an important property of a metal.

12.4 Degenerate Fermi Gases

In Fermi statistics, one frequently encounters integrals of the following form

$$I = \int_0^\infty \phi(\varepsilon) f(\varepsilon)\, d\varepsilon = \int_0^\infty \frac{\phi(\varepsilon)}{e^{(\varepsilon-\mu)/kT}+1}\, d\varepsilon, \tag{12.16}$$

where $\phi(\varepsilon)$ is some differentiable function of ε and $f(\varepsilon)$ is the Fermi function given in Eq. (12.9). At $T = 0$, $f(\varepsilon)$ is a step function, the integral is usually an easy one, as we discussed above. For $T \neq 0$, $f(\varepsilon)$ is no longer a step function, the integral is in general not easy to calculate. However at low temperatures the integral can be evaluated by a series expansion method. It is convenient to define a characteristic temperature T_{F}, known as the **Fermi temperature,** by the relation

$$T_{\mathrm{F}} \equiv \frac{\varepsilon_{\mathrm{F}}}{k} = \frac{h^2}{2mk}\left(\frac{3N}{g_s 4\pi V}\right)^{2/3}. \tag{12.17}$$

Thus T_{F} is the Fermi energy ε_{F} expressed in terms of temperature.

[1] The general definition of the Fermi surface is the surface spanned by the momentum (p_x, p_y, p_z) which satisfies $\varepsilon = \varepsilon(p_x, p_y, p_z) = \varepsilon_{\mathrm{F}} = \text{constant}$.

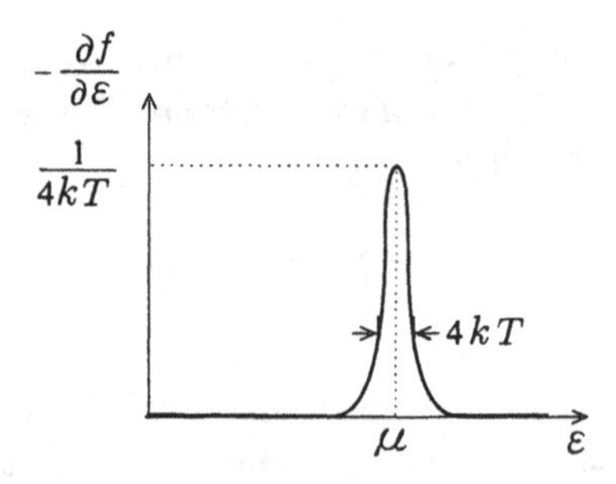

Figure 12.2: The plot of $(-\partial f/\partial \varepsilon)$ vs. ε for a degenerate Fermi gas.

Consider the integral I in Eq. (12.16), and do an integration by parts to obtain

$$I = [G(\varepsilon)f(\varepsilon)]_0^\infty - \int_0^\infty G(\varepsilon)\frac{\partial f(\varepsilon)}{\partial \varepsilon}\,d\varepsilon, \tag{12.18}$$

where

$$G(\varepsilon) = \int_0^\varepsilon \phi(\varepsilon)\,d\varepsilon. \tag{12.19}$$

The first term on the right hand side of Eq. (12.18) vanishes because $G(0) = 0$ and $f(\infty)$ approaches 0 exponentially. The integral in the second term can be evaluated by a series expansion if $T \ll T_F$. In this temperature regime the system is called a *highly degenerate Fermi gas* or simply a **degenerate Fermi gas.** We note that at $T = 0$ the Fermi function $f(\varepsilon)$ is a step function with the edge at $\varepsilon = \mu = \varepsilon_F$. Therefore $(-\partial f/\partial \varepsilon)$ is a delta function peaked at $\varepsilon = \varepsilon_F$ at $T = 0$. When $T \neq 0$ but $T \ll T_F$, the height of the delta-function peak is reduced to $1/4kT$, and the width is broadened to a few kT. Also the position of the peak is shifted by a small amount, i.e., $\mu \neq \varepsilon_F$. Hence $(-\partial f/\partial \varepsilon)$ is non-zero only for ε very close to μ (of order kT), where there is a sharp peak, as shown in Fig. 12.2. We may then expand $G(\varepsilon)$ around μ:

$$G(\varepsilon) = G(\mu) + G'(\mu)(\varepsilon - \mu) + \frac{1}{2}G''(\mu)(\varepsilon - \mu)^2 + \cdots. \tag{12.20}$$

By substituting the above result into Eq. (12.18), we obtain

$$\begin{aligned} I &= G(\mu)\int_0^\infty \left(-\frac{\partial f}{\partial \varepsilon}\right) d\varepsilon \\ &+\frac{1}{2}G''(\mu)\int_0^\infty (\varepsilon-\mu)^2\left(-\frac{\partial f}{\partial \varepsilon}\right) d\varepsilon + \cdots. \end{aligned} \tag{12.21}$$

This is known as the **Sommerfeld**[2] **expansion**. In obtaining the above equation we have used the fact that $(-\partial f/\partial\varepsilon)$ is non-zero only when ε is close to μ, the lower limit of the integral may therefore be replaced by $-\infty$ with negligible error, thus the integral vanishes for n =odd,

$$\int_0^\infty (\varepsilon-\mu)^n \left(-\frac{\partial f}{\partial\varepsilon}\right) d\varepsilon$$
$$\approx \int_{-\infty}^\infty (\varepsilon-\mu)^n \left(-\frac{\partial f}{\partial\varepsilon}\right) d\varepsilon = 0 \text{ for } n = \text{odd}.$$

The reason is that $(-\partial f/\partial\varepsilon)$ is an even function[3] with respect to $\varepsilon = \mu$, while $(\varepsilon-\mu)^n$ is an odd function if n is odd. Since

$$\int_0^\infty \left(-\frac{\partial f}{\partial\varepsilon}\right) d\varepsilon = f(0) - f(\infty) = 1,$$

the lowest order approximation is therefore

$$I \approx G(\mu) = \int_0^\mu \phi(\varepsilon) d\varepsilon, \tag{12.22}$$

which neglects all terms involving T. The above equation is an exact equation at $T = 0$ with $\mu = \varepsilon_F$.

Now consider the next lowest order non-zero term, $n = 2$,

$$\frac{1}{2}G''(\mu) \int_0^\infty (\varepsilon-\mu)^2 \left(-\frac{\partial f}{\partial\varepsilon}\right) d\varepsilon$$
$$= \frac{1}{2}G''(\mu)(kT)^2 \int_{-\infty}^\infty \frac{x^2\, e^x}{(e^x+1)^2} dx \quad \left(x = \frac{\varepsilon-\mu}{kT}\right)$$
$$= \frac{\pi^2}{6}(kT)^2 \left[\frac{d\phi(\varepsilon)}{d\varepsilon}\right]_{\varepsilon=\mu}, \tag{12.23}$$

where we have used the fact that the integral on the right-hand side of the second equality can be evaluated exactly to give the value $\pi^2/3$ (see Appendix E). Therefore, to the second order in T, we have

$$\int_0^\infty \phi(\varepsilon) f(\varepsilon)\, d\varepsilon \approx \int_0^\mu \phi(\varepsilon)\, d\varepsilon$$
$$+ \frac{\pi^2}{6}(kT)^2 \left[\frac{d\phi(\varepsilon)}{d\varepsilon}\right]_{\varepsilon=\mu} + \cdots. \tag{12.24}$$

[2] Arnold J. W. Sommerfeld, German physicist (1868–1951).

[3] Let $x = (\varepsilon-\mu)/kT$, then $-df/dx = 1/[(e^x+1)(e^{-x}+1)]$, which is even in x.

For many applications it is sufficient to expand to the T^2 term, so we will not go to the higher order terms here. We will apply this relation to obtain some important physical quantities by proper choices of the function $\phi(\varepsilon)$.

The most important T-dependent quantity in a Fermi gas is the chemical potential μ as a function of T, $\mu = \mu(T)$. In order to obtain this relation we choose $\phi(\varepsilon)$ to be the density of states $g(\varepsilon)$ given in Eq. (10.13). The number of particles N is independent of the temperature, therefore to the order of T^2 we have

$$\begin{aligned} N &= \int_0^{\varepsilon_F} g(\varepsilon)\, d\varepsilon = \int_0^\infty g(\varepsilon) f(\varepsilon)\, d\varepsilon \\ &= \int_0^\mu g(\varepsilon)\, d\varepsilon + \frac{\pi^2}{6}(kT)^2 \left[\frac{dg(\varepsilon)}{d\varepsilon}\right]_{\varepsilon=\mu} + \cdots . \end{aligned} \tag{12.25}$$

By noting that: (1) $g(\varepsilon) = c\varepsilon^{1/2}$ where c is a constant; and (2) every integral in the above equation contains $g(\varepsilon)$, the constant c will be cancelled; we can integrate the two integrals $\int g(\varepsilon)\, d\varepsilon$, and obtain

$$\frac{2}{3}\varepsilon_F^{3/2} = \frac{2}{3}\mu^{3/2} + \frac{\pi^2}{12}(kT)^2 \mu^{-1/2}. \tag{12.26}$$

Since $(kT/\mu)^2 \approx (kT/\varepsilon_F)^2 = (T/T_F)^2$, if T^4 and higher order terms are neglected, then it is straightforward to obtain

$$\mu = \varepsilon_F \left[1 - \frac{\pi^2}{12}\left(\frac{T}{T_F}\right)^2\right], \quad T \ll T_F. \tag{12.27}$$

We see that μ decreases as T increases, and the correction in μ due to non-zero T is less than 1% if $T/T_F < 0.1$. The relation given in Eq. (12.27) is essential for obtaining other thermodynamic properties of a Fermi system.

Next we consider the internal energy U of the system. In this case we choose $\phi(\varepsilon) = \varepsilon g(\varepsilon) = c\varepsilon^{3/2}$, then from Eq. (12.24) we obtain

$$U = \frac{2c}{5}\mu^{5/2} + \frac{c\pi^2}{4}(kT)^2 \mu^{1/2}.$$

By substituting μ from Eq. (12.27) we get

$$U = \frac{2c}{5}\varepsilon_{\mathrm{F}}^{5/2}\left[1+\frac{5\pi^2}{12}\left(\frac{T}{T_{\mathrm{F}}}\right)^2\right]$$
$$= \frac{3}{5}N\varepsilon_{\mathrm{F}}\left[1+\frac{5\pi^2}{12}\left(\frac{T}{T_{\mathrm{F}}}\right)^2\right], \quad T \ll T_{\mathrm{F}}, \qquad (12.28)$$

where the last equality is obtained because $U = \frac{3}{5}N\varepsilon_{\mathrm{F}}$ at $T = 0$. Alternately one may show explicitly that $N = \frac{2}{3}c\varepsilon_{\mathrm{F}}^{3/2}$ to obtain the above result. The constant volume heat capacity C_V is

$$C_V = \left(\frac{\partial U}{\partial T}\right)_V = \frac{\pi^2}{2}Nk\left(\frac{T}{T_{\mathrm{F}}}\right), \quad T \ll T_{\mathrm{F}}. \qquad (12.29)$$

Therefore C_V is proportional to T for a Fermi gas at low T, which satisfies the requirement of the third law, $C_V \to 0$ as $T \to 0$.

The entropy S of the system may be calculated by the following equation,

$$S = \int_0^T \frac{C_V}{T}\,dT = \frac{\pi^2}{2}Nk\left(\frac{T}{T_{\mathrm{F}}}\right), \quad T \ll T_{\mathrm{F}}. \qquad (12.30)$$

The Helmholtz free energy F may be obtained from the relation $F = U - TS$; then from the relation $P = -(\partial F/\partial V)_T$ we obtain the pressure P of the system:

$$P = \frac{2}{5}\frac{NkT_{\mathrm{F}}}{V}\left[1+\frac{5\pi^2}{12}\left(\frac{T}{T_{\mathrm{F}}}\right)^2\right], \quad T \ll T_{\mathrm{F}}. \qquad (12.31)$$

In obtaining the above equation we used the fact that the Fermi temperature T_{F} is a function of V. From the above equation and Eq. (12.28) we obtain the relation between PV and U:

$$PV = \frac{2}{3}U, \quad T \ll T_{\mathrm{F}}, \qquad (12.32)$$

for an ideal Fermi gas at very low T. Later we will show that this relation actually holds for all temperatures, and also it holds for an ideal Bose gas. Of course this implies that a classical ideal gas should also obey the same relation. Indeed, this is the case. For a classical ideal gas $PV = NkT$, and $U = 3NkT/2$ for a monatomic gas. Therefore $PV = 3U/2$. The proof that $PV = 3U/2$ holds for all T and for all kinds of ideal gases will be given in Sec. 12.10.

12.5 Application to Conduction Electrons

One of the most important Fermi systems is that of the conduction electrons in a metal. Even though in an open space there is a strong *Coulomb*[4] *interaction* between the electrons, the conduction electrons in a metal may be approximately considered as a **free electron gas**. The reason is that, because of the presence of the periodic positive ionic potentials and the screening of other electrons, the conduction electrons are weakly interacting. In the lowest order approximation we may consider the conduction electrons as an ideal Fermi gas. By using Eq. (12.13) we obtain the Fermi energy for an electron gas

$$\varepsilon_{\mathrm{F}} = \frac{h^2}{2m}\left(\frac{3N}{8\pi V}\right)^{2/3}, \tag{12.33}$$

where we have used $g_s = 2$ for electrons.

In most metals, the number density N/V of the conduction electrons is around 10^{28} m^{-3} to 10^{29} m^{-3}. From Eq. (12.33) the value of ε_{F} can be calculated to be a few eV, and the corresponding T_{F} is a few 10^4 K. Take copper as an example, $N/V = 8.45 \times 10^{28}$ m^{-3}, thus

$$\begin{aligned}\varepsilon_{\mathrm{F}} &= \frac{(6.63\times 10^{-34})^2}{2(9.11\times 10^{-31})}\left(\frac{3\times 8.45\times 10^{28}}{8\times 3.14}\right)^{2/3} \\ &\approx 11.2\times 10^{-19}\,\mathrm{J} \approx 7.00\,\mathrm{eV}\end{aligned}$$

$$T_{\mathrm{F}} \approx \frac{11.2\times 10^{-19}\,\mathrm{J}}{1.38\times 10^{-23}\,\mathrm{J\,K^{-1}}} \approx 81200\,\mathrm{K}.$$

This is a typical value of T_{F} for metals. Thus the conduction electrons form a *degenerate Fermi gas* at room temperature, because the condition $T/T_{\mathrm{F}} \approx 0.01 \ll 1$ is satisfied for $T = 300$ K. It is then sufficient to keep only the T^2 term in discussing thermodynamic properties of the conduction electrons. Therefore the formula for the chemical potential given in Eq. (12.27) and the constant volume heat capacity given in Eq. (12.29) can be applied to conduction electrons at room temperature or below.

We note that the constant volume heat capacity C_V given in Eq. (12.29) is much smaller than the classical result $\frac{3}{2}Nk$ even at room temperature. Apparently a large portion of the conduction electrons are not thermally

[4]Charles A. de Coulomb, French engineer (1736–1806).

excited. We can define an *effective number* of electrons N_{eff}, such that Eq. (12.29) can be written as

$$C_V = \frac{3}{2} N_{\text{eff}} k, \quad N_{\text{eff}} = \frac{\pi^2}{3} N \left(\frac{T}{T_{\text{F}}} \right). \tag{12.34}$$

The physical significance of N_{eff} is easily understood: Although there are in total N electrons, however *effectively* only N_{eff} electrons are thermally excited and contribute to C_V. The remaining electrons do not absorb thermal energy, so they make no contribution to C_V. For the conduction electrons in a metal at room temperature, typically $T/T_{\text{F}} \approx 0.01$, hence

$$\frac{N_{\text{eff}}}{N} \approx \frac{3.14^2}{3} \times 0.01 = 0.0329 = 3.29\%.$$

This means that at room temperature, typically only about 3% of the conduction electrons in a metal can absorb thermal energy, all other electrons are thermally ineffective. This can be understood from the Fermi function shown in Fig. 12.1(b). For $T/T_{\text{F}} = 0.01$, the distribution function is almost the same as at $T = 0$ for all energies except near the Fermi level $\varepsilon = \varepsilon_{\text{F}}$, where a small change occurs. Thus only electrons near the Fermi surface can absorb thermal energy and contribute to C_V. Electrons lying much below the Fermi surface remain in their original states and are thermally ineffective.

By using the effective electron concept which we just discussed, we may derive the T-dependence of C_V without using the rigorous derivation (to order of T^2) of the internal energy U given in Eq. (12.28). If only electrons with energy ε in the range $\varepsilon_{\text{F}} - \Delta\varepsilon \leq \varepsilon \leq \varepsilon_{\text{F}}$ are thermally excited, then *the number of effective electrons* N'_{eff} may approximately be written as $N'_{\text{eff}} \approx g(\varepsilon_{\text{F}})\Delta\varepsilon$ ($\Delta\varepsilon \ll \varepsilon_{\text{F}}$), where $g(\varepsilon_{\text{F}})$ is the density of states at the Fermi energy ε_{F}. From Eqs. (10.13) and (12.12) we have $g(\varepsilon_{\text{F}}) = 3N/2\varepsilon_{\text{F}}$. If we take $\Delta\varepsilon \approx kT$ and use the above equation, then $N'_{\text{eff}} \approx g(\varepsilon_{\text{F}})\, kT = \frac{3}{2}N(T/T_{\text{F}})$. The *approximated* internal energy U' and the heat capacity C'_V of the electron gas are then

$$U' = U_0 + N'_{\text{eff}} \cdot \frac{3}{2} kT \approx U_0 + \frac{9}{4} Nk \frac{T^2}{T_{\text{F}}},$$

$$C'_V \approx \frac{9}{2} Nk \frac{T}{T_{\text{F}}},$$

which gives the result that $C_V \propto T$. Moreover $C_V/C'_V = \pi^2/9 = 1.096$, the approximate result is almost equal to the rigorous result.

In metals, both the conduction electrons and the crystal vibrations contribute to the heat capacity. Hence at the low temperature limit we have

$$C_V^{\rm metal} = C_V^{\rm el} + C_V^{\rm ion} = aT + bT^3, \quad T \ll \theta_{\rm D}, T_{\rm F}, \tag{12.35}$$

where aT and bT^3 are the electron heat capacity and the lattice heat capacity, respectively. The theoretical expressions for the coefficients are $a = N^{\rm el}(\pi^2/2T_{\rm F})k$ and $b = N^{\rm ion}(12\pi^4/5\theta_{\rm D}^3)k$, where $N^{\rm el}$ and $N^{\rm ion}$ are, respectively, the number of the conduction electrons and the number of the ions. However in practice, the coefficients are determined experimentally. Measurement of $C_V^{\rm metal}$ for several different temperatures, plotted on a graph of $C_V^{\rm metal}/T$ vs. T^2 should give a straight line. The vertical intercept is a and the slope is b, because

$$\frac{C_V^{\rm metal}}{T} = a + bT^2. \tag{12.36}$$

From this we may determine $T_{\rm F}$ and $\theta_{\rm D}$ for a metal. It should be noted that the electron contribution aT is usually much less than the ion contribution bT^3, except at very low temperatures. For example, for copper $C_V^{\rm el}/C_V^{\rm ion}$ is negligibly small except for $T < 10$ K.

Finally, from Eq. (12.31) we see that at $T = 0$ the conduction electrons of copper ($N/V = 8.45 \times 10^{28}$ m^{-3}) have a pressure $P = 3.79 \times 10^{11}$ Pa on the surface of the metal; this is a very large pressure which is due to the Pauli exclusion principle.

12.6 Ideal Bose Gases Near $T = 0$

Now we consider an ideal Bose gas of N particles at temperature T in a volume V. The distribution function for bosons is given by Eq. (12.8)

$$f_{\rm BE}(\varepsilon) \equiv \overline{n}_{\rm BE}(\boldsymbol{p}) = \frac{1}{e^{(\varepsilon-\mu)/kT} - 1}, \tag{12.37}$$

where we have changed the state label from $\boldsymbol{p}$ to ε. The function $f_{\rm BE}(\varepsilon)$ is the **Bose distribution** (or *Bose-Einstein* distribution), which is the average number of particles in the energy state ε (assumed nondegenerate) at temperature T. As in the Fermi case, the chemical potential per particle μ is a function of T and N/V. Bosons do not obey the Pauli principle, therefore the only restriction of $f_{\rm BE}(\varepsilon)$ is that it must be non-negative. If we take $\varepsilon \geq 0$ then μ must be 0 or negative, otherwise $f_{\mathrm{BE}}(0) = 1/(e^{-\mu/kT} - 1)$ will be negative. Now we will study how μ varies with T when N/V is kept constant. The total number of particles N is given by (taking $s = 0$, $g_s = 2s + 1 = 1$)

$$\begin{aligned} N &= \int_0^\infty f_{\mathrm{BE}}(\varepsilon)\, g(\varepsilon)\, d\varepsilon = 2\pi V \left(\frac{2m}{h^2}\right)^{3/2} \int_0^\infty \frac{\varepsilon^{1/2}\, d\varepsilon}{e^{(\varepsilon-\mu)/kT} - 1}, \\ &= 2\pi V \left(\frac{2m}{h^2}\right)^{3/2} \int_0^\infty \frac{\varepsilon^{1/2}\, d\varepsilon}{e^{(\varepsilon+|\mu|)/kT} - 1}, \end{aligned} \tag{12.38}$$

where the condition $\mu \leq 0$ is used and $-\mu = |\mu|$.

Now if N/V is kept constant, then as T decreases $|\mu|$ must also decrease (i.e., μ increases) in order that Eq. (12.38) holds. Suppose that when the temperature is decreased to a certain temperature $T = T_c$ the chemical potential is 0, i.e., $|\mu| = \mu = 0$ at[5] $T = T_c$. When this happens Eq. (12.38) becomes

$$N = 2\pi V \left(\frac{2m}{h^2}\right)^{3/2} \int_0^\infty \frac{\varepsilon^{1/2}\, d\varepsilon}{e^{\varepsilon/kT_c} - 1} \tag{12.39}$$

$$= 2\pi V \left(\frac{2mkT_c}{h^2}\right)^{3/2} \int_0^\infty \frac{x^{1/2}\, dx}{e^x - 1}, \tag{12.40}$$

where in the last equality we let $\varepsilon/kT_c = x$. We note that the integral in the last equality is a definite integral, which can be evaluated to give the value $2.612\sqrt{\pi}/2$ (see Appendix E). Substituting this value into the above equation we can solve for T_c in terms of N/V to obtain

$$T_c = (2.612)^{-2/3} \frac{h^2}{2\pi mk} \left(\frac{N}{V}\right)^{2/3} = 0.527 \frac{h^2}{2\pi mk} \left(\frac{N}{V}\right)^{2/3}. \tag{12.41}$$

Thus T_c is proportional to $(N/V)^{2/3}$. The temperature T_c is a very special temperature whose physical significance can be understood from the following discussion.

We study what happens when T is decreased further, to values below T_c. Clearly we still must require $\mu = 0$. However we find that there is a difficulty such that Eq. (12.38) can not hold for $T < T_c$. We see that the left hand side of Eq. (12.38), which is N, a constant, is greater than the right hand side, which is a function of T ($T < T_c$), due to the relation

[5]More precisely, $\mu = \varepsilon_0$ (ε_0 =smallest value of ε) at $T = T_c$.

given in Eq. (12.39). For $T > T_c$ we can adjust μ such that the equality (12.38) holds, i.e., the right hand side is independent of T. However for $T < T_c$ we must set $\mu = 0$, thus we have nothing left to adjust so that the equality cannot hold, because the right hand side depends on T. It seems that there is something wrong in Eq. (12.38) when $T < T_c$. Indeed, we find that because the density of states $g(\varepsilon)$ is 0 for $\varepsilon = 0$, the particles in the state $\varepsilon = 0$ are not counted in the integral of (12.38). Therefore for $T < T_c$, Eq. (12.38) (with $\mu = 0$) should be replaced by

$$N = N_0 + 2\pi V \left(\frac{2m}{h^2}\right)^{3/2} \int_0^\infty \frac{\varepsilon^{1/2}\, d\varepsilon}{e^{\varepsilon/kT - 1}}, \quad T < T_c, \tag{12.42}$$

where N_0 is the number of particles in the ground state ($\varepsilon = 0$). We note that for $T \geq T_c$ there is no inconsistency in Eq. (12.38), because at these temperatures $N_0/N \approx 0$. However for $T < T_c$, N_0/N is no longer negligible; thus N_0 should be included in the counting of N. Hence T_c is a *critical temperature*, above which the number of particles occupying the ground state ($\varepsilon = 0$) can be approximated as 0, yet below T_c there are a large number (a finite fraction) of particles occupying the ground state. From Eqs. (12.39) and (12.42) we have

$$\frac{N_0}{N} = 1 - \left(\frac{T}{T_c}\right)^{3/2}, \quad T \leq T_c. \tag{12.43}$$

Therefore for $T \geq T_c$, $N_0/N = 0$, and for $T < T_c$, $N_0/N \neq 0$, as shown in Fig. 12.3. *The phenomenon that there is a finite fraction of particles occupying the ground state is known as the* **Bose-Einstein condensation** (for brevity, the *BE condensation*). The transition temperature is called the **condensation temperature**,[6] while the condensed particles are called the **condensate**. This is a condensation of particles in momentum space, which is quite different from a condensation in real space. For the latter case the particles are condensed into the liquid phase, while for the former case the condensate is still a gas but behaves differently from the normal gas. We will show in Sec. 12.8 that the BE condensation is a phase transition which satisfies the criterion for a first-order phase transition.

[6]This temperature is also known as the Bose temperature.

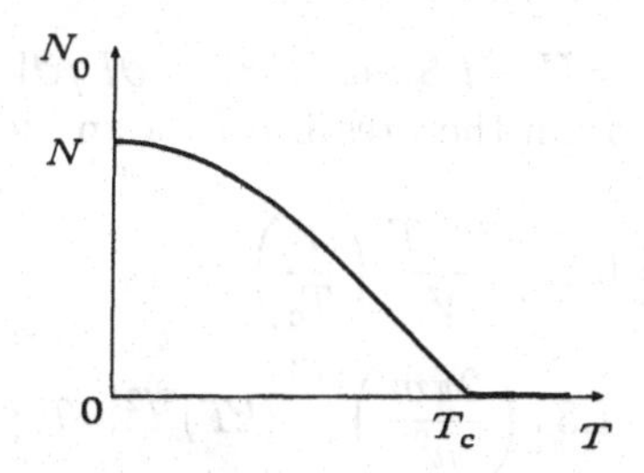

Figure 12.3: The plot of N_0 vs. T for an ideal Bose gas.

12.7 Ideal Bose Gases Below T_c

The internal energy U for an ideal Bose gas is given by

$$U = \int_0^\infty \varepsilon\, f_{\text{BE}}(\varepsilon)\, g(\varepsilon)\, d\varepsilon = 2\pi V \left(\frac{2m}{h^2}\right)^{3/2} \int_0^\infty \frac{\varepsilon^{3/2}}{e^{(\varepsilon-\mu)/kT} - 1}\, d\varepsilon.$$

This equation holds for both $T \geq T_c$ and $T < T_c$ because the energy for $\varepsilon = 0$ particles is 0. Thus the internal energy U is entirely due to the particles in the excited states. In the following we consider only the case $T < T_c$. In this case $\mu = 0$ and the above equation becomes

$$\begin{aligned} U &= 2\pi V \left(\frac{2m}{h^2}\right)^{3/2} \int_0^\infty \frac{\varepsilon^{3/2}}{e^{\varepsilon/kT} - 1}\, d\varepsilon \\ &= 2\pi kT\, V \left(\frac{2mkT}{h^2}\right)^{3/2} \int_0^\infty \frac{x^{3/2}}{e^x - 1}\, dx \\ &= 0.770\, NkT \left(\frac{T}{T_c}\right)^{3/2}, \quad T \leq T_c, \end{aligned} \tag{12.44}$$

where the integral on the right hand side of the second equality can be integrated to yield 1.78 (see Appendix E), and in the third equality the definition of T_c, Eq. (12.41), were used. Hence the constant volume heat capacity C_V is

$$C_V = \left(\frac{\partial U}{\partial T}\right)_V = 1.93\, Nk \left(\frac{T}{T_c}\right)^{3/2}, \quad T \leq T_c. \tag{12.45}$$

Thus $C_V \propto T^{3/2}$ for $T < T_c$. The entropy S of the system may be calculated from C_V,

$$S = \int_0^T \frac{C_V}{T}\, dT = 1.28 Nk \left(\frac{T}{T_c}\right)^{3/2}, \quad T \leq T_c. \tag{12.46}$$

From the relations $F = U - TS$ and $P = -(\partial F/\partial V)_T$, noting that T_c is a function of V, we obtain the pressure P of an ideal Bose gas:

$$\begin{aligned} P &= 0.513\,\frac{NkT}{V}\left(\frac{T}{T_c}\right)^{3/2} \\ &= 1.34\left(\frac{2\pi m}{h^2}\right)^{3/2}(kT)^{5/2}, \quad T \leq T_c. \end{aligned} \tag{12.47}$$

We note that P is independent of both N and V. This is because for $T < T_c$ excited state particles may condense to the ground state, which occupies no volume because these particles are motionless. From the above equation and Eq. (12.44) we also obtain

$$PV = \frac{2}{3}U, \quad T \leq T_c, \tag{12.48}$$

for an ideal Bose gas at low temperatures.

12.8 Phase Transition of the BE Condensation

Below T_c, an ideal Bose gas has two classes of particles, one class consists of particles in the ground state ($\boldsymbol{p} = 0$), the **condensate**, the other consists of particles in the excited states, the **gas particles** ($\boldsymbol{p} \neq 0$). We use the subscripts "0" and "1" to denote quantities in the condensate and the gas, respectively. Then $N = N_0 + N_1$ where N is the total number of particles of the system. From Eqs. (12.41)–(12.43) the number of particles in the gas is given by

$$\begin{aligned} N_1 &= 2.612V\left(\frac{2\pi mkT}{h^2}\right)^{3/2} \\ &= N\left(\frac{T}{T_c}\right)^{3/2}, \quad T \leq T_c. \end{aligned} \tag{12.49}$$

The condensate and the gas particles may be considered to be different phases. The condensate has no entropy, i.e., $S_0 = 0$, while the entropy of the gas S_1 is the S given in Eq. (12.46). $S_0 = 0$ can be understood from the fact that the multiplicity of the condensate is equal to 1, because it has only one state ($\boldsymbol{p} = 0$), therefore $S_0 = k \ln 1 = 0$. Also the volume of the condensate V_0 is 0 because the particles are motionless, while the volume of the gas V_1 is the volume of the system V.

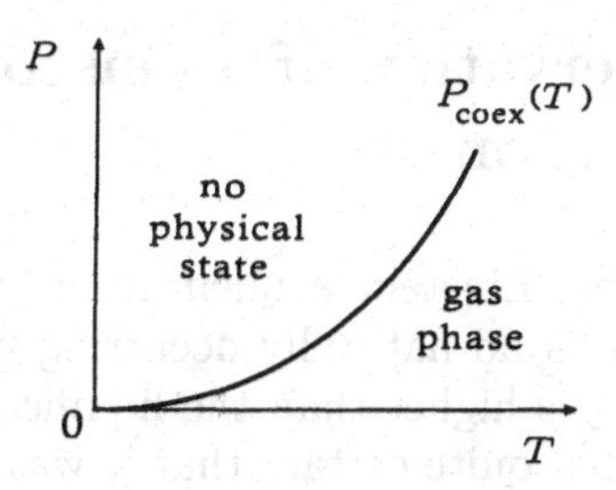

Figure 12.4: The P vs. T diagram for an ideal gas at low temperatures. The condensate-gas coexistence curve $P_{\text{coex}}(T)$ is given by Eq. (12.47). Note that there is no physical state above the coexistence curve.

That the phase transition is a first-order one can be understood from the following results. First, the chemical potential of the two phases are equal to each other,

$$\mu_0 = \mu_1 = 0. \tag{12.50}$$

This is the condition for two-phase coexistence. Second, the coexistence curve is given by Eq. (12.47), which can be used to obtain

$$\begin{aligned}\frac{dP}{dT} &= \frac{5}{2} \times 0.513 \frac{Nk}{V} \left(\frac{T}{T_c}\right)^{3/2} \\ &= 1.28 \frac{Nk}{V} \left(\frac{T}{T_c}\right)^{3/2} = \frac{S}{V},\end{aligned} \tag{12.51}$$

where $S = S_1$ is the entropy given in Eq. (12.46). Note that the above equation can be rewritten in the form of the Clausius-Clapeyron equation (7.9) for the first-order phase transitions:

$$\frac{dP}{dT} = \frac{S}{V} = \frac{S_1 - S_0}{V_1 - V_0}, \tag{12.52}$$

because $S = S_1$, $V = V_1$ and $S_0 = 0$, $V_0 = 0$. Therefore this is a first-order phase transition. The P-T diagram for an ideal gas at low temperature is shown in Fig. 12.4, where the condensate-gas coexistence curve $P_{\text{coex}}(T)$ is given by Eq. (12.47). Note that there is no physical state above the coexistence curve. The condensate exists only on the coexistence curve. The gas becomes 100% condensate only at $T = 0$ and $P = 0$.

12.9 The Observation of a Gaseous BE Condensation

The theory of the Bose-Einstein condensation was established in the 1920's. However there is no naturally occurring physical system whose critical temperature T_c is higher than the liquification temperature of a gas. Therefore it seemed quite certain that it was impossible to observe the BE condensation in nature. Helium-4 gas seems to be the best candidate. First, it is a Bose gas. Second, it has the lowest liquification temperature at 4.2 K among all Bose gases.[7] Third, a ^{4}He atom has the smallest mass among the bosons, except for the H_2 molecule, which has a much higher liquification temperature. A smaller mass means a higher T_c ($T_c \propto m^{-1}$). However if we substitute the mass for a ^{4}He atom $m = 6.65 \times 10^{-27}$ kg, and the number density at STP for an ideal gas $N/V = 2.69 \times 10^{25}$ m^{-3} into Eq. (12.41), we get $T_c \approx 0.03$ K, which is much lower than the liquification temperature 4.2 K. Therefore the BE condensation will not occur in ^{4}He gas, or any other natural gas.

However, even before the establishment of the theory of the BE condensation, physicists had already observed the phenomenon of superfluidity in liquid ^{4}He below 2.18 K. When the BE condensation theory was published, physicists had good reasons to believe that superfluidity has a strong connection with the BE condensation, because liquid ^{3}He (which obeys Fermi statistics) does not exhibit superfluidity. If we substitute the liquid ^{4}He density $N/V = 2.46 \times 10^{28}$ m^{-3} into Eq. (12.41), we get $T_c = 3.14$ K, which is quite close to the experimental value 2.18 K.

Nevertheless we cannot consider the superfluidity in liquid ^{4}He as a proof of the theory of the BE condensation, although the existence of a superfluid is strongly related to the BE condensation. The reason is that the BE condensation is a theory for an **ideal Bose gas**, while superfluidity occurs in the *liquid phase* in which the particle-particle interaction plays an important role. In the BE condensation, however, particles are non-interacting, the phase transition occurs *entirely* due to statistics. Moreover there are several differences between the gaseous BE condensation and the superfluid phase transition. These include the order of the phase transition and the behavior of the heat capacity C_V at very low temperatures. The BE condensation is a first-order phase transition, as we discussed above, while the superfluid phase transition

[7]Helium-3 gas has a lower liquification temperature at 3.2 K. However it is a Fermi gas.

is a second-order one. For a Bose gas below T_c C_V varies according to the $T^{3/2}$ law, while for superfluid C_V varies according to the T^3 law.

Therefore physicists had not given up looking for the gaseous BE condensation. Although it is impossible to observe this phenomenon in a natural physical system, it is possible to observe it in the laboratory. Since the system under consideration must remain in the gaseous phase down to extremely low temperatures, the number density N/V must be very small,[8] this will make T_c be lower than 10^{-5} K. It is rather difficult to cool a gaseous system down to such a low temperature, thus the limitation of the cooling technique is the main obstacle to observing the gaseous BE condensation. Due to the advancement of cooling techniques over the final decades of the twentieth century, physicists finally have been able to observe a gaseous BE condensation in the laboratory. The first observation was made in 1995 by Cornell, Ketterle, and Wieman, who by using the technique of laser cooling and the method of evaporation[9] cooled ^{87}Rb atoms to a temperature around 10^{-7} K and observed the BE condensation phenomenon. They trapped roughly 10^4 atoms in a volume of order 10^{-15} m^3 and observed an approximately Maxwell-Boltzmann velocity distribution when the temperature was around 2×10^{-7} K. However when T was down to 10^{-7} K, they observed a central sharp peak of the momentum distribution, i.e., a large fraction of the atoms were condensed into the ground state.[10] This is a signal of the BE condensation, because the energy of 10^{-7} K is about a hundred times greater than the ground state energy of an atom confined in a volume of 10^{-15} m^3 (see Problem 12.14). In subsequent years, Bose-Einstein condensations have also been observed in dilute gases of atomic sodium, lithium, and hydrogen.

12.10 Proof of $PV = \frac{2}{3}U$ for an Ideal Gas

We have shown that the relation $PV = \frac{2}{3}U$ holds for both an ideal Fermi and Bose gases at low temperatures. In this section we will show that,

[8] To remain in the gaseous phase, it is estimated that the average distance between particles must be greater than 10 times the particle size, thus N/V must be less than 10^{20} m^{-3}.

[9] See the explanation in Sec. 6.5.

[10] For the details of this experiment see Carl E. Wieman, "The Richtmyer Memorial Lecture: Bose-Einstein Condensation in an Ultracold Gas," *American Journal of Physics* **64**, 847–855 (1996).

in fact, this relation holds for all temperatures. From the definition of the grand potential Φ given in Eq. (10.96) and $\Phi = PV$ (Eq. (10.99)) we have

$$\frac{PV}{kT} = \ln \mathcal{Z} = \pm \sum_{p} \ln \left[1 \pm e^{(\mu-\varepsilon_p)/kT}\right]$$

$$= \pm \frac{V}{h^3} \int \ln \left[1 \pm e^{(\mu-\varepsilon_p)/kT}\right] d^3p \quad \begin{pmatrix} + & \text{Fermi} \\ - & \text{Bose} \end{pmatrix}. \quad (12.53)$$

We note that ε_p is independent of the direction of p, thus we may write $d^3p = 4\pi p^2 dp$. After an integration by parts, and noting that the integrated term vanishes, we obtain the following result (the complete derivation is left as an exercise for the reader, see Problem 12.17):

$$PV = \frac{4\pi V}{3h^3} \int_0^\infty \frac{p^3}{e^{(\varepsilon_p-\mu)/kT} \pm 1} \left(\frac{d\varepsilon_p}{dp}\right) dp$$

$$= \frac{2}{3}\frac{V}{h^3} \int \frac{\varepsilon_p \, d^3p}{e^{(\varepsilon_p-\mu)/kT} \pm 1}$$

$$= \frac{2}{3} \sum_{p} \overline{n}(p)\, \varepsilon_p = \frac{2}{3} U. \quad (12.54)$$

In the derivation we have used the fact that $p(d\varepsilon_p/dp) = 2\varepsilon_p$ for a real particle ($\varepsilon \propto p^2$). Note that the above derivation can also be applied to a photon gas, for which $\mu = 0$ and $\varepsilon = cp$. Therefore we obtain $PV = \frac{1}{3}U$ for a photon gas, which had been given in Eq. (8.75), obtained by using an argument in electromagnetic theory. Now we have found this relation using the grand potential method in statistical mechanics.

12.11 Problems

12.1. The definition of the grand partition function is $\mathcal{Z} = \sum_N e^{\beta\mu N} Z_N$, where Z_N is the N-particle partition function. Show that for a classical ideal gas which obeys Maxwell-Boltzmann statistics,

$$\ln \mathcal{Z} = \overline{N} = \frac{PV}{kT},$$

where $\overline{N}$ is the N which makes $e^{\beta\mu N} Z_N$ a maximum. [*Hint*: We may write $\mathcal{Z} = c\, e^{\beta\mu\overline{N}} Z_{\overline{N}}$, where c is a constant of order $\overline{N}^{\alpha}$.]

12.2. (a) Show that the entropy can be obtained from $S = \left[\frac{\partial(PV)}{\partial T}\right]_{\mu,V}$.

(b) For an ideal gas show that

$$S = \pm k \sum_{p} \{[1 \pm \overline{n}(p)] \ln [1 \pm \overline{n}(p)] \mp \overline{n}(p) \ln \overline{n}(p)\},$$

where the upper signs are for a Bose gas and the lower signs for a Fermi gas. [*Hint*: The result of (a) may be helpful.]

(c) Show that the result of (b) is equivalent to

$$S = -k \sum_{p} \left[\sum_{n} w_p(n) \ln w_p(n)\right],$$

where

$$w_p(n) = \frac{e^{n(\mu-\varepsilon_p)/kT}}{\sum_n e^{n(\mu-\varepsilon_p)/kT}}$$

is the probability that there are n particles in the state p.

12.3. A gas of molecules is in thermal equilibrium at temperature T with a surface which has N_0 adsorption centers. Each adsorption center can adsorb at most one molecule with an adsorption energy $-\varepsilon$ ($\varepsilon > 0$). Find the grand partition function of the adsorbed molecules and show that the chemical potential is given by

$$\mu = kT \ln \frac{\overline{N}}{N_0 - \overline{N}} - \varepsilon,$$

where $\overline{N}$ is the mean number of adsorbed molecules.

12.4. Consider a one-dimensional N electron gas confined in a length L at $T = 0$. (cf. Problem 10.1 for the density of states.)

(a) Find the number of particles with energy less than or equal to ε.

(b) Find the Fermi energy ε_F in terms of N/L.

(c) Find the average energy per electron in terms of ε_F.

12.5. Consider a two-dimensional N electron gas confined in an area A at $T = 0$. Repeat the previous problem with L replaced by A.

12.6. Consider an ideal Fermi gas of N particles in a three-dimensional volume V at $T = 0$. Suppose the energy-momentum relation for each particle is $\varepsilon = cp$ (i.e., a relativistic particle), where c is a constant (the speed of light).

(a) Find the density of states for the system $g(\varepsilon)$ if the spin degeneracy is $g_s = 2$.

(b) Calculate the Fermi energy ε_F in terms of N/V.

(c) Calculate the mean energy per particle in terms of ε_F.

12.7. Consider the conduction electrons in aluminum. The mass density of aluminum is 2.7×10^3 kg m^{-3} and its atomic weight is 27.

(a) Calculate the Fermi energy ε_F for the conduction electrons, assuming 3 electrons per atom.

(b) Calculate the chemical potential μ/ε_F at $T = 1000$ K.

(c) Calculate the electronic molar specific heat c_v^{el} at $T = 300$ K in terms of R.

12.8. Helium-3 atoms are spin-1/2 fermions, which liquefy at about 3.2 K under the pressure of 1 atm. However the interactions between liquid ^{3}He atoms are so weak that we may approximate liquid ^{3}He as a noninteracting ideal Fermi gas.

(a) Calculate the Fermi energy ε_F and the Fermi temperature T_F of liquid ^{3}He. The number density of liquid ^{3}He at 1 atm is 1.63×10^{28} m^{-3} and its atomic weight is 3.

(b) Use the ideal Fermi gas theory to calculate the heat capacity C_V for liquid ^{3}He for $T \ll T_F$. Compare the calculated result with the experimental value $C_V = 2.8\, NkT$ K^{-1}. How good is the agreement? Use the experimental value of C_V to calculate the liquid phase entropy S as a function of T.

(c) The entropy of solid ^{3}He below 1 K is mainly due to the spin entropy (each spin can have two directions). It is approximately a constant but drops to zero abruptly near $T = 0$. Sketch an S vs. T diagram for both liquid (from (b)) and solid ^{3}He for $T < 1$ K on the same graph. Estimate the temperature at which the two phases have the same entropy. Discuss how this will affect the shape of the solid-liquid coexistence curve (cf. Fig. 7.5(b)).

12.9. Complete the derivation of Eq. (12.31) and show that the relation $PV = \frac{2}{3}U$ holds for a degenerate Fermi gas ($T \ll T_F$).

12.10. Consider an ideal Bose gas at very low temperature. If the number of particles in the ground level $\varepsilon = 0$ is of the order N, show that the chemical potential μ varies as

$$\lim_{T\to 0} \mu \approx -\frac{kT}{N}.$$

12.11 Consider a hypothetical ideal Bose gas of ^{4}He atoms with a Bose condensation temperature $T_c = 0.02$ K. Find N/V for this gas. Calculate the percentage of the atoms which are in the ground state at $T = 0.01$ K.

12.12. Consider an ideal Bose gas below the condensation temperature. Show that (a) $C_V = \frac{5}{2}\frac{U}{T}$, (b) $S = \frac{5}{3}\frac{U}{T}$, (c) $F = -\frac{2}{3}U$, (d) $PV = \frac{2}{3}U$.

12.13. Consider an ideal Bose gas below the condensation temperature. If we write $S = N_1 s_1$, and $V = N_1 v_1$, find s_1 and v_1. Show that Eq. (12.52) satisfies the requirement of the third law, that $dP/dT \to 0$ as $T \to 0$.

12.14. (a) Consider an atom of mass m confined in a volume $V = L^3$. Use the rigid wall boundary condition to show that the ground state energy is $\varepsilon_0 = \dfrac{3h^2}{8mV^{2/3}}$.

(b) Find the ratios $\varepsilon_1/\varepsilon_0$ and $\varepsilon_2/\varepsilon_0$, where ε_1 and ε_2 are, respectively, the first and second excited state energies of the atom considered in (a).

(c) Consider a BE condensation experiment, in which 2×10^4 rubidium-87 atoms are confined in a volume of $(10^{-5}\ \text{m})^3$. Calculate the condensation temperature T_c for these atoms, assuming that the energy levels are quasi-continuous.

(d) Determine how many Rb atoms are in the ground state at $T = 10^{-7}$ K.

(e) Calculate the respective ratio of $kT/(\varepsilon_1 - \varepsilon_0)$ and $kT/(\varepsilon_2 - \varepsilon_0)$, where $T = 10^{-7}$ K and ε_0, ε_1 and ε_2 are given in (a) and (b).

(f) What is your comment on the results of (d) and (e)?

12.15. In a gaseous BE condensation experiment, the atoms are actually confined by a potential well rather than by a rigid wall container. We may assume that the potential well is harmonic and isotropic (in three dimensions). Energy levels in the well are $\varepsilon = nh\nu$, where ν is the classical oscillation frequency and $n = 0, 1, 2, \cdots$. (The zero point energy $\frac{3}{2}h\nu$ is unimportant and is neglected.)

(a) Show that the degeneracy of level n is $(n+1)(n+2)/2$.

(b) For $n \gg 1$, ε may be considered as quasi-continuous, show that the density of states may be approximated as $g(\varepsilon) = \varepsilon^2/2(h\nu)^3$.

(c) Show that the condensation temperature T_c is given by

$$T_c = \alpha \frac{h\nu}{k} N^{1/3},$$

where N is the total number of atoms in the well, and α is a numerical constant.

(d) By using the classical approximation for the oscillators; show that

$$\frac{3}{2}(2\pi\nu)^2 m(\Delta x)^2 = \varepsilon,$$

where m is the mass of an atom and Δx is the amplitude of the oscillation.

(e) If we take $\varepsilon \sim kT_c$ and $\Delta x \sim V^{1/3}$, where V is the volume of the system, show that

$$T_c \sim \frac{h^2}{mk}\left(\frac{N}{V}\right)^{2/3}.$$

Note that the above equation is exactly the same as Eq. (12.41) except that the numerical constant is different.

12.16. (a) Consider a two-dimensional ideal Bose gas of N particles confined in an area A. Show that there is no BE condensation except at $T = 0$. This implies that $T_c = 0$ for a two-dimensional Bose gas. What is $g(\varepsilon = 0)$ where $g(\varepsilon)$ is the density of states?

(b) Consider the same problem as in (a) for a one-dimensional ideal Bose gas.

12.17. Complete the derivation given in Sec. 12.10 that $PV = \frac{2}{3}U$ holds for all ideal gases independent of the statistics and temperature T. By using this approach, show that $PV = \frac{1}{3}U$ for a photon gas.

Appendix

In this appendix we evaluate the definite integrals and derive the approximations we have used in the text of this book. These include the Stirling approximation (Chapters 3 and 10), the volume of an N-dimensional sphere for large N (Chapter 3), the definite integrals involving classical statistics (Chapters 9 and 11), and the definite integrals involving quantum statistics (Chapters 8, 11 and 12). In addition we also introduce a useful function, the *gamma function*.

A. Integrals of Classical Statistics

The integrals, with a positive integer n,

$$I_n = \int_{-\infty}^{\infty} x^{2n} e^{-ax^2}\, dx, \quad a > 0, \tag{1}$$

are frequently encountered in Maxwell-Boltzmann statistics (Chapters 9 and 11). Since these integrals contain the function e^{-ax^2} (a Gaussian), they are also known as the **Gaussian integrals**. We first evaluate the case with $n = 0$ and $a = 1$. Denoting this integral by I_0', we have

$$\begin{aligned} I_0'^2 &= \left(\int_{-\infty}^{\infty} e^{-x^2}\, dx\right)\left(\int_{-\infty}^{\infty} e^{-y^2}\, dy\right) \\ &= \int_{-\infty}^{\infty}\int_{-\infty}^{\infty} e^{-x^2-y^2}\, dx\, dy \\ &= \int_0^{\infty}\int_0^{2\pi} e^{-r^2}\, r\, dr\, d\theta = \pi, \end{aligned} \tag{2}$$

where polar coordinates (r, θ) are used in the third equality. Hence

$$I_0' = \int_{-\infty}^{\infty} e^{-x^2}\, dx = \sqrt{\pi}. \tag{3}$$

By changing the variable x to $\sqrt{a}\,y$, we find

$$I_0 = \int_{-\infty}^{\infty} e^{-ax^2}\,dx = \sqrt{\frac{\pi}{a}}. \tag{4}$$

It is easy to see that

$$\int_{-\infty}^{\infty} x^{2n} e^{-ax^2}\,dx = (-1)^n \frac{d^n}{da^n} I_0 = (-1)^n \frac{d^n}{da^n}\sqrt{\frac{\pi}{a}}, \tag{5}$$

because the order of integration and differentiation can be exchanged. Since the integrand of I_n is even in x, we may change the lower limit of the integral from $-\infty$ to 0 and obtain

$$\int_0^{\infty} e^{-ax^2}\,dx = \frac{1}{2}\sqrt{\frac{\pi}{a}}, \tag{6}$$

$$\int_0^{\infty} x^{2n} e^{-ax^2}\,dx = (-1)^n \frac{1}{2}\frac{d^n}{da^n}\sqrt{\frac{\pi}{a}}. \tag{7}$$

B. The Gamma Function

Consider the integral

$$\int_0^{\infty} e^{-ax}\,dx = \frac{1}{a}. \tag{8}$$

Differentiating with respect to a n times we get

$$\int_0^{\infty} x^n e^{-ax}\,dx = \frac{1}{a^{n+1}}\,n!. \tag{9}$$

Setting $a = 1$, we get a formula for $n!$:

$$n! = \int_0^{\infty} x^n e^{-x}\,dx. \tag{10}$$

The integral on the right hand side can be evaluated numerically even when n is not an integer. The generalization of the above integral to *noninteger* n is called the **gamma function**, denoted as $\Gamma(n)$. It is more convenient to increase the argument by 1 and define

$$\Gamma(n+1) = n!, \tag{11}$$

thus we have

$$\Gamma(n+1) \equiv \int_0^\infty x^n e^{-x}\, dx. \tag{12}$$

From this we get the *recursion formula* for the gamma function

$$\Gamma(n+1) = n\,\Gamma(n). \tag{13}$$

It is easy to see that this formula holds for an integer n. For general values of n the above relation can be obtained by using an integration by parts from Eq. (12). This formula can be used to define the gamma function with a *negative argument*. It also defines the factorial of 0,

$$0! = \Gamma(1) = 1; \text{ and } \Gamma(2) = 1! = 1. \tag{14}$$

Another frequently used quantity is $\Gamma(1/2)$, which can easily be converted to a Gaussian integral with $n = 0$ and $a = 1$, thus

$$\Gamma\left(\frac{1}{2}\right) = \sqrt{\pi}; \text{ then } \Gamma(\frac{3}{2}) = \frac{1}{2}\Gamma(\frac{1}{2}) = \frac{\sqrt{\pi}}{2}. \tag{15}$$

The gamma function is useful in the evaluation of many definite integrals that occur in theoretical physics. We will see some of the examples in the following.

C. The Stirling Approximation

Consider

$$\begin{aligned}\ln N! &= \ln[N(N-1)(N-2)\cdots 2\times 1] \\ &= \ln N + \ln(N-1) + \ln(N-2) + \cdots + \ln 2 + \ln 1 \\ &= \sum_{n=1}^{N} \ln n. \end{aligned} \tag{16}$$

When N is large we can approximate the sum as an integral, thus, by an integration by parts, we get

$$\begin{aligned}\ln N! &\approx \int_0^N \ln x\, dx = x \ln x|_0^N - \int_0^N dx \\ &= N \ln N - N, \text{ for large } N. \end{aligned} \tag{17}$$

This is the Stirling approximation we used in Chapters 3 and 10.

D. The Volume for an n-Dimensional Sphere

Consider an n-dimensional space with rectangular coordinates $(x_1, x_2, \cdots, x_n)$. The "volume" element $d\mathcal{V}_n$ can be written as

$$d\mathcal{V}_n = dx_1\, dx_2 \cdots dx_n. \tag{18}$$

The "volume" of a sphere of radius R in this space is given by

$$\mathcal{V}_n(R) = \int_{0\le\sum_{i=1}^n x_i^2 \le R^2} dx_1\, dx_2 \cdots dx_n. \tag{19}$$

Obviously $\mathcal{V}_n$ will be proportional to R^n, thus we can write

$$\mathcal{V}_n(R) = C_n\, R^n, \quad R^2 = \sum_{i=1}^n x_i^2, \tag{20}$$

where C_n is a constant, which we would like to evaluate. From the above equation we can also write the volume element $d\mathcal{V}_n$, for the *spherically symmetric* case, as

$$d\mathcal{V}_n(R) = nC_nR^{n-1}dR. \tag{21}$$

Consider the Gaussian integral given in Eq. (3). Multiplying n such integrals, we obtain

$$\begin{aligned}
\pi^{n/2} &= \int_{-\infty}^{\infty} \cdots \int_{-\infty}^{\infty} \exp\left(-\sum_{i=1}^n x_i^2\right) \prod_{i=1}^n dx_i \\
&= \int_0^\infty e^{-R^2} nC_nR^{n-1}dR \\
&= \frac{1}{2}nC_n \int_0^\infty y^{n/2-1}e^{-y}dy \quad (y = R^2) \\
&= C_n\left[\frac{n}{2}\Gamma\left(\frac{n}{2}\right)\right] = \left(\frac{n}{2}\right)!\, C_n.
\end{aligned} \tag{22}$$

Therefore we have

$$C_n = \pi^{n/2} \Big/ \left(\frac{n}{2}\right)!\,. \tag{23}$$

For a large n we can use the Stirling approximation to obtain

$$\ln C_n \approx \frac{n}{2}\ln\pi - \frac{n}{2}\ln\frac{n}{2} + \frac{n}{2}. \tag{24}$$

This is, with $n = 3N$, the formula we used to obtain Eq. (3.26).

E. Integrals of Quantum Statistics

In quantum statistics we frequently encounter integrals of the form

$$\int_0^\infty \frac{x^s}{e^x \pm 1}\,dx, \tag{25}$$

where the plus (minus) sign is for Fermi (Bose) statistics, and s is a positive number, usually an integer, or a half-integer. Since $e^{-x} < 1$ for $x > 0$, we may substitute the following expansion,

$$\frac{1}{e^x \pm 1} = \frac{e^{-x}}{1 \pm e^{-x}} = e^{-x} \mp (e^{-x})^2 + (e^{-x})^3 \mp \cdots, \tag{26}$$

into Eq. (25) and integrate it term by term. For example, for $s = 1$, we have

$$\begin{aligned}\int_0^\infty \frac{x}{e^x \pm 1}\,dx &= \int_0^\infty (xe^{-x} \mp xe^{-2x} + xe^{-3x} \mp \cdots)\,dx \\ &= 1 \mp \frac{1}{2^2} + \frac{1}{3^2} \mp \frac{1}{4^2} + \cdots.\end{aligned} \tag{27}$$

For the Bose case the infinite series can be expressed as a Riemann[11] zeta function $\zeta(s)$ with $s = 2$. In general $\zeta(s)$ is defined as

$$\zeta(s) = 1 + \frac{1}{2^s} + \frac{1}{3^s} + \cdots = \sum_{n=1}^{\infty} n^{-s}, \tag{28}$$

with s a positive number. In fact it is required that $s > 1$, otherwise the series diverges. The evaluation of $\zeta(s)$ is beyond the scope of this book, so we just list the values of some of the frequently used zeta functions:

$$\begin{aligned}\zeta(2) &= \frac{\pi^2}{6} \approx 1.645, &\quad \zeta(4) &= \frac{\pi^4}{90} \approx 1.082, \\ \zeta(6) &= \frac{\pi^6}{945} \approx 1.017, &\quad \zeta(8) &= \frac{\pi^8}{9450} \approx 1.004,\end{aligned}$$

$$\zeta(3) \approx 1.202, \quad \zeta(5) \approx 1.037, \quad \zeta(7) \approx 1.008,$$

$$\zeta(\tfrac{3}{2}) \approx 2.612, \quad \zeta(\tfrac{5}{2}) \approx 1.341, \quad \zeta(\tfrac{7}{2}) \approx 1.127.$$

[11] Georg F. B. Riemann, German mathematician (1826–1866).

Therefore, from Eq. (27) we have

$$\int_0^\infty \frac{x}{e^x - 1}\, dx = \zeta(2) = \frac{\pi^2}{6}. \tag{29}$$

We need an extra effort to get the corresponding integral for Fermi statistics,

$$\begin{aligned}
\int_0^\infty \frac{x}{e^x + 1}\, dx &= \left(1 + \frac{1}{2^2} + \frac{1}{3^2} + \cdots\right) - 2\left(\frac{1}{2^2} + \frac{1}{4^2} + \frac{1}{6^2}\cdots\right) \\
&= \zeta(2) - \frac{2}{2^2}\left(1 + \frac{1}{2^2} + \frac{1}{3^2} + \cdots\right) \\
&= \zeta(2) - \frac{1}{2}\zeta(2) = \frac{1}{2}\zeta(2) = \frac{\pi^2}{12}.
\end{aligned} \tag{30}$$

By using the same approach, one can evaluate the following two types of integrals (for positive n):

$$\int_0^\infty \frac{x^n}{e^x - 1}\, dx = \Gamma(n+1)\,\zeta(n+1); \tag{31}$$

$$\int_0^\infty \frac{x^n}{e^x + 1}\, dx = \left(1 - \frac{1}{2^n}\right)\Gamma(n+1)\,\zeta(n+1). \tag{32}$$

From the above formulas we can get the integrals used in the text. For photons (Chapter 8) and phonons (Chapter 11):

$$\int_0^\infty \frac{x^3}{e^x - 1}\, dx = \Gamma(4)\,\zeta(4) = \frac{\pi^4}{15}. \tag{33}$$

For an ideal Bose gas below T_c (Chapter 12):

$$\int_0^\infty \frac{x^{1/2}}{e^x - 1}\, dx = \Gamma(\frac{3}{2})\,\zeta(\frac{3}{2}) \approx 2.612\,\frac{\sqrt{\pi}}{2}; \tag{34}$$

$$\int_0^\infty \frac{x^{3/2}}{e^x - 1}\, dx = \Gamma(\frac{5}{2})\,\zeta(\frac{5}{2}) \approx \frac{3}{4}\sqrt{\pi} \times 1.341 \approx 1.783. \tag{35}$$

For a degenerate ideal Fermi gas (Chapter 12):

$$\begin{aligned}
\int_0^\infty \frac{x^2\, e^x}{(e^x + 1)^2}\, dx &= -\left.\frac{x^2}{e^x + 1}\right|_0^\infty + 2\int_0^\infty \frac{x}{e^x + 1}\, dx \\
&= 2\left(1 - \frac{1}{2}\right)\Gamma(2)\,\zeta(2) = \frac{\pi^2}{6},
\end{aligned} \tag{36}$$

where an integration by parts is used to obtain the the first equality. Some other useful integrals are

$$\int_0^\infty \frac{e^x}{(e^x+1)^2}\,dx = 1; \quad \int_0^\infty \frac{x^4 e^x}{(e^x-1)^2}\,dx = \frac{4\pi}{15}. \tag{37}$$

References

C.J. Adkins, *Equilibrium Thermodynamics,* 3rd edition, Cambridge University Press, Cambridge, UK, 1983.

R. Baierlein, *Thermal Physics,* Cambridge University Press, Cambridge, 1999.

R. Bauman, *Modern Thermodynamics with Statistical Mechanics,* Macmillan, New York, 1992.

R. Bowley and M. Sanchez, *Introductory Statistical Mechanics,* Clarendon Press, Oxford, 1996.

H.B. Callen, *Thermodynamics and an Introduction to Thermostatistics,* 2nd edition, John Wiley and Sons, New York, 1985.

A.H. Carter, *Classical and Statistical Thermodynamics,* Prentice Hall, Upper Saddle River, New Jersey, USA, 2001.

D. Chandler, *Introduction to Modern Statistical Mechanics,* Oxford University Press, Oxford, 1987.

C.B.P. Finn, *Thermal Physics,* 2nd edition, Chapman and Hall, London, 1993.

C. Garrod, *Statistical Mechanics and Thermodynamics,* Oxford University Press, Oxford, 1995.

D.R. Gaskell, *Introduction to the Thermodynamics of Materials,* 3rd edition, Taylor & Francis, Washington, DC, 1995.

W. Greiner, L. Neise and H. Stöcker, *Thermodynamics and Statistical Mechanics,* Springer-Verlag, New York, 1995.

T. Grenault, *Statistical Physics,* 2nd edition, Chapman and Hall, London, 1995.

E. Guha, *Basic Thermodynamics,* Alpha Science, Pangbourne, UK, 2000.

T.L. Hill, *An Introduction to Statistical Thermodynamics,* Dover, New York, 1986.

K. Huang, *Statistical Mechanics,* 2nd edition, John Wiley and Sons, New York, 1987.

C. Kittel and H. Kroemer, *Thermal Physics,* 2nd edition, Freeman, San Francisco, 1980.

R. Kubo, *Thermodynamics,* John Wiley and Sons, New York, 1960.

R. Kubo, *Statistical Mechanics,* North-Holland, Amsterdam, Netherlands, 1965.

S.K. Ma, *Statistical Mechanics,* World Scientific, Singapore, 1985.

A.P. Pippard, *The Elements of Classical Thermodynamics,* Cambridge University Press, Cambridge, 1987.

M. Plischke and B. Bergersen, *Equilibrium Statistical Physics,* 2nd edition, World Scientific, Singapore, 1994.

L.E. Reichl, *A Modern Course in Statistical Physics,* 2nd edition, John Wiley and Sons, New York, 1998.

F. Reif, *Fundamentals of Statistical and Thermal Physics,* McGraw-Hill, New York, 1965.

B.N. Roy, *Principles of Modern Thermodynamics,* Institute of Physics, London, 1995.

D.V. Schroeder, *An Introduction to Thermal Physics,* Addison-Wesley, Reading, Massachusetts, USA, 2000.

F.W. Sears and G.L. Salinger, *Thermodynamics, Kinetic Theory, and Statistical Thermodynamics,* 3rd edition, Addison-Wesley, Reading, Massachusetts, USA, 1975.

M. Sprankling, *Heat and Thermodynamics,* MacMillan, London, 1993.

K. Stowe, *Introduction to Statistical Mechanics and Thermodynamics,* John Wiley and Sons, New York, 1984.

M.W. Zemansky and R.H. Dittman, *Heat and Thermodynamics,* 7th edition, McGraw-Hill, New York, 1997.

Answers to Selected Problems

1.2. (a) $A = P_1^2/nRT_1$; (c) $T_2 = T_1/4$; (d) $3nRT_1/8$.

1.3. (b) 0, 5.62×10^3 J, -5.05×10^3 J, 1.57×10^3 J.

1.4. (a) 0.2 m^3; (b) 150 K; (c) 200 K, 0.13 m^3; (d) 225 K, 0.15 m^3.

1.5. 1.52×10^{-3} m^3, 9.5 J.

1.6. (a) $V_f = 30.6 \times 10^{-3}$ m^3, $T_f = 373$ K; (b) 2168 J, 3000 J;
(c) $c_v = 21.7$ J mole^{-1} K^{-1}, $c_P = 30$ J mole^{-1} K^{-1}.

1.7. 2/7, 1.

1.9. (a) a.

1.10. c, $c + R/[1 - 2a(v-b)^2/RTv^3]$.

1.11. (a) $Q = n[a(T_2 - T_1) + b(T_2^2 - T_1^2) + c(1/T_2 - 1/T_1)]$;
(b) $\bar{c}_P = a + b(T_2 + T_1) - c/T_1T_2$.

1.16. (a) 0.25 liters; (b) 188 J; (c) 520 K.

1.17. 0.82.

1.22. $v = v_0 \exp(aT^3/P)$, $a/b = 1/3$.

2.1. (a) 0°C; (b) 1.06 kg; (c) 15 kcal.

2.3. (a) 4×10^6 J; (b) 4×10^6 J; (c) 4×10^6 J, 2×10^6 J; (d) 50%, 1.

2.5. $W_{\text{dia}}/W_{\text{mon}} = 1/3$.

2.6. $T_{\text{A}} = 1200$ K, $T_{\text{C}} = 300$ K.

2.7. $Q_2 = 1000$ J, $Q_1 = 600$ K, $\eta \approx 44\%$.

2.11. (a) 1.22 kJ/K; (b) 6.06 kJ/K.

2.12. (a) 1300 J/K, -1121 J/K, 179 J/K; (b) 1300 J/K, -1207 J/K,
93 J/K.

2.13. $\Delta s_{\text{AB}} = 0$, $\Delta s_{\text{BC}} = -(15R/2)\ln 2$, $\Delta s_{\text{CA}} = (15R/2)\ln 2$,
$\Delta s_{\text{tot}} = 0$.

2.14. (a) Heat: 132 J, 0, −109.6 J; work: 0, 132 J, −109.6 J;
(b) 0.547 J/K, 0, −0.547 J/K; (c) 17%.

2.15. (a) $Q_{AB} = 12$ kJ, $Q_{BC} = -7.2$ kJ, $Q_{CD} = -6$ kJ, $Q_{DA} = 2.4$ kJ;
(b) 0.6×10^5 N m^{-2};
(c) $S_{AB} = 11.0$ J/K, $S_{BC} = -5.54$ J/K, $S_{CD} = -11.0$ J/K,
$S_{DA} = 5.54$ J/K.

2.16. (b) $Q_{AB} = Q_{CD} = 0$, $Q_{BC} = 2.08 \times 10^6$ J, $Q_{DA} = -8.31 \times 10^5$ J;
(c) 0.60; (d) 0.667.

2.17. 293 J/K.

2.19. (a) $T_2/(T_1 - T_2)$;
(b) $T_2 = T_1 + \{W/\alpha - [(W/\alpha)^2 + 4T_1W/\alpha]^{1/2}\}/2$; (c) 2°C.

2.20. 0.33 kcal vs. 0.26 kcal.

2.26. $c[(T_2 - T_1) - T_1 \ln(T_2/T_1)]$.

3.1. (a) 750 Ne, 75 He; and 250 Ne, 25 He; (b) $(0.75)^{1000}(0.25)^{100}$.

3.3. (a) (i) 0; (ii) $1.91nR$.

3.5. (a) $(M + N - 1)!/M!(N - 1)!$;
(b) $T^{-1} = (k/h\nu)\ln[(U/N + h\nu/2)/(U/N - h\nu/2)]$.

3.6. (a) $S = Nk[\ln N - (N_0/N)\ln N_0 - (N_1/N)\ln N_1]$;
(c) $N_0 = N/(1 + e^{-\varepsilon/kT})$, $N_1 = Ne^{-\varepsilon/kT}/(1 + e^{-\varepsilon/kT})$.

3.7. (a) $S = 2k[N\ln N - n\ln n - (N - n)\ln(N - n)]$;
(b) $n = N/(1 + e^{\varepsilon/2kT})$.

3.8. 1.02 J/K.

4.1. (a) $\Delta S_{\text{body}} = 6.93$ J/K, $\Delta S_{\text{res}} = -5.0$ J/K, $\Delta S_{\text{univ}} = 1.93$ J/K;
(b) 11.0 J/K, −6.67 J/K, 4.33 J/K;
(c) −6.93 J/K, 10.0 J/K, 3.07 J/K.

4.2. (a) 4.25 J/K; (b) 0; (c) −6.37 J/K.

4.3. (a) $(5R/2)\ln(v_2/v_1)$; (b) $R\ln(v_2/v_1)$; (c) 0.

4.4. (a) 7.0×10^4 cal; (b) 215 cal/K; (c) 7.0×10^3 cal.

4.5. (a) $T_f = 600$ K, $T_i = 300$ K; (b) 300 K.

4.16. (a) $c_v = a + bT - R$; (b) $s = a\ln T + bT - R\ln P + s_0$;
(c) $h = aT + bT^2/2 + h_0$.

4.21. (a) 4.56 J/kg; (b) 153.5 J/K.

4.22. (b) 24.7 J/K.

4.23. $\Delta S = Mq_0/T_0 - M(c_l - c_s)\ln(T_0/T_1)$.

5.1. $\Delta S = -38.3$ J/K; $\Delta F = \Delta G = 1.15 \times 10^4$ J/K.

5.2. (a) (i) $\Delta U = \Delta H = \Delta F = \Delta G = 0$;

(ii) $\Delta U = \Delta H = 0$, $\Delta F = \Delta G = -1.91nRT$.

(b) $\Delta U = \Delta H = 0$, $\Delta S = R\ln 2$, $\Delta F = \Delta G = -RT\ln 2$.

(c) $\Delta U = C_V(T_2 - T_1)$, $\Delta H = C_P(T_2 - T_1)$,

$\Delta F = C_V(T_2 - T_1) - S(T_1, P_1)(T_2 - T_1)$,

$\Delta G = C_P(T_2 - T_1) - S(T_1, P_1)(T_2 - T_1)$.

5.3. $W = 33$ J/mole, $\Delta U = -125$ J/mole.

5.4. $\Delta S = -[a(P_B - P_A) + b(P_B^2 - P_A^2)/2 + c(P_B^3 - P_A^3)/3.]$

5.9. (a) $P(v + A) = RT$; (b) $s = -R\ln(P/P_0) + (dA/dT)P$;

(d) $h = -AP + (dA/dT)TP$; (e) $u = (dA/dT)TP - RT$.

6.3. -2.1 K.

6.4. (a) 3.72×10^6 J; (b) 1.15×10^4 J/K.

6.5. (a) $h = u_0 + c_vT + RTv/(v - b) - 2a/v$;

(c) $\kappa = v^2(v - b)^2/[RTv^3 - 2a(v - b)^2]$.

6.6. (a) $\eta = 0$, $\mu = -b/c_p$; (b) $\eta = -b/c_v$, $\mu = 0$.

6.7. 35.4 K.

6.8. 1.24×10^{-7} K.

6.14. $S = 2aT^{1/2} + bT^3/3$. No violation of the third law.

7.1. -2900 J mole^{-1}.

7.2. (a) -1.35×10^7 Pa/K; (b) 1.35×10^7 Pa.

7.7. 64.1 atm.

7.8. (a) $T = 200$ K, $P = 1.01$ atm;

(b) $\ell_{31} = 49.9$ J/mole, $\ell_{32} = 33.3$ J/mole, $\ell_{21} = 16.6$ J/mole.

7.9. 0.0075°C.

7.14. (a) $v_{\text{solid}} < v_{\text{liquid}}$; (c) T decreases.

7.15. $-3.2°C/km$.

7.19. (a) 2.

7.20. (c) $T_c = \lambda/2R$.

7.21. (a) $f(T, P) = \mu_B^0(T, P) + \lambda(T, P)$.

8.2 $a = C_c B^2/T^2$, $b = -C_c B/T$.

8.4. $-(M_2B_2 - M_1B_1)/2$.

8.14. -1.03×10^3 J.

8.15. (a) $s = 3.51 \times 10^{-4}$ J m^{-2} K^{-1}, $c_A = -8.75 \times 10^{-5}$ J m^{-2} K^{-1}.
(b) -2.4 K.

8.16. (a) 0.0128 torr; (b) 5.8×10^4 torr.

8.18. (d) 6.5 K/km.

8.20. 2.99×10^3 N/m^2.

8.21. 6.9×10^5 N/m^2.

8.24. $4bT^4V/3$.

8.25. (b) $du(V_b - V_a)/3$; (c) $4u(V_b - V_a)/3$.

9.1. (a) $2v_0/3$; (d) $0.707v_0$.

9.2. $v_m \approx 422$ m/s; $\overline{v} \approx 476$ m/s; $v_{\rm rms} \approx 519$ m/s.

9.3. (a) 6667 collisions; (b) 1.226 atm.

9.5. (a) $f(v) = (m/kT)e^{-mv^2/2kT}$; (b) $\overline{v} = (\pi kT/m)^{1/2}$, $\overline{v^2} = 2kT/m$;
(c) $PA = NkT$; (d) $(N/A)(kT/\pi m)^{1/2}$

9.7. $P \approx 0.36$ N/m^2; $\tau \approx 1.01 \times 10^{-4}$ s.

9.8. $\ell \propto P^{-1}$; $\tau^{-1} \propto P$.

9.9. 4.4×10^{-5} s.

9.10. (a) 3679; (b) 821; (c) 2386; (d) 184.

9.11. $\sqrt{2}\ell$; 0.

9.12. $\ell_1 = [4\pi a_1^2 n_1 + \pi(a_1 + a_2)^2 n_2]^{-1}$, $\ell_2 = [4\pi a_2^2 n_2 + \pi(a_1 + a_2)^2 n_1]^{-1}$.

9.13. $2.77V/(\overline{v}A)$.

9.14. (a) 0.707; (b) 0.5.

9.15. $P_1(t) = P_0[1 + e^{-A\overline{v}t/2V}]/2$.

9.16. $0.485\,P$.

9.17. (a) $4I$; (b) $I/2$; (c) $4I$; (d) $I/2$.

9.18. $(c_2/c_1)(m_1/m_2)^{1/2}$.

9.19. (a) $m^{1/2}$; (c) $T^{1/2}$.

9.20. (a) $(8kT/\pi m)^{1/2}k/(2\pi d_0^2)$; (b) 0.069 J k^{-1} m^{-1} s^{-1}.

9.21. 1.18.

9.22. $\eta = 1.24 \times 10^{-5}$ N s m^{-2}; $\lambda = 8.93 \times 10^{-3}$ J m^{-1}s^{-1}K^{-1};
$D = 9.62 \times 10^{-6}$ m^2/s.

10.1. $c_2 = g_s 2\pi mA/h^2$; $c_1 = g_s(m/2)^{1/2}L/h$.

10.2. (a) $\delta n_1 = \delta n_3 = 1$;
(b) $W_{\text{before}} = \frac{60!}{30!\,20!\,10!}$, $W_{\text{after}} = \frac{60!}{31!\,18!\,11!}$.

10.3. (a) 9; (b) $\Omega = 84$; (c) $\overline{n}_j$ =1.33, 1, 0.71, 0.48, 0.29, 0.14, 0.05.

10.4. (a) $\Omega = 9$, $\overline{n}_j$ =1.33, 1, 0.78, 0.44, 0.22, 0.11, 0.11.
(b) $\Omega = 1$, $\overline{n}_j$ =1, 1, 1, 1, 0, 0, 0.

10.5. (a) $k \ln 3$.

10.6. (a) 1716; (b) 28.

10.12. (b) 1.58×10^4.

10.13. (a) 0.98; (b) 0.60; (c) 0.40; (d) 0.40; (e) 0.02.

10.15. (a) $\overline{n} = 25.3$, $p(0) = 0.038$, $p(1) = 0.037$, $p(2) = 0.035$;
(b) $\overline{n} = 2.13$, $p(n)$ =0.32, 0.218, 0.148.
(c) $\overline{n} = 0.0218$, $p(n)$ =0.979, 0.0209, 0.00044.

10.16. $(\varepsilon - \mu)/kT > \ln 2000 \approx 7.6$.

10.17 (a) $Z = 1 + 2e^{-\beta\varepsilon} + 2e^{-2\beta\varepsilon} + e^{-3\beta\varepsilon}$;
(b) $Z = 3 + 2e^{-\beta\varepsilon} + 3e^{-2\beta\varepsilon} + e^{-3\beta\varepsilon} + e^{-4\beta\varepsilon}$;
(c) $Z = (4 + 4e^{-\beta\varepsilon} + 5e^{-2\beta\varepsilon} + 2e^{-3\beta\varepsilon} + e^{-4\beta\varepsilon})/2!$.

10.21. (a) $1/\sqrt{2\pi}\sigma$; (b) x_0.

11.1. $Z_1 = A(2\pi mkT)/h^2$, $C_A = Nk$, $C_P = 2Nk$.

11.2. $Z_1 = 8\pi V(kT)^3/h^3c^3$, $C_V = 3Nk$, $C_P = 4Nk$.

11.3. $Z_1 = 2\pi VkT/h^3\omega$, $u = kT$, $c_v = k$.

11.4. (a) x_0; (b) $x_0 - 3bkT/4a^2$;
(c) $Z_1 = (L/h)(2m/a)^{1/2}(\pi kT)[1 + (15b^2/16a^3)kT]$;
(d) $\bar{\varepsilon} = kT + (15b^2/16a^3)(kT)^2$, $c_v = k + (15b^2/8a^3)k^2T$.

11.7. $7/2 + \ln[(2\pi mk/h^2)^{3/2}/\theta_{\rm rot}]$ for CO,
$7/2 + \ln[(2\pi mk/h^2)^{3/2}/(2\theta_{\rm rot})]$ for O_2.

11.8. $c_v = 9k/2$; $\overline{|\boldsymbol{r}_1 - \boldsymbol{r}_2|^2} = 3kT/K$. .

11.11. 1.11×10^3 J kilomole^{-1} K^{-1}; 2.67×10^3 J kilomole^{-1} K^{-1}.

11.14. (a) 2.24×10^{16} atoms, 2.08×10^{-7} J T^{-1};
(b) 6.67×10^{17} atoms, 6.18×10^{-6} J T^{-1}.

11.15. (a) $Z_1 = e^{-\beta\varepsilon_0} + e^{-\beta\varepsilon_1}$; (e) $c_v \to 0$ $(T \to 0)$, $c_v \to 0$ $(T \to \infty)$.

11.16. (a) $Z_1 = 1 + 2e^{-\beta\varepsilon} + e^{-2\beta\varepsilon} = (1 + e^{-\beta\varepsilon})^2$.

11.18. (a) $C_c = NJ(J+1)(g\mu_B)^2/3k$.

11.19. (c) $\alpha = Np^2/3k$.

11.23. (c) $\lambda_{1,2} = e^{\beta\varepsilon}[\cosh\beta B \pm (\sinh^2\beta B + e^{-4\beta\varepsilon})^{1/2}]$.

12.4. (a) $N = (L/h)(8m\varepsilon)^{1/2}$; (b) $\varepsilon_F = (h^2/8m)(N/L)^2$; (c) $\varepsilon_F/3$.

12.5. (a) $N = 4\pi mA\varepsilon/h^2$; (b) $\varepsilon_F = (h^2/4\pi m)(N/A)$; (c) $\varepsilon_F/2$.

12.6. (a) $g(\varepsilon) = (4\pi V/h^3c^3)\varepsilon^2$; (b) $\varepsilon_F = (3/8\pi)^{1/3}hc(N/V)^{1/3}$;
(c) $3\varepsilon_F/4$.

12.7. (a) 1.87×10^{-18} J; (b) $\mu/\varepsilon_F = 1 - 4.5 \times 10^{-5}$; (c) $1.10 \times 10^{-2}R$.

12.8. (a) $\varepsilon_F = 6.9 \times 10^{-23}$ J, $T_F = 5.0$ K;
(b) $C_V = 0.99\,NkT$; (c) 0.25 K.

12.11. $N/V = 1.11 \times 10^{24}$ m^{-3}; $N_0/N = 65\%$.

12.14. (b) 2, 3; (c) 1.37×10^{-7} K; (d) 38%; (e) 121, 60.5.

Index

Physics Constants

Boltzmann constant	$k = 1.381 \times 10^{-23}$ J K^{-1}
	$= 8.617 \times 10^{-5}$ eV K^{-1}
Avogadro's number	$N_A = 6.022 \times 10^{26}$ kilomole^{-1}
gas constant	$R = 8.314 \times 10^{3}$ kilomole^{-1} K^{-1}
Planck constant	$h = 6.626 \times 10^{-34}$ J s
	$\hbar = h/2\pi = 1.054 \times 10^{-34}$ J s
speed of light	$c = 2.998 \times 10^{8}$ m s^{-1}
gravitational constant	$G = 6.673 \times 10^{-11}$ m^3 kg^{-1} s^{-2}
atomic mass units	1 amu $= 1.661 \times 10^{-27}$ kg
electronic charge	$e = 1.602 \times 10^{-19}$ C
electron mass	$m_e = 9.109 \times 10^{-31}$ kg
proton mass	$m_p = 1.673 \times 10^{-27}$ kg
Bohr magneton	$\mu_B = 9.274 \times 10^{-24}$ J T^{-1}
	$=9.274 \times 10^{-21}$ erg Oe^{-1}

Conversion of Units

volume	$1\ m^3 = 10^6\ \mathrm{cm}^3$
	$= 10^3$ liter
force	$1\ \mathrm{N} = 1\ \mathrm{kg\ m\ s}^{-2}$
	$= 10^5$ dyne
pressure	$1\ \mathrm{Pa} = 1\ \mathrm{N\ m}^{-2}$
	$= 10\ \mathrm{dyne\ cm}^{-2}$
	$= 9.872 \times 10^{-6}$ atm
	$= 10^{-5}$ bar
	$= 7.502 \times 10^{-3}$ torr
	$1\ \mathrm{atm} = 1.013 \times 10^5\ \mathrm{Pa}$
	$= 1.013$ bar
	$= 760$ torr
energy	$1\ \mathrm{J} = 1\ \mathrm{N\ m}$
	$= 10^7$ erg
	$= 2.390$ kcal
	$= 6.242 \times 10^{18}$ eV
	$1\ \mathrm{kcal} \doteq 4186\ \mathrm{J}$
	$1\ \mathrm{eV} = 1.602 \times 10^{-19}\ \mathrm{J}$
magnetic induction	$1\ \mathrm{T} = 10^4$ gauss